P9-AFZ-492

KOROLEV

KOROLEV

How One Man Masterminded the Soviet Drive to Beat America to the Moon

James Harford

John Wiley & Sons, Inc.

New York • Chichester • Weinheim • Brisbane • Singapore • Toronto

This text is printed on acid-free paper.

Published by John Wiley & Sons, Inc.

Library of Congress Cataloging-in-Publication Data
Harford, James
Korolev : how one man masterminded the Soviet drive to beat America to the moon / James Harford.
p. cm.
Includes bibliographical references and index.
ISBN 0-471-14853-9 (cloth : alk. paper)
1. Korolev, Sergei Pavlovich, 1907–1966. 2. Astronautics—Soviet Union—Biography. I. Title.
TL789.85.K62H37 1997
629.4′092-dc20
[B] 96-35311

Printed in the United States of America

10 9 8 7 6 5 4 3 2

To my son Jim, whose patience and skill enabled me to get into the U.S. and Russian electronic information network, without which I would still be working on this book.

CONTENTS

Preface			ix
A Note About Russian Names			xvi
Abbreviations			xvii
Prologue		The Anonymous Chief Designer	1
Chapter	1	Kibalchich to Korolev	8
	2	Growing Up in Ukraine: Broken Family, Bolsheviks, Gliders	16
	3	To Moscow: Tupolev, Tukhachevsky, First Rockets	29
	4	The Gulag and Sharaga Years	49
	5	The German V-2: Bedrock Technology	64
	6	The World's First ICBM: Aimed at the USA	91
	7	Sputnik: No Big Deal to Khrushchev–At First	121
	8	Unmanned Firsts: Hitting the Moon and Venus	139
	9	Gagarin First, Shepard an Anti-climactic Second	159
	10	*Voskhod:* A "Circus Act"	180
	11	Spy Sats and Com Sats	190
	12	The Organization: Korolev Up and Down	201
	13	The Technology: Simple but Reliable	219
	14	The Party, the Paranoia	230
	15	Racing Apollo: The Odds Were Enormous	246
	16	The End of Anonymity: Burial in the Kremlin Wall	276
	17	After Korolev: Demise of the Circumlunar and Lunar Landing Missions	286
	18	Ironic Epilogue	313
Appendix		A Chronology of the Moon Race, 1957–1974	324
Notes			340
Bibliography			372
Index			377

Preface

IN ONE OF my first interviews in Moscow I was told by Gyorgi Vetrov, who worked for Sergei Pavlovich Korolev and who is now one of Russia's most eminent space historians, "Your book on Korolev could play a great role. . . . It could have international importance and must be done well. The trouble is that in Russia there are no good publications focusing on the space program. Everything that has been printed about the lunar program . . . has contained distortions. There are people still alive who are withholding the truth, and they are still influential."

The Korolev story is very Russian, and remarkable for its manifestation of ingenuity and resourcefulness in what was surely one of the most paranoid and ruthless societies in history. A survivor of the gulag system, Sergei Pavlovich Korolev, "SP" to his team, in the 1940s personally persuaded the Soviet political apparatus to provide support for the development of intercontinental ballistic rockets and then the space program, fighting off rivals, working with relatively unsophisticated technology, and—by order of Stalin, Khrushchev, and Brezhnev—remaining unknown to his own countrymen until his death and burial in the Kremlin wall.

When he was taken in the middle of the night to prison in 1938, Korolev, only thirty-one years old, had already done pioneering work on liquid propellant rockets. Like millions of Soviet citizens spirited off by the NKVD during the Stalin purges, he was forced to confess to trumped-up charges. Unlike the others, however, he was released after perhaps a year, peremptorily transferred from one of the Soviet Union's most infamous prisons, the Kolyma gulag in far eastern Siberia, to a different kind of prison, near Moscow, where he went back to work on rockets.

The tale of how he not only got out of Kolyma, but gained full freedom and eventually led a gigantic enterprise that built the first nuclear bomb delivery system, then shocked the world with the first satellite and the first man in space, lets some light into the black hole of the Soviet social, political, and technological system in the pre-glasnost decades.

A dose of humility potion was needed before undertaking this project. I realized that, even with the help of the many people in Russia and the United States who are credited in these pages, it would be very difficult to review and assess objectively and accurately the life of this legendary person I had never met, and who was anonymous throughout his career in one of history's most secretive and paranoid societies.

Another complicating factor: Sergei Pavlovich Korolev died at age fifty-nine, at the peak of his effort to lead the USSR in its race to beat the Americans to the Moon. Leonid Brezhnev's decision to then uncover his identity, personally helping to carry his remains to the Kremlin wall, led to a national outpouring of testimony to his accomplishments. What he had done was, indeed, remarkable. But, as with the too-early passing of other great men, many of the memoirs about Korolev were so unmitigatedly worshipful and sentimental that they called for cooler analysis.

Knowledgeable readers in Russia may disagree with some of my assessments and may find errors of fact as well. I hope there are few of the latter. Although the former Soviet Union began to open its records to history in the Gorbachev era, and now allows the publication of memoirs dealing with many long-classified subjects, the fact is that policy has not caught up with practice. Some archives are still closed to scholars, and others are still buried in unsorted files. And, alas, there were some important figures in Korolev's life, fortunately only a handful, who would not allow me to interview them. Most of this complex man's former colleagues, however, were generous with their time and were quite willing to open up freely. The record had been obscured long enough, and it was time, as well, for them to get credit for their own singular contributions.

My interest in doing this book hatched very early on—not long after the Soviets shocked the Western world with Sputnik in 1957. It was Korolev, of course, known then only as the "Chief Designer," who led that project. At the time I was the youthful Executive Secretary of the American Rocket Society (ARS). Some Americans, besides CIA specialists, I reasoned, ought to follow the Soviet space program. But my ARS duties left precious little time for extracurricular research. Others, notably the late Charles Sheldon at the Library of Congress, did outstanding work sleuthing out the Soviet capabilities and achievements. I owe a big debt to him, and to his successors at the Library's Congressional Research Service. One of them, Marcia Smith, has been particularly helpful to me over the years.

In 1959 I enrolled in my first Russian language class—a once-a-week, ten-week course at the Princeton Adult Education School. Over the years I took several others, usually missing half the classes because of the rigorous travel schedule I had at ARS. Fortunately, some of that travel was to the annual congresses of the International Astronautical Federation (IAF)—usually in Europe—where I got to meet the small delegations of KGB-escorted Soviet space scientists permitted to attend. No engineers were allowed out of the USSR, as the Soviets were totally close-mouthed about their technology. Occasionally some of these same Soviet scientists would attend ARS meetings in the United States, and in spite of the restrictions on what they were able to say, a few became, and still are, my friends. Among them are Yuri Zonov, Gorimir Chernyi, Vadim Gogossov, Oleg Gazenko, and Vladlen Vereshchetin, each of whom has been helpful in enabling me to identify Korolev colleagues.

In 1963 ARS merged with the Institute of Aeronautical Sciences to form the American Institute of Aeronautics and Astronautics, and after a year as deputy to the late S. Paul Johnston, who encouraged my interest in the Soviets, I became Executive Director of that organization. AIAA is the world's largest professional society in aerospace, with some 35,000 members—engineers and scientists, as well as students.

In 1973 I made my first four trips to the USSR when AIAA cosponsered an aeronautical conference in Moscow with the Soviet Ministry of Aviation, and, in the same year, the Soviets hosted the IAF Congress in Baku near the Caspian Sea. The latter is only a couple of hours by air from the main Russian launch center, Baikonur, but there was no way that the Soviets would let us visit that highly secret facility. It would not be until twenty-one years later, in 1994, that I would get there.

Over the two-plus decades I made many trips to Russia, most of them devoted primarily to interviews and other research for this book. I had begun early on to accumulate published material on the Soviet space program, and on Korolev, from USSR sources. A much-valued source was the now-suspended (regrettably) special section on Russian space published in English translation by the National Technical Information Service. I improved my language capability with cassette tapes and weekly tutorial sessions with four other students taught by the patient Tatyana Ermolaeva in our homes. Of most importance, however, was the Russian 102 course at Princeton University, which Veronika Dolenko kindly allowed me to audit. It was rigorous—I was the only oldster in a class of mostly very bright freshmen—but it enabled me to bolster my knowledge of the grammar of this rich language, and to build a modest vocabulary. Fluency was not mastered, however, and so on my Russian trips I needed good interpreters—familiar with space technology history as well as English idioms. Back home, I spent many hours, with dictionary handy, translating Russian documents. Fortunately, I was able to consult with two wonderfully helpful Russian-born scientists at Princeton University: Oleg Bukharin and Nick Kukharkin.

When I retired from AIAA in 1989 I decided to work intensively on the book, with the intent to cover broadly the development of the

Russian program, but with a humanizing focus on its dominant figure—Sergei Korolev. I worked on the book half time for the next three years, devoting the other half, as a consultant to AIAA, to what became the largest space meeting ever held—the World Space Congress, commemorating the 1992 International Space Year. I am deeply grateful to the AIAA presidents and vice presidents who indulged my Russian interests over the years, including Larry Adams, Holt Ashley, Norm Augustine, Bill Ballhaus, Burt Edelson, Dan Fink, Joe Gavin, Alan Lovelace, Art Mager, John McLucas, Rene Miller, George Mueller, Court Perkins, Bill Pickering, John Swihart, and Mike Yarymovych. My gratitude goes also to AIAA staff members Cort Durocher, my successor as Executive Director, and Mireille Gerard, head of the Institute's international programs.

With the World Space Congress behind me I was able to devote full-time to the book, first as 1992–1993 Verville Fellow at the National Air and Space Museum, and after that on my own. Gregg Herken, Cathleen Lewis, and Frank Winter of the NASM space history department, and Michael Neufeld and Von Hardesty of the aeronautical department, provided good pragmatic advice and criticism of specific chapters, as well as moral support.

At the Russian end, I have Boris Rauschenbakh, Viktor Sokolsky, Alexei Drozhilov, and Alexander Gurshtein of the space history branch of the Academy of Sciences to thank on many counts—advice, a place to hang my hat in Moscow, hotel and interview arrangements, and their own excellent scholarly works.

Two other eminent Russian space historians, however, were so important to this project that I doubt I would have been able to carry it out successfully without their sustained help. They are Maxim Tarasenko, Moscow Physical Technical Institute, and Gyorgi Vetrov, who had a long career at the Korolev design bureau, himself. Tarasenko is the leading researcher on Russian military space, historical as well as contemporary. Vetrov's own three-volume work on Korolev is slated for publication in Moscow, soon I trust.

Also in a special category is Vladimir Syromiatnikov, who has worked at the Korolev bureau for some forty years and who is a key designer of the system that will permit American and Russian vehicles to dock with the forthcoming International Space Station. Syromiatnikov not only advised, critiqued, and arranged for me, but also—with his wonderful wife Svetlana—fed me, transported me, fur-capped me, and otherwise acclimated me to Russia's unique environment.

Other Russians who imparted their wisdom and hospitality were Vladimir Agapov, Viktor Legostaev, Mikhail Marov, Alexei Leonov,

Kyril Kondratyev, Vladimir Solovyev, Vitaly Sevastyanov, Oleg Alifanov, and Vassily Mishin, who served as Korolev's deputy, succeeded him as Chief Designer, and who still—in his eighties—lectures at Moscow Aviation Institute.

Special gratitude goes to Natalya "Natasha" Koroleva, daughter of Sergei Pavlovich, who not only allowed me to interview her at length on three occasions, but who gave me copies of precious photographs from her own collection. Sergei Khrushchev, a very able engineer-veteran of the Soviet space program who is also Nikita's son, gave me a long interview at his current post at Brown University, and critiqued several chapters dealing with our talks.

The bibliography and footnotes record important Russian publications that I have referenced, but I must cite the works of Yaroslav Golovanov as having been particularly fruitful and reliable sources for this book. Golovanov, whose 800-page *Korolev Fakti i Miti* (*Facts and Myths*), published in 1994, is the definitive Russian biography, generously allowed this potential American competitor to interview him twice, and suggested others who should be interviewed.

I must not forget the remarkable Lenin Library in Moscow, which allowed me to take out a regular card so I could pore over books, speeches, and newspaper clips from the Korolev era the old-fashioned way.

Back in the USA were many who helped me. Chris Faranetta of the Washington office of the S. P. Korolev Rocket Space Corporation Energia (RSC Energia for short) was always ready to respond to questions. The NASA history department I visited almost as often as I phoned either the director, Roger Launius, or archivist Lee Saegesser of encyclopedic memory. Others who made my work more comprehensive: the late Nancy Doyle of the Princeton University Engineering Library, and the entire research staff of the Princeton Public Library, but especially Mary Lou Hartman—who located books for me all over the United States.

Jerry Grey, space policy expert and longtime friend, offered candid criticism of several chapters. Charles Vick sent me his line drawings of Soviet vehicles—some of them, amazingly accurate, made when only meager facts were available, and read part of the manuscript. For hard information as well as criticism of some chapters I am very grateful to Nicholas Johnson, whose personal lore of Soviet space may be unmatched in the United States and who published annually for a decade or so fact-laden summaries of Soviet space developments; Dick Martin and William Patterson, formerly of Convair Astronautics; Steve Zaloga, probably the most knowledgeable American authority on Soviet ICBMs; and John Logsdon, who runs the Space Policy Research Institute at George Washington University, and whose own writings, as well as those of his staff, in particular Dwayne Day and Lorenza Sebesta, filled in several important holes in my research. Bill Pickering, former director of the Jet Propulsion Laboratory; James Van Allen of the State University of Iowa; Herbert Friedman, Naval Research Laboratory; and Roald Sagdeev, former director of the Soviet Institute of Space Research, now at the University of Maryland, were especially helpful in reviewing and commenting on the chapters dealing with the Sputniks and the early lunar and planetary missions.

Anyone who follows Russian space history will be dependent, as I was continually, on the marvelous articles in the publications of the British Interplanetary Society, dating back to when Ken Gatland got them started in the 1950s. Another faithful chronicler has been the History Committee of the International Academy of Astronautics, with annual sessions at congresses of the International Astronautical Federation.

Thanks to a small grant from the John F. Kennedy Library in Boston I was able to dig into the trove of JFK space material from the early 1960s in that wonderful place, with the help of William Johnson and staff. I was helped as well to get material on Lyndon Johnson's crucial efforts on behalf of congressional support for space exploration from the LBJ Library in Austin, thanks to an intercession from a longtime Senate Space Committee staff expert, Glen Wilson.

I spent a lot of mostly unrewarding effort trying to get information on what our Central Intelligence Agency knew about Soviet space developments in the 1950s and 1960s. After I got in touch with an old friend, Albert Wheelon, a space industry executive who worked, at one time, for "The Agency," I got the information I needed on our spy satellites and what they told us about Russian space capabilities. Jeff Richelson of the private National Security Archive later sent me a batch of good stuff on that subject, and so did Major Patricia Wilkerson of the Department of Defense's National Reconnaissance Office.

A much-appreciated award for winning the Robert H. Goddard Historical Essay contest of the National Space Club, for Chapter 5 of this book, dealing with what the Soviets learned about rocketry from the Germans, helped cover some of my Moscow travel costs, and also generated useful inputs from Frederick C. Durant, Frederick I. Ordway, Professor Logsdon, and R. Cargill Hall.

I got solid help on the German V-2 chapter from Ordway, and from a longtime friend and colleague, Ernst Stuhlinger, University of Alabama at Huntsville; from Mitchell Sharpe; and from Kurt Magnus, one of the few surviving members of the German contingent who were commandeered to the USSR at the end of World War II.

Advice on how to penetrate the sensitive Soviet archives, although it led to only modest success, came from Librarian of Congress James Billington and his Moscow office staff; Vladislav Zubok of the Kennan Institute for Advanced Russian Studies; James Hershberg of the Woodrow Wilson Center's Cold War International History Project; and from Patricia Kennedy Grimsted, Harvard Ukrainian Research Institute, who has spent many long months in Moscow working that problem more productively than I did. I can furnish anyone trying to assault the Russian archives battlements with a long list of people who turned me away.

My agent, Julian Bach, not only told me—ten years ago—that I "must write this book" but encouraged me through the tough periods, and made cogent editorial suggestions.

Yvonne Brill, herself an important contributor to U.S. rocket technology, was willing to read and critique the entire manuscript, a chore for which I will be everlastingly grateful. My editor, Hana Lane, at John Wiley, spotted awkward phrases and redundancies and offered wise recommendations on how to make the story flow better.

A host of friends, three brothers and a sister, four children, and most important of all, my wife Millie, kept telling me I could do it—and here it is.

A Note About Russian Names

I HAVE FOLLOWED the method used by the Library of Congress in transliterating Russian names, even though it results in mispronunciation. A transliteration that would approximate the correct pronunciation of Korolev, for example, would be Kuralyov because the unaccented "o" in Russian gets an "uh" sound and the "e" in the third syllable has two dots over it (like the German umlaut), giving the "e" a "yo" sound. Using the same rules, Khrushchev—a name that also figures prominently in this book—should be Khrushchyov. But the Library of Congress and the American Library Association have chosen to transliterate for potential reversibility. That is, someone—or something, like a computer—can take the romanized script and convert it easily back to the original script. So, although I've gotten used to saying Kuralyov in order to be understood on my many trips to Russia for this book, the power of the Library of Congress is so strong that Korolev and Khrushchev have become widely used, and I've gone along.

ABBREVIATIONS

USSR

CPCC	Communist Party Central Committee
CPSU	Communist Party of the Soviet Union
GDL	Gas Dynamics Laboratory
GIRD	Group for Studying Reaction Propulsion
GSKB	State Union Design Bureau
IKI	Institute of Space Research
IMBP	Institute of Medical and Biological Problems
KGB	Committee of State Security
MAI	Moscow Aviation Institute
MGTU	Moscow State Technical University (also known as Bauman Institute)
MGU	Moscow State University
MOM	Ministry of General Machine Building
MPTI	Moscow Physics & Technology Institute
MVTU	Moscow Higher Technical Professional School (former name of MGTU)
NII-1	Scientific Research Institute-1 (successor to RNII)
NII-88	Scientific Research Institute-88
NIITP	Scientific Research Institute for Thermal Processes (successor to NII-1)
NITSKD	Scientific Research Center of Space Documentation
NKVD	Predecessor of KGB
NPO	Scientific Production Association
OAVUK	Society of Aviation and Aerial Navigation of Ukraine and the Crimea
OKB	Experimental Design Bureau
RKA	Russian Space Agency
RNII	Reaction Scientific Research Institute
RSC Energia	Rocket Space Corporation Energia (RKK Energia in Russian)
SKB	Special Design Bureau
TsAGI	Central Aerohydrodynamic Institute
TsIAM	Central Institute of Aviation Motors
TsKBEM	Central Design Bureau of Experimental Machine Building
TsNIIMash	Central Research Institute of Machine Building
TsSKB	Central Specialized Design Bureau

USSR

TsUP	Center of Flight Control
VPK	Military Industrial Commission

U.S.

ABMA	Army Ballistic Missile Agency
AF SAB	Air Force Scientific Advisory Board
AF BMD	Air Force Ballistic Missile Division
CSM	Command and Service Module
EOR	Earth Orbit Rendezvous
EVA	Extra-Vehicular Activity
LEM	Lunar Excursion Module (later just LM for Lunar Module)
LEO	Low Earth Orbit
LM	Lunar Module
LOR	Lunar Orbit Rendezvous
MSFC	Marshall Space Flight Center (NASA)
NACA	National Advisory Committee for Aeronautics
NASA	National Aeronautics and Space Administration

PROLOGUE

The Anonymous Chief Designer

ONE DAY in the early 1960s, Sergei Pavlovich Korolev was looking at a newspaper photograph of Wernher von Braun, then being lionized in the United States for his part in the upcoming Apollo program. His comment, recalled Antonina Otrieshka, a staff assistant: "We should be friends."[1]

On the surface the remark is astounding. But what it reveals is Korolev's feeling of identity with von Braun's space prowess, as well as a certain envy over attention he wasn't enjoying himself. He, after all, was the number one Chief Designer of the Soviet space program, the head of the Council of Chief Designers, the principal architect of the spectacular Russian achievements in space which, at that time, had decisively outclassed those of the Americans. Yet no one except insiders even knew his name. Under a policy started by Stalin and continued through the reigns of Khrushchev and Brezhnev, Korolev remained officially anonymous to the outside world until his death in 1966, at the peak of his effort to beat the Americans to the Moon. The fear, apparently, was that he might be done in by U.S. espionage agents.

It's not so strange that Korolev should link himself to von Braun, even though the two became virtual icons of violently opposed Cold War powers. They were alike in a lot of ways.[2] Both were charismatic leaders who inspired extraordinary dedication and loyalty from their people. Both began experimenting with rockets as amateurs in the 1930s. Both maintained that space flight to the Moon and planets was their goal. Both, nonetheless, got their early impetus and funding from the development of military missiles. Both were arrested for alleged subversion of military projects, although von Braun's detention by the SS was brief, only two weeks' imprisonment. Korolev got terrible treatment—about seven years in prison, from 1938 to 1945. The first part of his incarceration, maybe as long as a year, was spent in one of the infamous gulag camps in Kolyma, far northeastern Siberia, and the last six years or so in what was called a *sharaga*—a prison for engineers and scientists. Both died of cancer-related causes while at their career summits, Korolev at fifty-nine, von Braun at sixty-five.

Boris Rauschenbakh, who worked with Korolev for many years, recalls other similarities between the two men:

> Korolev was not principally a designer. He was, as von Braun was, a good manager, not only a manager but a commander. Like Napoleon, he sometimes made decisions without full information. If you work in a field where there is no experience you are forced to make decisions without full information. Therefore you are liable to make mistakes. The good commander makes the right decisions instinctively.
>
> . . . Both found it very dangerous to discuss ideology. Von Braun had to be a Nazi. Korolev had to be a Communist, although he rarely attended Party meetings. Both went directly to high political figures for their support.[3]

But, aside from Korolev's anonymity and von Braun's visibility, there is a big difference between the two. Von Braun's contributions to American missile and space achievements, while very significant, were matched, or nearly so, by many U. S. space leaders—James Webb, Hugh Dryden, Karel Bossart, George Low, Robert Gilruth, Robert Seamans, William Pickering, Max Faget, C. Stark Draper, Sam Phillips, George Mueller, Chris Kraft, Kurt Debus, Eberhard Rees, and others.

Korolev, however, was by far the dominant figure in the Soviet rocket and space exploration effort. While there were other able Russian contributors who figure in this book—for example, Friedrikh Tsander, Valentin Glushko, Vladimir Chelomei, Mikhail Yangel, Mikhail Tikhonravov, Konstantin Feoktistov, Konstantin Bushuyev, Viktor Legostaev, Sergei Kryukov, Boris Chertok, Vassily Mishin, Vladimir Barmin, Nikolai Pilyugin, Mikhail Reshetnev, Boris Rauschenbakh, Leonid Voskresensky—they did not permeate the programs to the extent that Korolev did.

Compare the records. Von Braun, after his voluntary surrender to the U.S. military at the end of World War II with 126 of his Peenemunde colleagues, and a period of detention in Fort Bliss, Texas, led the teams that developed the Jupiter intermediate-range ballistic missile; the Redstone rocket that launched America's first satellite and first American space traveler, Alan Shepard; and—his crowning achievement—the giant Saturn V rocket that enabled twelve astronauts to walk on the Moon. An eloquent and stirring speaker, he was a media star, made numerous Congressional appearances, was on the covers of *Time* and *Life,* was awarded honorary degrees, authored magazine articles, even had an unmemorable B movie made on his life, with Curt Jurgens playing him.

But Korolev not only developed the first Soviet ICBM and launched the Space Age with Sputnik, he put into space the first dog, first man, first two men, first woman, first three men; he directed the first "walk in space"; he created the Soviet Union's first spy satellite and first communications satellite; he built the launch vehicles and the spacecraft that first reached the Moon and Venus and passed by Mars. He was frustrated, though, from achieving his biggest goal: sending a man to the Moon before the United States did. Whether he could have done it if he had not died is an open question. The answer is very likely not, because the Soviets simply failed to match the gigantic mobilization of manpower and money that the United States put into the Apollo program. But this extraordinary man gave the effort a magnificent go.

Like a football coach, he would whip up his team for the competition with the United States, especially during the Sputnik preparations in 1957. One of his designers, Oleg Ivanovsky, remembered:

> He would tell us that "the Americans are at our heels, and the Americans are serious people." He wouldn't use the word "Amerikantsi" but "Amerikan-ye" as if these weren't just American residents but the entire American culture we were competing with. He didn't mean this as an insult but as a show of respect for the competition.[4]

Korolev's performance was the more remarkable not only because it was achieved with much less funding, but also because he had sometimes to contend with rivals who were out to do him in, and with capricious government superiors who wanted to control him or quash him.

It must have been supremely galling for Korolev, an engineer-manager of tremendous achievement and high ego, to have to reconcile himself, because of Soviet paranoia, to career-long obscurity. He not only received no public kudos, in a country where such recognition is a staple for the big shots, he was never even allowed proper visibility on the many occasions when cosmonauts were feted in the Kremlin or Red Square. He was never bestowed openly with Kremlin honors, and he wasn't even allowed to wear the medals indicating that he was twice a "Hero of Socialist Labor." The only way he got into the newspapers was through occasional articles on space in Pravda under the pseudonym "Professor K. Sergeev."

What kind of a person was he?

In many interviews with his family and colleagues over a period of several years, and in the reading of his papers, letters, and the memoirs of his fellow workers, I was to learn much about his complexities. Most of his colleagues and fellow workers revered him, although some tempered their praise, like one of his deputies, Boris Chertok, who said, "He was a wonderful organizer [but] there has been a tendency to canonize Korolev."[5]

It's not strange that the man took on an extraordinary aura. His ability to inspire large teams as well as individual engineers is proverbial. He had a roaring temper, was prone to shout and use expletives, but he was quick to forgive and forget. His consuming passion was work, work, work for space exploration and for the defense of his country. One wonders how he maintained such an unswerving loyalty to a system that had treated him so cruelly. But he is far from the only Russian who suffered and stayed loyal to the cause.

It seems clear that his treatment by the Soviet system imbued him with a degree of opportunism, and sometimes Machiavellianism, that served him well in his efforts to promote his programs. Some who worked with Korolev remember him for being, at times, devious and scheming. Andrei Sakharov, who helped design the H-bomb payload carried on Korolev's rockets, described him as "a brilliant engineer and organizer and a colorful personality" who "dreamed of the cosmos" but was "a bit more cunning, ruthless, and cynical than Kurchatov [Igor, the head of the Soviet nuclear weapons program]."[6]

On the other hand, Yaroslav Golovanov, Korolev's principal Russian biographer who spent twenty years researching his subject, and who worked in the Soviet space program himself, expressed the view that "He was not devious, but he was skilled in reaching his objectives and at making necessary compromises."[7]

Indeed, Sergei Pavlovich apparently followed whatever zigzag path got him to his goals. One of his most candid characterizations came from Oleg Gazenko, a distinguished doctor who became the leader of the Soviet Union's space medicine program.

> He was a typical product of the Soviet system of that time. He had been imprisoned but still he worked hard for his country.
>
> The vectors of his objectives and some of the goals of the regime coincided at that time. They had the same goal to create a powerful missile shield for the country. Korolev adapted well in various conditions. He was clever in dealing with various political personalities.
>
> I agree with Sakharov's description. Korolev was typical of his generation. He adapted very well. It was very difficult for me to deal directly with Korolev so I tried to avoid contact with him. He had no respect for people. He used them. I don't want to give names but, for example, on one occasion a rocket was being prepared for launch. The final test of electrical circuits was being run and a short circuit occurred. Korolev began to shout at the electrical crew, making very serious accusations, using the lowest swear words.
>
> For doing good work he would take money from his safe without any State authorization. He could use money as he wished.
>
> History, according to a Russian saying, knows no subjunctive mood. I can't say if Korolev was the only person who could have run such a

successful program . . . but he was the best person for the times and the situation.[8]

Another outstanding scientist who remembered Korolev's easy access to funds, as well as his earthiness, but nevertheless admired him, is Josef Gitelson, who made major contributions to the development of ecological systems for Soviet spacecraft. Still in a senior position at the Institute of Biophysics in Krasnoyarsk in Siberia, Gitelson recalled a meeting with Korolev in 1961 at the latter's design bureau in Kaliningrad, when he and the Institute Director at the time, Academician Leonid Kirensky, were explaining the results of recent studies, apparently in too much detail:

> He [Korolev] interrupted,
>
> "Don't deliver a lecture . . . Too many academicians give lectures that are not worth a damn." He used a more vulgar expression, actually, which translates roughly as "not a cock's worth."
>
> "What *results* do you have?"
>
> In effect, he despised the superior attitude of the prestigious metropolitan academics who usually spoke condescendingly to people from the provinces. We had results and we described them. He was impressed and made a fast decision.
>
> "How much money do you need?"
>
> Kirensky wasn't prepared to give an answer.
>
> "Is one million rubles enough?"
>
> It was plenty and we got the money in new rubles, not old ones (new ones were 10 times the value of old ones) by bank transfer from Korolev's design bureau.
>
> Indeed, the money began coming very quickly—without the usual bureaucratic procrastination—directly from Korolev's design office, which wasn't the usual practice. The established procedure would have taken years of planning and coordination to allot such considerable monies. To breach—break—such a tradition, to do business quickly, this is Korolev all over.
>
> We went to work developing a bioreactor for making algae to consume carbon dioxide and develop oxygen to produce pure water from the cosmonaut's urine and perspiration.
>
> We later visited Korolev frequently to update him on our work. In the five years until his death we had only a few meetings to solve key problems. Still, every meeting is memorable both in pithiness and bright utterances.[9]

The man's accomplishments were truly amazing when one recognizes the severe limitations on his freedom of action, the penalties that the system meted out to those who failed, and the fact that he was in poor physical condition in his later years, apparently the legacy of the privations he experienced during his gulag days. Yet his underlings remember how merciless he was on himself—and on his staff—working long days, traveling at the end of the day to the Baikonur launch complex to minimize the time away from the job.

Otrieshka said that, "When a ukase [edict] came out in 1953 stating that people should work eight hours a day he laughed and said, 'That's not for us.' He usually worked himself from 8 A.M. to 10 P.M. Then after 10 P.M. he'd have private conversations with various people. I would only get his attention after 10 P.M."[10]

Having spent more than forty years in the aerospace field myself, I am well aware that working long hours, weekends, and flying the "red eye" overnight were common practice for U.S. engineers and scientists as well. But at least the Americans had decent accommodations in clean hotels on Florida beaches, and hearty restaurant meals, before their launches. Not so Korolev and colleagues. I've visited the simple Baikonur cabin where Korolev stayed during launch preparations. It is virtually monastic: a bed, a bureau, a desk, a table, a fridge, a toilet, and a sink. That accommodation was luxurious compared to what his staff endured—tents on a frigid or roasting, dust-swept desert in the early days, and in later years barracks that were unheated in winter and not air-conditioned in summer. Baikonur's temperatures were often below zero in snowy winter, and soared over 100°F in summer.

Americans used to postponement after postponement of Cape Canaveral launches due to "cloud conditions" or "high winds aloft" can't help but be impressed by the record of Soviet on-time launches even in the worst conditions, especially during the early days. In good weather and bad, sun and snow, they rolled their rockets to the pad, raised them vertically, and sent them off with a remarkable success rate.

Perhaps above all, Korolev was a man who had assigned himself a destiny that he was determined to fulfill to the end, and impatient to see it come to fruition. Gitelson illustrated the impatience, sustained to his last days, in this recollection.

> My last meeting with him was in late 1965, about three months before he died. We talked about the future use of our systems. At the time the technology for controlled ecological systems was not developed sufficiently to apply our research results. I said to him,
>
> "Sergei Pavlovich, do you suppose our results came too early?"
>
> "No, too late," he answered. "I have a short time before me, maybe ten years, and I want to send humans to the nearest planet. I will pay you twice as much if you cut the time for developing the technology in half."[11]

According to Pavel Vorobiev, who still works at the Korolev design bureau, Korolev really meant the "nearest planet." His original plan, even before concentrating on the Moon project, was to launch a cosmonaut around Mars. Vorobiev told me that he had in mind putting the space traveler in a closed-loop life support system and an artificial low gravity environment for the lengthy—probably two- to three-year—round trip.[12,13]

Korolev did not get the ten years, nor even ten months. But whenever a man, or a woman, or more likely a mix of men and women, reaches Mars—possibly thirty or forty years from now—the Chief Designer will have helped them get there.

KIBALCHICH TO KOROLEV 1

"I AM WRITING this project in prison, several days before my death . . . I believe in the feasibility of my idea and this faith supports me in my terrible situation. If learned specialists find my idea realistic I shall be happy to be able to render service to my country and mankind. I shall meet death calmly then, knowing that my idea will not perish with me."[1] Thus wrote one Nikolai Kibalchich (1853–1881) from Sts. Peter and Paul Fortress in St. Petersburg, just before being executed for the assassination of Tsar Alexander II.

Kibalchich's idea was a clever scheme for using the principle of jet propulsion to achieve human flight. His papers weren't discovered for decades, but when they surfaced they established him as the first person to show scientifically that an engine using the energy of "slow-burning explosives" would be logical for propelling a flying machine.[2] The principle he defined was meant to be applied to a vehicle flying in Earth's atmosphere, but the extrapolation to space became clearly evident to subsequent thinkers. One of the most prolific popularizers of space flight in Russia, Y. I. Perelman, later described Kibalchich's invention as "the first step in the history of space flight."[3]

It may seem bizarre, but the fact is, as historian Walter A. McDougall put it, "Modern rocketry and social revolution grew up together in tsarist Russia. There is no anomaly in the fact that the most 'backward' of the Great Powers before World War I was the one that fostered violent rebellion against the chains of human authority and the chains of nature."[4]

While Sergei Pavlovich Korolev, who would lead the Soviet space program to spectacular firsts, never rebelled against the government, he spent a lifetime suffering its restrictions and he certainly belongs prominently in that group that rebelled against nature. Korolev was one of a number of young Soviets who, in the early Bolshevik years, were literally levitated from their drab and difficult existences by the writings of Konstantin Edvardovich Tsiolkovsky (1857–1935), considered the father of space travel not only by the Russians, but by most space professionals.

For years science fiction writers from many countries had fantasized about visiting the Moon and other planets. A recognized classic is *From the Earth to the Moon,* written by Jules Verne (1828–1905) and published in 1865, but it is technically naive, depending on a cannon to propel humans. It was the Russians, and prominently Tsiolkovsky, who first showed the way to the practical accomplishment of space travel with rockets, although the world hardly knew about their ideas for years.

There was, in fact, a sufficient body of material extant—much of it Russian—to enable Nikolai Rynin (1877–1942), of Leningrad, to write a nine-volume encyclopedia of space travel between 1928 and 1932.[5] The Soviet Army general Gregory A. Tokaty-Tokaev, who defected to England in 1948, wrote that, "During only the second half of the 19th century at least twenty projects, or project-like ideas, of rocket-propelled flying machines were suggested in Russia by Konstantinov, Ivanin, Kibalchich, Evald, Gashvend, Tretesky, Sokovnin, Teleshev, Nezhdanovsky, Fyodorov, and others."[6] The ideas of some of those thinkers, and others who wrote about rockets as far back as the 1880s and 1890s—for example, Lebedev and Meshchersky—are documented in a comprehensive paper by the space historian Viktor Sokolsky published in English in 1986.[7] Nezhdanovsky, writes Sokolsky, suggested in the 1880s a liquid rocket using bipropellants later to become common—a liquid hydrocarbon as fuel and nitric acid as oxidizer.[8] Lebedev, a prominent physicist, "produced a schematic diagramme of a rocket engine for aircraft in 1892 and calculated the necessary amount of fuel."[9] There was also Yuri Kondratyuk, who anticipated, in a 1929 book, the idea of John Houbolt of NASA for using lunar orbit rendezvous to shortcut man's voyage to the moon.

The aforementioned Kibalchich was perhaps the quintessential bomb-maker cum scientist. He had been arrested first at age twenty-two, for revolutionary activity, and spent three years in prison. He had tried several times to kill Alexander II (1818–1881) who, ironically, had been one of the more socially sensitive of the tsars, having tried earnestly to introduce reforms during his reign. "At the beginning of the reform period," according to Sir Donald Mackenzie Wallace, "there had been much enthusiasm for scientific as opposed to classical education. Russia required, it was said, not classical scholars, but practical scientific men capable of developing her natural resources. The government, in accordance with that view, had encouraged scientific studies until it discovered to its astonishment that there was some mysterious connection between natural science and revolutionary tendencies."[10] Kibalchich, apparently, was one of those "Nihilists," as they became known, who was determined to do in the emperor. He succeeded on March 13, 1881, was apprehended on March 17 and put to death on April 3, at age twenty-seven. At his trial the chairman of the court described him as having "a remarkable mind, unusual stamina, hellish energy and amazing calmness."[11]

Friends wrote reminiscences of him, but the details of Kibalchich's ideas did not reach a broad audience until Rynin published them in 1918, in the journal *Byloe* [The Past].[12] Rynin later wrote in his encyclopedia, that " . . . Kibalchich's priority should be established in advancing the idea of the use of jet engines for flying, the idea which, it is true, has not been put to practice yet but which is basically correct and promises a great deal for the future, particularly if thinking of interplanetary flights."[13]

It is not known whether Robert H. Goddard (1882–1945), the Massachusetts physicist generally called the father of American space flight, knew of the ideas of Kibalchich or the other Russians. It is unlikely that he did before he, himself, developed independently the mathematics of reaching Earth-escape velocity with rockets in 1912, six years before Rynin wrote in *Byloe*. Goddard, however, who launched the world's first successful liquid fuel rocket in 1926, is accorded very high stature as one of the pioneer space flight thinkers. His ideas, like Tsiolkovsky's, encompass far more than liquid rockets.[14] Also deserving of credit are the Transylvanian Hermann Oberth (1894–1989), although his writings did not begin to appear until 1923, and Robert Esnault-Pelterie (1881–1957), the Frenchman whose works were entirely theoretical.

Tsiolkovsky, however, was the most influential conceptualizer. His story, like Kibalchich's and Korolev's, is riven with the pain, privation, and sacrifice that seem classically Russian. Growing up in the small town of Kaluga, two hours south of Moscow, he lost most of his hearing after a bout with scarlet fever at age nine. His mother died when he was thirteen. He taught himself mathematics, and went to Moscow, perhaps on a pitifully small subsidy from his father, at sixteen. In his autobiography, written in 1926 for the Rynin encyclopedia, he writes, "My parents were poor. My father (an itinerant forester) was a failure—an inventor and a philosopher. My mother, as father used to say, possessed the spark of talent."[15] He found a mentor in Moscow, a visionary named Nikolai Fyodorov (whose admirers were said to include Tolstoy, Dostoyevsky, and Leonid Pasternak, Boris' father).[16] Fyodorov tutored the young Tsiolkovsky in the library daily for some three years, introducing him to books on mathematics and science and discoursing with him on, among other subjects, the philosophical imperative leading humankind towards space exploration. From the autobiography in Rynin:

> I read only whatever could help me solve the problems that interested me and which I thought important. I thus got interested in the theory of the centrifugal force, because I believed it could be used to lift a spacecraft from the earth. There was a time when I thought I had found a solution to the problem (I was 16 years old then). I was so worked up

that I couldn't sleep all night—I wandered about the streets of Moscow, pondering the profound implications of my discovery. But by morning I saw my invention had a basic flaw. My disappointment was as strong as my exhilaration had been. That night had a lasting impact on me. Now, thirty years later, I still have dreams in which I fly up to the stars in my machine, and I feel as excited as on that memorable night.[17]

Tsiolkovsky went back to Kaluga and became a high school teacher of mathematics and physics. But he worked constantly on his scientific ideas. For a number of years he was preoccupied with subjects other than space travel—steam engines, stellar radiation, and then dirigibles.

In 1885, at the age of 28, I definitely decided to devote myself to aeronautics and the theoretical design of a metal-made dirigible. I worked for two years almost continuously. I had always been a dedicated worker and used to come home from school exhausted, having spent most of my energy there. Only in the evening could I make my calculations and conduct my experiments. What could I do? I did not have enough time, and I spent all my energy on my students. So I decided to get up at dawn; and after having worked on my treatise, I would go to school.[18]

The rigors of teaching all day and writing through the dawn hours eventually sapped Tsiolkovsky's energies, but new ideas about the possibility of space flight became his preoccupation.

In the twenties I stopped teaching because of ill health . . . it absorbed all my strength and left me very little energy for my own studies and projects. I wrote, made calculations, and built [projects] mostly on holidays and during the vacation.

I worked out certain aspects of the problem of going into space by means of a reaction-driven device such as a rocket. Mathematical calculations, based on scientific knowledge and repeatedly checked, show that it should be possible to go into space with such devices, and perhaps to set up living facilities beyond the atmosphere of the earth. It is likely to be hundreds of years before this is achieved and man spreads out not only over the face of the earth, but over the whole universe.[19]

Tsiolkovsky closed his autobiography, written at age sixty-eight, with the sentence, "Will I have enough strength, will I be able to complete my projects?" He would be dead nine years later, just as his ideas were being converted to working technology. But today in Kaluga there is an entire museum, built by the Soviet government, which houses the great man's work—models and reproductions of his dirigibles and space concepts right next to their modern consequences—satellites, spacecraft, space stations. The place has become a ritual stop for space engineers, cosmonauts, American astronauts, and especially for young students from all over Russia.

I went there several times myself, and also visited the home in which Konstantin Edvardovich worked in virtual solitude for some forty years. Walking through the simple brown clapboard structure—its tiny bedrooms where the old man lived with his wife and daughters; the study, which still has his ear trumpets and his telescope on the desk; the workshop with its turning lathe and models of dirigible shapes that Tsiolkovsky tested in what might have been the world's first wind tunnel—it is difficult not to be impressed by the fact that in this place, a mostly deaf teacher, living in a small town cut off from communications with world science, could have made such substantial contributions.

Tsiolkovsky wrote prolifically, starting with *Free Space* (1883), which dealt with the use of jet propulsion in the vacuum of space; then *Investigation of Cosmic Space by Reactive Vehicles* (1903). This latter was especially important because it developed his theory into a possible vehicle—"an elongated configuration of least aerodynamic drag; liquid hydrogen and liquid oxygen propellant supplied to the combustion chamber by mixer valves; divergent nozzle; cosmonaut's compartment. . . . "[20] The suggestion of using liquid hydrogen and liquid oxygen anticipated the use of those propellant components in the powerful upper stages of the Saturn V vehicle that sent the Apollo 11 astronauts to the Moon. "I do not know of any group of bodies [liquid oxygen and liquid hydrogen]," Tsiolkovsky wrote, "which when combined chemically would yield per unit mass of resultant product such an enormous amount of energy."[21]

His 1903 work, in the words of Korolev himself, "contained a whole range of extremely important and radically new technical propositions in the field of rocketry." "He did not merely theorize," wrote Korolev, "but consistently supported his theoretical conclusions, which were sometimes very unusual, with serious and detailed considerations, a vast majority of which are still widely applied in all countries of the world engaged in rocket technology."[22] Korolev then mentions the effort Tsiolkovsky put into choices of propellants still in use today—liquid oxygen, liquid hydrogen, and various hydrocarbons.[23]

Korolev cites other Tsiolkovsky design practicalities, including the idea of the rocket nozzle, which Korolev calls "a special explosion tube in the form of a flared cone," and "a combustion chamber to which fuel was supplied by pumps, suggesting automated control of combustion in the engine, to gear its operation to the different conditions encountered during the rocket's flight along its trajectory." Tsiolkovsky also proposed, again according to Korolev, "Grids with slanted vents to be installed at the entrance to the explosion tube in order to create the most favourable conditions for combustion" and he anticipated the concept of regenerative cooling of the rocket by circulating the cold fuel around the hot combustion chamber. He went on to suggest the use of refractory materials to deal with the rocket engine's high temperatures and he examined the ways in which the fuel is fed, atomized, and ignited.[24]

Among Tsiolkovsky's particularly advanced possibilities Korolev calls out his speculations on the possible use of "atomic energy, solar radiation and the energy of space radiation for propelling rockets."

In other writings this remarkable loner proposed such innovations as the gyroscopic stabilization of rocket ships in space; the use of a centrifuge to test the effect of gravity on living organisms; the control of flight by insertion of tilting surfaces in the rocket exhaust; the protection of space travelers from the effect of rapid acceleration during takeoff by, in Korolev's description, "immersing them in special costumes in tubs filled with a liquid whose density approaches that of the human organism";[25] and the creation of an artificial gravity field in the space ship to counter the effects of prolonged weightlessness.

Especially perceptive, wrote Korolev, was Tsiolkovsky's idea for building an "interplanetary station" that would be a "combination of several rockets linked to each other after being placed in orbit. . . . The station's orbit could be changed by burning an additional small quantity of fuel . . . communication between the station and the earth could be maintained by special rockets." This idea presaged the USSR's first space station, whose design was originally undertaken during Korolev's last years, and which was first launched in 1971. Almost continuously since then the Russians have had a space station in orbit.

In the 1920s and early 1930s, Tsiolkovsky continued his productivity with *The reaction engine,* 1927–1928; *A new aeroplane,* 1928; *Jet-propelled aeroplane,* 1929; *The theory of the jet-engine,* 1930–1934; *The maximum speed of a rocket,* 1931–1933. One of his classics, and his final work before his death, was *Space rocket trains,* which he worked on from 1924 to 1934.[26] This huge, mathematically dense work outlined the technique for designing multi-stage rockets. He would tell a group of students at the Zhukovsky Academy in 1934 that " . . . I am not at all sure, of course, that my 'space rocket train' will be appreciated and accepted readily, at this time. For this is a new conception reaching far beyond the present ability of man to make such things. However, time ripens everything; therefore I am hopeful that some of you will see a space rocket train in action."[27]

Indeed, they would. Multi-staging would eventually become the accepted mode for delivering payloads to the planets, and, starting in

1969, for landing twelve astronauts on the Moon and returning them to Earth in the Apollo spacecraft.

Although unrecognized for the first sixty years of his life, Tsiolkovsky's work was eventually sponsored by Soviet officialdom. During the ten years before his death, from 1925 to 1935, reports Tokaty, some fifty-eight of his books and articles were published by the State, which financed his work after he was elected to the Academy of Sciences in 1919. He became a consultant to rocket and aircraft companies and was invited regularly to important conferences.[28, 29]

One of Korolev's biographers, Alexander Romanov, describes in some detail a visit that he alleges Korolev made to Kaluga to see Tsiolkovsky in 1929,[30] but Yaroslav Golovanov concludes that the two met only in 1932, in Moscow, in connection with Tsiolkovsky's award of the Order of the Red Banner of Labour from the Kremlin.[31] Even that meeting may not have taken place, however. Gyorgi Vetrov, an eminent space historian who once worked as a designer under Korolev, says that Sergei Pavlovich was ambiguous on the matter himself, using phrases like, "After my acquaintance with Tsiolkovsky's work . . . I started rocket development."[32]

There is no doubt, though, about the degree to which Tsiolkovsky's works gave direction to Sergei Korolev. In simple fact, Korolev began to build what Tsiolkovsky had conceived.

There are those in Russia today who feel that Korolev's plaudits have been overblown. "He belongs in a large circle which includes also Valentin Glushko [1908–1989], Mikhail Yangel [1911–1971], Vladimir Chelomei [1914–1984], others. It is a stereotype that it was mostly Korolev who built the Soviet space program. Glushko certainly never got enough credit," were the words of journalist Mikhail Rudenko.[33] Boris Gubanov, an outstanding rocket designer who worked with Korolev, Yangel, and Glushko, put it this way: "Korolev was the older man, the pathfinder, and so Yangel, Chelomei, and others were not given enough credit for their contributions, which are only now becoming known."[34]

It is certainly true that Glushko, Chelomei, and Yangel—whose stories are intertwined with each other as well as with Korolev—deserve high ranking in the pantheon of Russian missile and space technologists. All three, as well as Korolev, were accorded similar honors by the Party hierarchy. Each became members of the Communist Party, each was twice decorated as Hero of Socialist Labor, each was elected to the Academy of Sciences.

By and large, however, the technical people regard Korolev as clearly number one. Vsyevolod Avduyevsky, who had numerous involvements with Sergei Pavlovich dating back to the R-5 missile in the 1940s, but who worked directly for Chelomei at one time, spoke with conviction: "It is absolutely correct that Korolev was the founder of cosmonautics in the Soviet Union. After [his death] in 1966 there was no equivalent to Korolev."[35]

The fact is that the range of Korolev's work was much more sweeping than any of the others, covering not just rocket engines, as Glushko's did; or cruise missiles, launch vehicles, and space stations, as Chelomei's did; or mostly ballistic missiles, as Yangel's did, but the whole gamut of rocket engines, cruise and ballistic missiles, satellites, spacecraft, lunar and planetary probes, and launch vehicles. Furthermore, Korolev did much more than anyone else to gain support from the Soviet citizenry and the political leaders for initiating ballistic missile development and space exploration, even though he had to remain anonymous personally.

Glushko, Chelomei, and Yangel may well have resented the huge swell of postmortem glory accorded Korolev—burial in the Kremlin wall, statues in various space centers, museums, streets in his name, publications galore.

But this book provides persuasive testimony that the dominant place given Korolev in Russian space history is richly deserved.

2

Growing Up in Ukraine: Broken Family, Bolsheviks, Gliders

THE MAN who would lead the world into the Space Age was born in modest circumstances, into a family that would be called dysfunctional today. Sergei Pavlovich Korolev's birth took place in Zhitomir, a small city in Ukraine near Kiev, on December 30, 1906 by the old Russian calendar, which is January 12, 1907 in the new calendar. His mother, Maria Nikolaevna Moskalenko, from an old Cossack family, had contracted, through parental pressure, an unhappy marriage with Sergei's father, Pavel Yakovlevich Korolev, a teacher of literature in the local gymnasium.

"Serezha was going on three years old when our family fell apart . . .," wrote his mother decades later. "I moved him from Kiev to my parents in Nezhin [about 100 miles northeast]."[1] Apparently Maria Nikolaevna was far from forthright with her son about the vital facts, because for years Korolev thought that his biological father had died at the time of the move. But many years later he received a letter from a heretofore unknown stepsister, Yevgeniya Zenchenko-Rshevskaya, the child of his father's second marriage, telling him that his father had lived until 1929, when Korolev was a grown man of twenty-two. In fact, Pavel Yakovlevich wrote to his former wife just before he died, requesting an opportunity to meet his son, but the letter never reached Sergei. Gyorgi Vetrov says that Korolev's second wife, Nina Ivanovna, once told him that her husband said, "If I had known [about the letter], I would have crawled on all fours [to him]."[2]

One of the reasons for the marital breakup, according to Golovanov, had been Maria Nikolaevna's determination to enroll in a course of education for women, which her sister Anna was taking. She finally decided to live with Anna, sending Sergei to his grandparents in Nezhin. This was very upsetting to her husband Pavel. Golovanov writes of an incident, which may not have been recalled objectively by Maria Nikolaevna, that Pavel "was beside himself. He begged, entreated, and suddenly began shouting. Once he ran into her room utterly beside himself, the whites of his eyes shining, and threatened and demanded that she return. 'Try and understand,' she said in a quiet, almost tender voice, 'I shall never return.' "[3]

Zenchenko-Rshevskaya has another view. She argues that the real cause for the breakup was that after the death of Pavel's parents he had adopted two of his young sisters and the whole family sank into financial difficulties, thereby severely stressing the environment, which had not been a loving one in the first place.[4]

In any case, Maria Nikolaevna and her son moved to Nezhin. There, as his mother remembered, little Serezha was "surrounded by the attention of his grandmother and grandfather. He was everyone's favorite, this little fellow with the long golden locks and the dark big eyes, which one time evoked an unexpected remark from an oculist professor: 'What clear eyes you have, little boy! You will be a good student . . . yes, you will learn well!' "[5] She continued:

> He liked my stories. . . . We "flew" together on a fairy-tale magic carpet and saw Konek-Gorbunek [the humpback pony], the grey wolf, and many other miracles beneath us.
>
> This was enthralling and wonderful. My son would press close to me, would look wide-eyed into the sky, while the silvery Moon peeked out from amidst the small clouds.
>
> . . . aside from toys he received many books. He memorized things well. There were no other children his age around. The neighbors had no children, and he didn't know the noise and worry of a group of children. Playing alone, he had to show initiative, inventiveness, and learn to think independently. His failures tempered his force of will. He was not supposed to cry.[6]

Serezha learned to read at an early age from his grandfather's newspaper, and his preschool teacher said that he had an excellent memory and was very good at arithmetic.

But besides being fatherless, he was often motherless as well, because after a year Maria Nikolaevna resumed her courses in Kiev, and returned home only on weekends. "Everyone loved him," Golovanov writes, "but he missed the love of his parents at a time when he needed it most. He was always neatly dressed, always excellently fed, and always lonely, and nearly always sad . . . but in time he grew used to it and did not suffer from loneliness . . . He spent hours sitting before a big box of bricks that Uncle Yuri had brought him from Lodz, and a whole town with a tall cathedral, big-pillared houses, and shops and bridges would rise in grandfather's bedroom."[7]

When he was six years old it was announced in Nezhin that the famous pilot Sergei Utochkin would perform an airplane flight at the local fairgrounds.[8] It is impressive to realize how quickly the technology of flight had reached the hinterlands of Russia. This was 1913, only ten years after the Wright Brothers flew at Kitty Hawk, and only five years after their first public European flights in France in 1908. But Utochkin by then had already become renowned by making barnstorming flights all over Russia. Golovanov described the scene vividly:

> The town seethed with excitement . . . nobody grumbled . . . about the cost of tickets—one whole rouble. By three o'clock in the afternoon on the day of the flight the fairground, where the biplane had been brought from the railway station, was surrounded by people without tickets. Urchins clustered on the trees and on the roofs; soldiers of [the] 44th Artillery Brigade, which was stationed in Nezhin, sealed off the grounds for decent folk.
>
> Sergei arrived at the fairground with his grandfather and grandmother. . . . Sitting on his grandfather's shoulders Sergei saw a short, resolute, ginger-headed man walk toward the biplane, and climb into it, and heard him shout something loudly and sharply to a soldier standing by the propeller. The soldier turned the blade, the aeroplane started rattling and shaking with twenty or so soldiers holding its wings and tail. The cloud of yellow dust raised by the propeller swept toward the boaters and parasols of the well-dressed public. . . .
>
> The engine was warmed up for about half-an-hour. At last Utochkin waved his arm, the plane roared wildly, the dust rose in a whirlwind so that Sergei could hardly make out the shapes of the soldiers in the yellow cloud. Then the aeroplane began rolling jerkily over the ground at ever increasing speed. The soldiers ran with it for a while holding the wings. . . . And then the plane flew! Sergei saw how it hopped at first, hitting the ground lightly with its wheels, then rose . . . and flew! . . . Listing a little the plane gained height and was already some fifteen metres up.
>
> Utochkin flew a kilometre or two and landed in a field near the retreat buildings of the nunnery. The crowd rushed to cheer the hero, and Sergei and his grandfather and grandmother went home.[9]

The impact of the early days of airplane flight was affecting people—especially young boys—all over the world. I remember my own father telling me of the excitement that enveloped Jersey City when the newspapers announced that an airplane would be flying down the Hudson River on such and such a day. It might have been in 1908, when my father was twelve. The route was chosen because authorities wanted to be sure that a crash, if it were to occur, would be in the water and not on the land. Thousands of spectators lined the New York and New Jersey sides of the river to see the new phenomenon. In my father's case it was an unforgettable event. In Sergei Korolev's case the impact was career-forming. As a young man he would become a pilot himself.

It would be some time, however, before his next exposure to aviation. In the meantime, he was often deep in loneliness. "I did not have any childhood," one of his relatives remembers Sergei complaining.[10] He lived in Nezhin until he was almost eight and, as his first teacher,

Lidia Mavrikievna Grinfeld, recalls, "He didn't have any child acquaintances of his own age, and never knew child's games with little friends. He was often completely alone at home, and when I arrived from the gymnasium he would cry: 'Is that you, Lidia Mavrikievna? I'm glad that you've come.' . . . They didn't allow him to run in the street. The gate was always bolted. He would sit a long while on the upper cellar door and watch what was happening in the street."[11]

Little Serezha was a good student from the beginning. Grinfeld recalls that "He accepted instruction avidly. . . . He was attentive, industrious, and capable. He especially loved arithmetic, and could already count to a million. He quickly mastered the four rules, and the multiplication tables, first in rows of one hundred, and then in thousands. . . . He soon mastered writing, already rather skilfully, and wrote short dictations spoken to him. He really loved reading, and would avidly recall what he had read. I remember that he liked the fables."[12]

But difficult times were ahead for the seven-year-old. In the summer of 1914 the tsar's army mobilized to fight Germany as World War I began. At the same time, there was great social unrest in Russia, including in the Kiev area. Workers went on strike and, as a result, were conscripted into the army. Their places were taken by German prisoners of war. The war fared badly in the summer of 1915 and it was difficult to believe that " . . . not so long ago, [there had] been quiet green Nezhin, tea-parties behind closed shutters, and the excitement of Utochkin's flight. . . . No one had much time for Sergei."[13]

One consequence for young Sergei was the emergence of a streak of obstreperousness. Margarita Ivanova Rudomino, a niece by marriage of Sergei's uncle Yuri Moskalenko, recalls a summer interlude at a camp operated by Uncle Yuri on the Dnieper in 1916. " . . . since I was the oldest [of the children, at sixteen] I had to look after the boy, but I wanted to be with . . . people my own age . . . Serezha seemed a difficult charge to me. He was stubborn, persistent, and argumentative. I was no authority for him, and in general he was not distinguished by his good behavior."[14]

The same year, 1916, however, brought one Grigory Mikhailovich Balanin into the picture, and Sergei's behavior improved substantially. Maria Nikolaevna had met Balanin, a young electrical engineer, while she was studying in Kiev. He had received his diploma in Germany but needed a Russian engineering degree, so he had enrolled in Kiev Polytechnic Institute. After divorcing Pavel Korolev, Maria Nikolaevna married Balanin in November 1916, when Sergei was nine. Balanin would turn out to be an excellent stepfather, and a considerable influence on his new son, counseling him often on his manners and study habits.

In 1917 Balanin got a job in Odessa with the Southwestern Railway and later was appointed to an important position as head of the port power station. The family occupied a nice two-story house with a balcony overlooking the sea. The stability of the new family was especially important because Odessa became a very turbulent place.[15]

As 1918 began, arguments raged between factions representing the Bolsheviks, the Provisional Government, the Odessa town government, and a group of Ukrainian counterrevolutionary separatists, with soldiers and sailors supporting the Bolsheviks. Sergei experienced the shutdown of his school and he was kept at home when shooting broke out in the streets. Eventually the new Soviets prevailed and the schools were opened. But the calm was short-lived. In a few weeks Austrian and German troops moved in and took over the city.

One wonders how the eleven-year-old Serezha could concentrate on books during that summer of 1918, but he did, "drinking in everything, and taking everything to heart. . . ." His appetite was eclectic, as he consumed a geometry textbook, manuals on the strength of materials, some Chekhov, and volumes by the German poet-novelist Wilhelm Hauff and the French anarchist-balloonist Élisée Reclus.[16]

The Germans cleared out in November, but in the same month a British destroyer showed up. Then came a contingent of Serbs, then a French landing party. Balanin characterized the situation as "New bandits come to take the place of the old ones." There were shootings between the Bolsheviks and the counterrevolutionaries, and the city experienced a severe food shortage during the winter. The Balanin family suffered along with everyone else, and to make matters worse Sergei got typhus, although he was better by spring 1919. The French sailors mutinied, then all of the French left Odessa. During the summer and fall the counterrevolutionaries continued to fight with the Bolsheviks but the Soviets prevailed by February 1920.[17]

The subsequent period was "a time of hunger, rationing, and wild winter nights on the streets of Odessa," in the words of a young newspaper reporter quoted by Golovanov, who then speculates that Korolev's character might have been profoundly formed by his first-hand experiences. "His concepts of good and evil, of courage and cowardice, of justice and injustice . . . grew from what he saw before his eyes." He could have seen personally the exploits of the heroes of the civil war as they were brutalized and beaten by the "White-guard butchers," says Golovanov, so that in later years "people marvelled at Korolev's exceptional faculty to see the essence of a man. Was it not there, in Odessa, that the roots of this rare talent are to be found?"[18]

Well, maybe, but Sergei Pavlovich would have plenty of opportunity in later life to have his character formed by his own brutalization at the hands of such "heroes" of Bolshevism as Nikolai Yezhov, Lavrenti Beria, and Josef Stalin himself, albeit indirectly in the last case.

It took some time for Odessa to get back to a state of normalcy in the early 1920s. N. Y. Abezguz, a schoolmate during Sergei's teenage years, recalled that

> This was a menacing time . . . the hunger of 1921 had not yet been forgotten. Apartments were heated with cast iron stoves. Often they were fueled simply with household furniture. Electrical lighting was still greatly limited. Kerosene lamps seemed a luxury, and we had to make do with homemade carbide or oil lamps. The water lines were always breaking. On the streets you could hear the click of the "woodies" of the Odessa residents—sandals with soles made out of two boards held together with strips of leather. But we, the young people, knew that all of these things were "life's trifles"—trivia next to the bright future. After all, the Land of the Soviets was changing over to a peacetime mode. Reform in the system of public education was being implemented. Pay and private gymnasiums [high schools] and commercial schools were eliminated, and the separate education of boys and girls was abolished. The schools were undergoing polytechnical transformation and vocational instruction was being introduced.[19]

Abezguz and Korolev were among the first students to undergo this new vocational training. Their school, Stroyprofshkola (Professional Building Construction School) No. 1, was housed in what had been a private girls academy, the Department of Empress Mariya. It must have been an unusual vocational institution, indeed, because, although Abezguz trained to be a plasterer, and Korolev a roofer and tile layer, they also had as faculty "the city's best pedagogues . . . which to a certain degree attracted young people to our school. I can recall the physicists A. G. Aleksandrov and V. P. Tverdyy, the mathematician F. A. Temtsunik, the brilliant erudition of the writer B. V. Lupanov. The instructor of art and drafting, artist A. N. Stilianudi,[20] took us to the land of the wonderful."[21]

That hardly sounds like a vocational school, a fact supported by Abezguz's statement that the "vocational training had no clearcut guidelines" and even included courses by Aleksandrov on resistance of materials and on mathematics and physics "studied to a greater depth than in the former gymnasium." Certainly there is no counterpart for such an educational institution in the United States, nor, for that matter, in the Russia of today.

Although schooled in the academics, practical skills were strongly stressed, with Korolev doing especially well at carpentry. Groups of students even got "on-the-job training at repair work sites. We painted," says Abezguz, "with iron oxide the iron sheet metal roof of the main building of the Odessa Medical Institute and the intricate portholes on the roof of the building of one of its faculties . . . This was not enough for Korolev, and he at his own initiative finished 'in walnut' the wooden door of the institute's main building, obviously after consulting with the master-instructors."[22]

Abezguz says that Korolev was "the most mature . . . energetic and independent among us" and seemed to be guided "by a clear goal in life." He resented, moreover, those school activities that "distracted him from working in the port's glider club and in the province aviation sports section, and from independent study of the theory of flight."[23]

Sergei's interest in flight, it seems, was already full blown, and he was still only fifteen or sixteen. One factor that had fanned his excitement about aviation was the proximity of a military seaplane detachment in the port of Odessa, "stationed there," as his mother wrote, "to protect the coastal boundaries. We really could watch the seaplanes flying close by." Sergei, an excellent swimmer, would swim the considerable distance over to the jetty next to the seaplane operations and, as Maria Nikolaevna remembers, "hang onto the barbed wire for hours, as if mesmerized, watching with interest what was going on there. Once a mechanic shouted to him: 'Well, what are you hanging around for? Why don't you give me a hand? Can't you see I'm having trouble with this motor?' That is all Sergei needed. He quickly crawled under the wire. Soon they got used to seeing him around in the detachment. . . . "[24]

A major outlet for young Sergei's aviation zeal was created in 1923: the Society of Aviation and Aerial Navigation of Ukraine and the Crimea (OAVUK). Golovanov describes OAVUK as the "republican" branch of a much larger organization—the Society of Friends of the Air Force. It is interesting to note how thoroughly party-oriented even such an organization became in the early years of the USSR. SFAF's national council, says Golovanov, included such party stalwarts as Mikhail Frunze, M. I. Kalinin, Anastas Mikoyan, G. K. Ordzhonikidze, Kliment Voroshilov, and Stalin himself. Its formation was recognized by the Thirteenth Party Congress, one of its squadrons even being named for Lenin. It grew to over a million members in just one year.[25]

Golovanov seems to indulge in outright Soviet indoctrination as he interprets the significance of the new "passion for flying." It was not, he writes, " . . . simply youthful enthusiasm. It originated from the admirable conviction that a free people must and can overcome immemorial and humiliating backwardness in every field, and do it quickly. And this feeling was reinforced by a clear understanding of the need to strengthen the defences of the young republic. On posters were blazoned: 'Let us answer the imperialists' mad development of armaments by new squadrons, built by the workers and peasants of the Soviet Union, friends of the Air Force.' "[26]

Whether Sergei needed that kind of pumping up is unlikely. In any case, his involvement with the Odessa hydroplane squadron became a passion. There were eight patched-up plywood M-9 biplanes, designed by D. P. Grigorovich. The squadron's commander, Alexander Shlyapnikov, a veteran of the storming of the Winter Palace, gave young Korolev his first flight. "They flew out beyond the breakwater, climbing against the wind, the engine roaring . . . Sergei saw tiny people, toy ships. . . ."[27] He said nothing about the experience to his mother until one day, when she remarked about the beauty of the clouds, he responded with, "Oh, mother, if you could only see them from the top. . . . It was clear that he had already been flying."[28]

In the summer of 1923 he flew often as a passenger. On one occasion he got a full measure of thrills when, flying in a biplane with the pilot Brzhezovsky, the engine cut out. While Korolev was doing a wing walk to check the oil supply, the plane shook violently and he fell into the sea, fortunately from an altitude of only about 10 meters. Brzhezovsky landed, picked him up, and got him to shore.[29]

Such physical feats were not unnatural to the young Sergei, who was an accomplished gymnast. One of his classmates, L. A. Aleksandrova, especially remembers, "in the years 1923–24: this figure of a stocky guy, charging the length of the school corridor at recess, on his hands, feet up. He practiced this with surprising stubbornness, and before long he could go on his hands the whole corridor, a distance of a few tens of meters. He fashioned some wooden grips for his hands as a convenience. Then, shifting his center of gravity from one hand to the other in order, he could travel along the school corridor to the amazement of all his comrades and teachers."[30]

Besides his physical prowess, Sergei was cool in a crisis. One demonstration of that characteristic is recalled by Margarita Rudomino. In the summer of 1923 she was again vacationing with the Moskalenko family, including the sixteen-year-old Serezha, on the Dnieper River. Unlike her earlier experience with the difficult nine-year-old, "I was interesting to him at that time," she wrote, "as a Moscow resident [during] the first years of Soviet rule. . . . He never left my side, asking about everything. . . ."

Margarita, Sergei, and Sergei's youngest uncle, Vassily Moskalenko, whom Margarita would marry the following year, would sometimes go for a boat ride on the Dnieper in the evenings. Margarita recalls that

> It was in Kanev, where Shevchenko[31] is buried. At that time the wonderful monument marking his grave had not been put up yet, but . . . there was a huge and uniquely magnificent wooden cross on the hill where the remains of the poet had been laid to rest. . . . On that evening there was only one old boat on the shore, possibly one which wasn't being maintained, and as always (Serezha at the oars and I at the helm), we set off for the opposite shore. It was a wonderful evening—warm, quiet, beautiful, calm, and a good feeling in the soul. We floated along, laid on the sand, and set out on the return trip. The Dnieper was wide. It had gotten dark . . . and there wasn't a single other boat on the river. . . . We were a little tired. . . . Suddenly we noticed that there was water in the bottom of the boat, and it was filling up more and more. . . . Vassily Nikolayevich took the wheel, and I began to bail water with some rusty bucket . . . but despite all my efforts the water continued to rise. At first we laughed and tried to scare each other that we were sinking. . . . But with each minute the water . . . was rising. We took off our shoes and got quiet . . . it was clear that this could come to a bad end. . . . (I) knew how to swim (but) couldn't tread water for a long time. . . . Vassily . . . couldn't swim at all. And suddenly Serezha took command . . . he yelled: "Don't panic! . . . Listen to me . . . I'll take Uncle Vassily and deliver him safely to shore . . . no one grab me by the neck, arms or legs. You, Margarity, if you get too tired, can hold on to my hand lightly. In any case, we'll swim side by side. . . . The main thing is to be calm. Don't do anything without my signal." . . . So we continued to float and with faint heart waited for his signal to jump in the water. I continued to shout . . . at the top of my voice: "Help, help, save us!" And suddenly, as if by magic . . . appeared from around the bend . . . a buoy keeper who was lighting fires on the river. He quickly floated over to us. Just as quickly, we got into his boat, and no sooner had we gone off 5–6 meters than our poor little boat sank before our very eyes, forming a funnel in the water. . . . That is how Serezha saved us from a catastrophe . . . with his calm manner and decisiveness.[32]

It is not surprising that Rudomino would remember the incident many years later, since she would certainly have observed how the Chief Designer called on the same qualities in his high-pressure job.

During these years not all was harmonious in Sergei's life as his relationship with his stepfather Balanin became discordant. Balanin was concerned about the young fellow's preoccupation with flying and was strict about monitoring his academics. When Korolev's mathematics teacher complained about his lack of concentration on study, Balanin lectured him, "Education and a profession will make you independent. . . . Remember that. Handstands, yachts, aeroplanes are all rubbish. . . ."[33]

But Sergei's interest in airplanes was deeply serious. He began to design a glider plane, as an independent project, while at the same time studying for his graduation exams at the vocational school. He had become adept at using drafting instruments and slide rule, and got his design of what he called the K-5 glider accepted for construction by the OAVUK commission. It was 1924 and he was just seventeen years old.

In the same year he became attracted to a classmate who would one day become his wife, Xenia Vincentini. Golovanov describes her as "slender, very pretty, with a plait that hung below her waist, and big eyes." The talk was that the Vincentini forebears, years earlier, had come from Italy to grow grapes. The Vincentini household was a veritable social center led by a pleasant father whom everyone, including his own children, called "Max," and mother Sophia, who liked all the boys and girls who were constantly in her home. Xenia, nicknamed Lyalya, had several talented, witty beaus, so it was not going to be easy for the relatively shy Sergei to win her hand.[34]

The courtship was made more difficult by his intense program at the vocational school, from which he graduated in 1924. His certificate credits him with completion of studies in the elements of political knowledge, Russian, mathematics, strength of materials, physics, labor hygiene, history of culture, Ukrainian, German, drawing, and manual training.[35] And he could lay tiles, a craft which any visitor to today's Russian bathrooms will acknowledge has been sorely neglected. He and nine of his graduating classmates, including Lyalya Vincentini, were assigned after graduation to repair the roof of a medical institute, apprenticed to a professional tiler. Alas, as Golovanov reports, they "worked badly, breaking expensive Marseille tiles, doing things in a slip-shod way, in order to get sacked. They were not in a mood for work. . . . "[36]

In fact, Korolev's mind was on aviation, and his hope was to go to Moscow to enroll in the Zhukovsky Academy, where, among other Russian aeronautical greats, Andrei Tupolev was teaching while designing the Soviet Union's military aircraft. But Sergei was not a qualified military pilot, and the academy delayed in responding to his application for a waiver of that requirement. In the meantime, he heard that the Kiev Polytechnical Institute had started an aviation section, so he decided to enroll there.

He asked Lyalya to marry him but she said no, she wanted to study and she had been accepted at the Chemical and Pharmaceutical Institute in Odessa.[37] So off he went to Kiev. He filled out his application in Ukrainian, and identified himself as Ukrainian, although his father had probably been Russian, and the family spoke Russian at home. Golovanov writes an interesting characterization of the Polytech student body, which for its variety smacks of the post–World War II students educated under the GI bill in American universities:

> Korolev enrolled at the same time as a group of Rabfakists [students from a special evening course of young workers given a crash course at the secondary school level to prepare them for higher education]. . . . Many of them not only had not had the opportunity of studying the history of Greek drama or strength of materials, like Korolev at his Odessa school, but two years earlier had not even been able to read. They were workers and peasants, former soldiers, who had come to students' desks straight from the flames of the Civil War. Here at the Poly they were all mixed together—the "professional" students in the worn student uniform of the tsarist days, wearing pince-nez, ironical, supercilious, and exceptionally lazy; the Rabfak intake, healthy, gauche, still very much in the dark, but seizing books in an iron grip with an indestructible love, or rather passion for knowledge; and . . . callow youths and quite good, already established skilled workers—engine drivers, mechanics . . . also the young sons of Nepmen [those who had profited from the capitalism permitted under the New Economic Program], with the way of Russian merchants, which they fobbed off as aristocratic, a little gang of easy-going, well-fed young men, slicked with brilliantine.[38]

Although Kiev Polytech's aeronautical heritage was not as impressive as that of the Zhukovsky Academy in Moscow, it was not at all insignificant. In fact one of Zhukovsky's own students, a professor named N. B. Delone, had built a biplane glider at Kiev Poly in 1909, only a year after the Wright Brothers had demonstrated their airplane in Paris. Another professor, Alexander Kudashev, who occupied the chair of structural stability, built no less than four airplanes powered by engines of 25 to 50 hp. Among the students turned on by this activity during those years was one Igor Sikorsky. Sikorsky, with two other students—Bylinkin and Iordan—started a workshop where they built airplanes for sale and did experimental work, unsuccessfully, on autogiros. These were still pre-Bolshevik times and it is interesting to learn that the sons of several wealthy sugar refiners also took part in what became an airplane-building craze in the 1909–1911 period.

By the time Korolev had enrolled in Kiev Polytech in 1924 there was, therefore, a sizeable aeronautical activity, including not only professors but students who had been pilots during the war. Korolev, as a freshman, was low on the totem pole, so he was unable at first to join the action, which was focused on building a glider for the Second All-Union Glider Rally in Koktebel, near Feodosia in the Crimea.

In Kiev Sergei lived first with his uncle Yuri, then got his own "nook," more distant from the Institute. He earned rubles delivering newspapers, repairing the Institute's roof, and with other jobs.[39] Although he had little money from home, Sergei, coming from the intelligentsia, had to pay for his studies, although the fees were not large. Workers and peasants, or their children, paid nothing. The children of Nepmen paid high fees. But almost all of the students worked at one job or another and the school's hours were geared to that fact. Classes started at four in the afternoon and finished at ten o'clock at night. There were no exams, although there were numerous pass or fail tests. The subjects covered by the twenty-seven tests that Korolev took in his two years at Kiev Polytech—math, physics, chemistry, mechanics, strength of materials, thermodynamics, machine parts, electricity, architecture and building, statics, and political economy—are not greatly different from what this author studied in the first two years of the mechanical engineering curriculum at Yale in the 1940s. But while the author's nonclass activity was, by Soviet standards, dilettantish—baseball and basketball (he did have to wait on tables for room and board)—Korolev had required workshops in such subjects as "lubrication," and on top of that got practical experience as the fireman on a locomotive.[40]

In the beginning of 1925 still another activity was added to Korolev's load at Polytech—a course on glider construction. Many students tried to join but only sixty were chosen, Sergei among them. Funding for the course was stopped after the first few months but the momentum was so great to continue that the students kept going on their own. They decided to work on four gliders during the summer, one of them a rebuilt craft from the previous year's Koktebel competition, the second an improved version of that, the third a trainer, and the fourth would be a glider that aimed for the record at the next competition. Sergei Pavlovich, as a first-year student, was not picked to work on the record-aiming craft, so he concentrated on the trainer.[41]

The high point of that experience was actually flying the finished aircraft. Sergei's first flights were short, merely tens of yards after launch from a catapult, and on one of them he crashed into a water pipe, cracked his ribs, and spent two days in bed. But his elation at actually rising into the air was intoxicating. His big disappointment, however, was in not being chosen to go to two upcoming competitions, first Germany, then Koktebel. Sergei's fellow pilots won three silver cups for the

Soviets in Germany. Then at Koktebel in October one of them, Konstantin Yakovchuk, set a Soviet record by staying up for more than 9 1/2 hours in the rebuilt glider. Soon after the graduation of Yakovchuk and several of the other leaders, the glider activity fell apart.

Back to the books went Korolev, with classes sometimes going until midnight. His loneliness was mitigated somewhat by letters from Lyalya and from his mother, who had moved with Balanin, in the fall of 1925, to Moscow. He finished his tests at the end of his second academic year, in June 1926, and decided that he, too, should head for Moscow.

On July 29, 1926, he notified the rector of Kiev Polytech that he had accepted admission to the Moscow Higher Technical School (MVTU).

3

To Moscow: Tupolev, Tukhachevsky, First Rockets

All over Moscow there are tsarist-era buildings that have been converted to Soviet, now Russian, institutions. They have not all aged gracefully. On many of them the exterior stone has blackened and the entry steps have crumbled, compromising their dignity. Inside, the corridors are poorly illuminated, and there is a pervading aroma that seems to come from lavatories not faithfully tended. But many of these buildings have rich histories and function in important roles. One of them is the prestigious Bauman Institute, formally known as Moscow State Technical School (MGTU), and before that the Moscow Higher Technical School (MVTU). This was Sergei Pavlovich Korolev's next stop in the educational process, starting in the fall of 1926. The structure was created in the middle of the eighteenth century as the palace of one A. Bestuzhev-Ryumin. Golovanov writes that a visiting member of the Polish royal family, one Stanislas Poniatowski, once said of the building that, " . . . in all of Europe there is none like it for luxury and decoration." In the 1870s, as Russia realized its need for engineers, it had become the Imperial Technical School, but, the name Bauman Institute came thirty-five years later, and stuck, after an incident described by Golovanov:

> . . . however thick the walls of this house were, they could not shelter its inmates from the mutinous winds of the twentieth century. Here, in 1905, a crowd seethed and boiled with anger, and vows of wrath sounded in the speeches over the body of a handsome man, still quite young, with a pointed ginger beard. They had brought him here, to the draughtsman's hall, already dead, with his head smashed in, and the feet of the hundreds of people who came smeared the drops of his blood on the tiles, grey with age. This was where his funeral began, a political demonstration thousands strong, never before known in Czar Nicholas' empire, a procession terrible in its undisguised fury. His name—Nikolai Bauman—rang across Russia like a tocsin summoning people to battle.[1]

It took some time during the post-Revolution years before the new Moscow Tech was integrated with the sons of workers. One of the professors had refused to recognize the Bolsheviks, regarding them as vandals. Golovanov describes the student body as a mix of "thirty-five-year-old men in black greatcoats" and "former Rabfak evening students, grim fellows in faded Russian blouses and greasy caps, some of whom had only learned to read a few months before."[2] Korolev was in neither group, but entered the words "middle class" on his application where it called for "Former Social Status of Parents." In 1927, the year after his enrollment, only 13 percent of the graduates were workers' children and only 4 percent members of either the Communist Party or of Komsomol, the Young Communist League.[3]

Korolev's interest at Moscow Tech, however, was not politics but aviation, and he plunged into his studies with vigor. A strong aeronautical heritage had been started with the arrival at Tech of one Nikolai Yegorovich Zhukovsky (1847–1921) in 1872. Zhukovsky built one of the world's first wind tunnels there in 1902, a year before the Wright Brothers flew at Kitty Hawk, and in 1910 he set up an aeronautical laboratory.[4,5] Andrei Tupolev, who would become an important figure in Korolev's personal future, had graduated from MVTU himself in 1918, only eight years before Korolev's arrival. Tupolev eventually designed many of the Soviet Union's military and civil aircraft over a period of more than fifty years. He was one of Sergei's lecturers at MVTU, as was V. P. Vetchinkin, who received Russia's first degree in aircraft construction. Vetchinkin's "simplicity and spontaneity" particularly impressed Sergei, writes Golovanov, citing an instance when he encouraged a student who had been tossing paper airplanes to keep at it in order to " . . . see what makes them fly."[6]

As a third-year student, the nineteen-year-old Sergei Pavlovich was no longer studying fundamental subjects like physics and mathematics, but practical aeronautical engineering—flight mechanics, aerodynamics, aircraft and engine design and construction. This is more specialized than what a student at the same level in the United States would have been studying. The syllabus at Massachusetts Institute of Technology for 1929 shows that aeronautical engineering students took a single aeronautical course—it was called "Theoretical Aeronautics"—in junior year, and their first airplane design course in the second semester of that year. The other courses occupying MIT juniors were either fundamental—applied mechanics, thermodynamics, structures and materials—or they were in the humanities—political economy, plus one language. The fourth-year curricula at both Moscow Tech and MIT were more focused on airplanes and engines, although MIT seniors were required to take physical chemistry and electrical engineering courses as well.

The lectures at MVTU were not compulsory (they weren't at MIT either) but attending them was so obviously advantageous that most students were present anyway. Moscow Tech, however, offered a practical day-to-day exposure to aeronautics that was not available to MIT students.[7] MVTU's nearby neighbor was the Central Aerohydrodynamic Institute (TsAGI), where many of Korolev's professors were also directing active programs. The students, therefore, inevitably became involved in their projects. "None of them," said Golovanov, "could draw a line between their work at the Institute and teaching at the Moscow Tech. . . . From the first year almost everyone was designing and constructing something."[8]

Sergei lived with his parents on 38 Alexandrovskaya (now Oktyabarskaya) Street. His room was also the family dining room, and that, of course, was not an optimum arrangement for a student who not only had to study hard but who wanted to work on aircraft design at the same time. Margarita Rudomino, by then married to Sergei's uncle Vassily and living in Moscow as well, remembers how difficult it was for all concerned.

> In the winter when we came to visit the Balanins, we often found young people drafting something on boards on the big dining room table in Serezha's room, which doubled at first as the family's dining room. They would discuss and argue, paying no attention to us . . . when dinner time approached, Maria Nikolaevna would ask Serezha to clear off the table. "Just a minute, another minute" . . . but this "minute" lasted too long, and Maria Nikolaevna would lose patience and set the small collapsible table in the room where we sat. Among Serezha's friends at that time I remember Yuri Pobedonostsev and Petya Flerov, whom I met at Serezha's many times later. . . . Serezha often complained to me that not many people know or understand his work, that few people support or value it, and that many do not believe in its future or in its success. Often, he said, it is necessary to hit one's head against a brick wall, and it is not always possible to achieve a positive result. His enthusiasm, his devotion to his ideas, and his constant struggle for bringing them to life were dear to me.[9]

Even today in Russia, university students often have to live at home with their parents, or if they do have dormitory accommodations they are hardly comparable to what American students have. A few years ago I hosted in Princeton, where I live, a group of professors from Moscow Aviation Institute. On a tour of the Princeton University campus we encountered a young graduate student who happened to be studying aeronautical engineering. The Gothic buildings, with bell tower and courtyards, and their setting next to the beautifully landscaped University golf course, suggest Hollywood's version of what a college should look like. I asked the student if I could show my Russian guests his room. "Of course." It was impressive—decent furniture, an orderly bedroom, a large living room that featured a big fireplace and a modern computer terminal.

"Is it a university computer?" asked one of the Russian professors.

"No, it's mine."

"Yours alone? How many students live here?"

"Just me."

"Only one? In Moscow these rooms would take at least six!"

Korolev added to his academic load a commitment to work with the members of a local group of amateur gliding devotees under the aegis of an aeronautical society[10] called Osoaviakhim (an organization with a spectacularly omnibus name that translates roughly into Society for the Promotion of Defense, the Furthering of Aviation and the Chemical Industry).[11,12] There was gliding all through the winter of 1926–1927, even when the temperature dropped as low as –26°C. In the spring of 1927, Sergei, now all of twenty, began to design a light airplane with a partner, Sergei Krichevsky, under the auspices of an organization called AKNEZH, the acronym for the Zhukovsky Academic Circle, a student scientific society. But the two argued and the partnership split up after a short time.[13]

Sergei finally got to his first Koktebel glider competition in the Crimea in the fall of 1927, and found the actual flying of the gliders a heady experience.

> . . . he got satisfaction not only from the feeling of oneness with the machine . . . but also . . . understanding its behaviour . . . why it tilted to the right, why its wings lurched, why it nosed down. The span and profile of the wing, the lift coefficient, the density of the air, all the symbols in the formulae, all the figures in the calculations became realities in the air instead of abstractions and from being dead on paper came alive here up in the clouds.[14]

It was around this time that the more mature Sergei switched rooms with his parents, getting the bedroom while they occupied the dining room-living room. He was now mustachioed, and began wearing a suit and tie. He was to live in this room for some ten years.

As hundreds of trade union members and party members were enrolled in the school in a national effort to speed up the country's output of engineers, 1928 saw a politicization of MVTU. Towards the end of the year Korolev became determined to design his own glider and have it ready for the Koktebel competition scheduled for the following October. He got as a partner a classmate, Savva Lyushin, who insisted that they go to work that very evening.[15] Korolev was determined that the glider would be reliable even if the aerodynamics were not optimum.

It would be named *Koktebel,* in honor of the competition. During that summer he and Lyushin worked intensively on their design, interrupted only by a momentous event in early August when Korolev soloed for the first time in an ancient Avro 504K biplane. He continued to fly all that month.[16]

At the October 1929 competition, he flew his own glider, the one that he and Lyushin had designed. He wrote to his mother ecstatically describing the experience:

> I feel a colossal sense of satisfaction and want to shout something into the wind that kisses my face, and makes my red bird tremble, with its gusts. . . .
>
> It's hard to believe that such a heavy piece of metal and wood can fly. But it's enough to leave the ground to feel how the machine becomes alive and flies whistling, answering to the least movement of the controls. Isn't it the greatest satisfaction and reward to fly by oneself in one's own machine.[17]

Korolev's *Koktebel* glider got outstanding praise, evoking this comment from the journal *Aeroplane*: "The intention of the designers, Lyushin and Korolev, was to build a machine very stable in the longitudinal direction that would not fatigue the pilot in long flights. In this they have been fully successful."[18]

Earlier that year, in April 1929, the sixteenth All-Union Conference of the Communist Party had responded to the country's urgent need for engineers by recommending that their education be accelerated. Moscow Tech, consequently, acted to reduce the number of examinations needed for diplomas. Students who had accomplished practical designs could qualify for their diplomas. Korolev was quick to propose the light plane that he had started to design with Krichevsky two years earlier. The project was accepted, with Andrei Tupolev designated as his advisor. Tupolev later described his student:

> Korolev was one of the "easiest" diploma students. I saw at once what he was after, and it was sufficient just to help him a little, to make some minor corrections here and there. . . . He already made an excellent impression on me then as regards both his personality and his talent for design. I would say that he was a man with unlimited devotion to his job and his ideas.[19]

It was a relief to Tupolev that Korolev required little supervision because he, himself, besides teaching at MVTU, was working all hours at TsAGI on the ANT-9, called *Soviet Wings,* a plane designed for Mikhail Gromov, one of Russia's most famous pilots. The aircraft, due to make its maiden flight at the May 1 parade in Red Square, would then be flown by Gromov to Berlin, Paris, Rome, London, and Warsaw.

Sergei defended his light plane design in December 1929, and a few weeks later got his certification as a qualified aero-mechanical engineer.

In late 1929 Korolev went to work full time at Factory 22 in Fili. This was an extraordinary enterprise—an aircraft design bureau headed not by a Russian but by a Frenchman, one Paul Richard. Richard, out of work in France, had been invited to the USSR to take over the failed hydroplane projects of Dimitri Grigorovich. "It was a custom of long standing in Russia to regard foreigners as more clever than the folk at home . . . " writes Golovanov. Renamed OPO-4—the Fourth Experimental Section—the company had been given an extremely ambitious charge. It was to design no less than ten varieties of seaplanes, including a fighter, a reconnaissance plane, and a torpedo bomber. However, the latter, designated TOM-1, became the single focus, and the project occupied a number of engineers who would be among Russia's most brilliant designers, including S. A. Lavochkin, whose LA-7 would be built in a quantity of some 22,000 and become one of Russia's top fighter planes in the Great Patriotic War. Korolev, though, did not shine in the group. His capabilities, manifested throughout his career, were primarily in synthesizing the ideas of others into a sound overall system, not in working on a particular element like a wing or a rocket nozzle. Besides, he continued to be occupied with not one but two personal projects—a glider and a light plane.

The glider, called *Red Star,* he took on with the help of his stepfather, Grigory Balanin. Their objective was to build not a simple sail plane like his earlier *Koktebel* but one which could do aerobatics. They wanted the craft to do loops, a feat performed at that time only in America. Further, the plane should perform the loops unassisted, that is, without being towed to altitude by an airplane. With modest funding from Osoaviakhim, and the help of a few colleagues, the SK-3 *Red Star* was completed, its design having been approved by no less than S. V. Ilyushin, designer of many of the Soviet Union's most famous military and commercial aircraft. The first test in Moscow, however, was a failure, as there was insufficient wind for a takeoff. Korolev was sure that there would be a wind in the Crimea, so he disassembled the glider and took it to the 1930 competition in Koktebel. There the veteran pilot Vassily Stepanchonok electrified the onlookers by doing three loops in the craft, evoking this comment from Ilyushin:

> The loops made by pilot Stepanchonok V. A. on the SK-3 glider deserve recognition as this year's great achievement. The glider is very

important from the point of view of introducing aerobatics into the training program of glider flying and also of equipping gliders with instruments, assessing their qualities. . . .[20]

Alas, designer Korolev could not accept the plaudits personally. He was in a hospital in Feodosia very ill with typhoid fever. The fever led to a severe ear infection and so his mother, Maria Balanina, rushed to the Crimea and took him back to Moscow where he was operated on. He spent months recuperating, so long that he had to be given a temporary disability stipend, but worked the whole time at the drafting board, his head bandaged.[21]

By March 1931, he had recovered and was ready to have his two-seater light plane, the SK-4, tested. It had problems from the start with insulation of some of the leads, with leaks, and with poorly machined parts. The experience may well have fostered the meticulous attention to manufacturing detail that characterized his later designs. After several somewhat successful test flights the veteran pilot, Dimitri Koshits, crashed the airplane when the engine stalled. Although he was not injured, it was the airplane's demise.[22]

Now came a shift in Korolev's technical direction. It is probable that for some time he had been aware of Tsiolkovsky's writings on the possibilities of space travel. In April 1927, Professor Vetchinkin of Moscow Tech had recommended that his students attend lectures on "From Human Flight in the Air to Flights in Universal Ether" by the inventor Gyorgi Polevoi and an investigator of rocket-powered automobiles, Alexander Fedorov. It is likely that Korolev attended, and he might also have participated, in the same month, in the "First World Exhibition of Interplanetary Apparatus and Devices" on what is now Tverskaya Street in the middle of Moscow. The organizers had made a large display of a lunar landscape, with planet Earth on the horizon, anticipating by more than four decades the famous photo of Earth taken from lunar orbit by the Apollo 8 astronauts in 1968. The exhibition, which attracted large crowds, featured models, drawings, and photos illustrating the concepts of Tsiolkovsky, as well as of the American Robert H. Goddard, the German Max Valier, and the Transylvanian Hermann Oberth.

A year earlier, on March 16, 1926, Goddard had launched the world's first successful liquid-fuel rocket, and during 1926 had exchanged some of his publications with Nikolai Rynin, the Leningrad space historian, so it is likely that some of Rynin's copies of the Goddard writings had reached Fedorov for the 1927 exhibition.

Although Korolev surely must have followed all of this, he did not become an impassioned space flight enthusiast overnight. More likely he envisioned the rocket as a means of improving the performance of aircraft. After all, his work had been on gliders and light planes and he would get his pilot's license in 1930. Aeronautics was still his main field of interest. But the idea of putting a rocket on an airplane was both challenging and logical. He would soon get his chance to do just that.

This period, beginning in 1930, saw the first steps toward development of working liquid rocket engines in the USSR—in two different places. One was in Moscow, where the OR-1 engine would be conceived by an organization of rocket amateurs called the Group for Studying Reaction Propulsion (GIRD). It was led by Friedrikh Arturovich Tsander (1887–1933), a man with extraordinarily visionary ideas about the possibilities of interplanetary flight.

The other was the Gas Dynamics Laboratory (GDL) in Leningrad, where a young engineer named Valentin Glushko (1908–1989), working under Nikolai Tikhomirov (1859–1930) on the grounds of Sts. Peter and Paul Fortress, began to build the ORM-1 and ORM-2 (experimental rocket motors) that had electrical ignition systems, expandable nozzles, and various liquid fuels.[23]

The rocket group at Moscow GIRD had been one of four research groups formed within the Osoaviakhim organization, the other three being devoted to sports planes, airplane production, and stratospheric balloons. Tsander, GIRD's leader, rates a very high place in the pantheon of Soviet rocket pioneers. He was born in Riga, Latvia, into an intellectual German family. He lost his mother when he was two. He is said to have been inspired as a child by the stories told by his father, a doctor, about the stars and planets.

He had begun his work on jet propulsion theory in 1908. In that year, at age twenty-one, he had ideas for combining aircraft and rocket capabilities to produce a vehicle not dissimilar to today's space shuttle. He graduated with honors from Riga Polytechnic Institute in 1914, moved to Moscow in 1915, and went to work in the Motor Aircraft Works. However, his interest in space flight was so all-consuming that he left his job and worked for thirteen months independently on the concept of an interplanetary space ship. Evidently his passion was infectious, because his fellow workers at the Motor Works even collected two months' pay to subsidize him. Tsander called it " . . . the first donation to the cause of interplanetary communications." He began writing articles about the possibilities of rocket engines and interplanetary flight. Tsander was one of the first to consider how the minute force of sunlight might be used to accelerate a spacecraft to very high velocity over interplanetary distances.[24]

Although dreaming of a flight to Mars—he wrote *Flights to Other Planets* in 1924—he had begun, in 1929, to actually build a much more basic piece of technology: a rocket engine. He experimented with the little device all during the summer of 1930 in an abandoned German church. Golovanov calls him "the first man in the USSR to take the first steps to turn astronautics into an applied science."[25] The design of the engine, which led to the OR-1, while truly pioneering, was very simple. Golovanov, in fact, calls it "a reconstructed soldering lamp."[26] Gasoline burned in air to achieve a thrust of only about eleven pounds.

Tsander and Korolev had been working together at TsAGI during this period, but they were devoting their after hours and weekends to Moscow GIRD. Korolev's job at TsAGI was lead engineer on development of an autopilot for Andrei Tupolev's TB-3, a heavy bomber. But GIRD became his passion, as it was Tsander's.

Over fifty tests were run on OR-1, and it led to a much more mature engine, the OR-2, developed in coordination with another TsAGI friend of Tsander and Korolev, Yuri Pobedonostsev. OR-2 used gasoline and liquid oxygen as fuels, developed 110 pounds of thrust, and had such sophisticated features as a combustion chamber cooled by the liquid oxygen, and a nozzle cooled by a closed-circuit water circulating system.[27]

Meanwhile, GDL in Leningrad was also working intensively on rocket development. The organization had actually had its roots in 1921 in Moscow, where Tikhomirov, demonstrating an ego that seems characteristic of many of the rocket pioneers—Tsiolkovsky excepted— created the "Laboratory for the Development of N. I. Tikhomirov's Inventions."[28] Glushko writes that Tikhomirov was "another of the remarkably foresighted Russians who was active in rockets before the turn of the century." As early as 1894 he had experimented with small solid propellant devices.[29] Somehow he managed to keep his enterprise working, presumably on an amateur basis, for the next two decades. He apparently knew which revolutionary button to push, appealing directly to Lenin, in 1919, for support, because, as Glushko wrote, "it would promote the build-up and welfare of the young republic of workers and peasants." That initiative resulted in a "go-ahead," although Glushko's report does not explain where the financing came from before 1921. After that date it came from the government, which seems remarkable in the light of the financial stringency of the early days of the Revolution. Not much was offered, however, since times were difficult and significant money was hard to come by.[30] In the 1920s the Laboratory carried out testing of its rocket projectiles at the Main Artillery Range in Leningrad, and in 1925 the whole establishment moved to that city from Moscow.

In 1928 the Laboratory test-fired the first ever smokeless-powder rocket leading to the famous Katyusha rockets that became an important weapon in the Great Patriotic War. It was after those successful firings that the Laboratory was expanded and renamed the Gas Dynamics Laboratory.[31]

This was about time that Glushko came to GDL. Another future Soviet space hero with a Ukrainian connection, Glushko was the son of a Ukrainian father and a Russian mother who was a nurse. Born in 1908 in Odessa, he went to a trade school in that city, learning to be a sheet metal worker while Korolev, by coincidence, was learning to be a tiler in the same city at around the same time although the two did not know each other.

"They gave me a diploma . . . after I had worked as an apprentice for six months at a hydraulic fittings plant, first as a fitter, and then as a lathe operator."[32] While Korolev then became intensely involved in airplanes and gliders, Glushko had already developed a passion for space travel—he had written a letter to Konstantin Tsiolkovsky in 1923, when he was 15 years old.

Tsiolkovsky wrote back asking the youngster how serious was his interest in space travel. Glushko replied, "All I can say is that this is my ideal and my life's goal—I want to devote my life to this great cause . . ."[33]

Decades later, he wrote, "We must remember that in those years rocket technology was not generally accepted. Only amateur enthusiasts engaged in this sphere of technology and they were called 'lunatics.'"[34] In his Odessa years, Glushko experimented with explosives that had been recovered from unexploded artillery shells left by the retreating White Guards. In 1924 and 1925, he wrote magazine articles on the possibilities of exploring the moon, and on the use of the reaction engines proposed by Tsiolkovsky for space travel. Then he went to Leningrad State University.

According to Vetrov, the specialties offered there—optics and mechanics—were not interesting to Glushko, so in April, 1929, at age 20, he left the University, without getting a diploma, to pursue rocket research at GDL.

Tikhomirov had been asked by the "Military Department of the Committee for Inventions for an expert opinion on my proposal for a rocket engine. The proposal was accepted, and a section was set up within the GDL for the development of electric and liquid-propellant rocket engines and rockets, which went to work under my guidance. . . ."[35] Glushko wrote.

Electric rockets more than sixty years ago? Right. Glushko claims that his work at GDL in 1929–1930 produced the "first ever electrothermal rocket engine." Not known for his modesty, he wrote that,

"It antedated the state of science and technology by some thirty years."[36] It is a fact that electric rocket thrusters—the first U.S. versions were developed in the 1960s—are commonly used today to control spacecraft in orbit and in the future may be used to accelerate space ships to the very high velocities required for interplanetary and interstellar voyages.

Meanwhile, at Moscow GIRD a period of intensive experimentation was interrupted by an important event in Korolev's personal life. For several years he had been courting Lyalya Vincentini by mail and by short visits to Kharkov, where she had graduated from the Kharkov Medical Institute, and during brief trips on her part to Moscow. Although she hoped to become a surgeon, as her daughter Natasha would one day, she had gotten a job as manager of the district hygiene center and then as deputy inspector of the district health department. On one occasion Korolev visited her there for several days, although it was in the middle of an epidemic of dysentery and typhoid fever and so she was working constantly. Lyalya, it seemed, was in no hurry to get married, and this made Sergei very angry. Finally, however, she came to Moscow for just two days and they were married on August 6, 1931. Golovanov describes what must have been a singularly unromantic event:

> What he yearned for all these years suddenly came true, quickly and very simply. There was no wedding, or rather there was a very small and short wedding party. Their only guests were Mikhail Gromov and Dmitri Koshits, who congratulated the newly-weds, drank a quick bottle of champagne with them, helped the bride into a cab and rode with her to Kursk station; Lyalya went to Kharkov and from there to the Donbas to make arrangements for her transfer to Moscow. Everything was done rather clumsily and hurriedly, and sadly.[37]

Marriage, however, did not diminish Korolev's dedication to his rocket work. The decision was made, in January 1932, to put the OR-2 on a tail-less glider with a trapezoidal wing, the BICH-11, designed by one Boris Cheranovsky. Even Tsander was bedazzled by the idea of installing it on a real flying machine. Unfortunately, in numerous tests throughout the year and into 1933, the engine was plagued with problems—gas leaks, tank deformation, a faulty feed system, burnt-out nozzles. The intrepid inventor, who had never been robust and who had pushed himself mercilessly during the development, got so sick he was sent to a sanatorium.

Leonid Korneev, one of the GIRD team, described the furious pace of the project in those days.

> I remember that for three days running, we couldn't get a vital test ready. All the members of the group were younger than Tsander, and it was much easier for them to stand such a strain. Seeing that Tsander was very tired and asleep on his feet we presented him with an "ultimatum:" if he didn't go home immediately everyone would stop work. . . . Tsander disappeared. . . . Five or six hours passed, and one of the mechanics shouted . . . "Everything's ready. Raise the pressure, we're off to Mars!" And suddenly everyone was stupefied. A couch that stood at the far end of the basement toppled with a crash, and from behind it appeared Tsander.[38]

Alas, shortly thereafter, both the OR-2 and its designer died. On March 28, 1933, as Tsander lay on his deathbed in the Caucasus spa of Kislovodsk, 1,000 miles away, his OR-2 engine made a final effort in Moscow. It actually fired for twenty seconds but then the nozzle burned out, and the engine became technological history.

It would be another Moscow GIRD engine, the GIRD-09, designed by Mikhail Tikhonravov, that would power the first liquid-fueled rocket to be launched successfully in the USSR. The rocket took off from a test stand in the woods of Nakhabino, about twenty miles west of Moscow, on August 17, 1933, at 7:00 P.M. The rocket weighed 18 kilograms (40 pounds) and had a fuel load of 1 kilogram of "solid gasoline" and 3.45 kilograms of liquid oxygen. Pressure in the oxygen tank was 13.5 atmospheres. The flight lasted eighteen seconds. An excerpt from a GIRD report of the event:

> Takeoff occurred slowly, at the maximum altitude (about 400 meters "by rule of thumb") the rocket went horizontal and hit the adjacent forest on a flat trajectory. The engine operated throughout the flight. The shell crumpled on impact. The change in arc from vertical to horizontal and rotation to a downward direction was the result of a gas puncture [hot spot] in the flange which created a lateral force.[39]

Korolev gave the event momentous perspective on a poster he placed on the wall at GIRD. The poster was placed proudly next to a message of official congratulations for achieving "the first practical results in the mastery of reactive motion techniques" from the "administration of military inventions of the technical staff of the chief of armaments of the Workers' and Peasants' Red Army." Characteristically, Korolev couldn't resist including a dig at his colleagues on inattention to quality control:

> The first Soviet liquid-propellant rocket has been launched. The day of August 17th will doubtlessly be significant in the life of GIRD, and from this time on Soviet rockets must fly above the union of republics.

> The GIRD collective must exert all our efforts so that this year will be reached the performance calculated for the rockets necessary for operation by the Workers' and Peasants' Red Army.
>
> In particular, special attention must be paid to the quality of work at the range, where, as a rule, there are always many discrepancies, incompletions, etcetera.
>
> It is necessary also to master and release into the air other types of rockets as soon as possible in order to thoroughly study and attain adequate mastery of reactive techniques.
>
> Soviet rockets must conquer space![40]

There is a stone marker in the Nakhabino woods commemorating the event. It pays tribute not only to Korolev and Tikhonravov, but to the departed Tsander. Only a few weeks after the flight of the GIRD-09, an upgraded version of Tsander's OR-2, designated GIRD-X, exerting 150 pounds of thrust, was launched from the same spot.

Like most of the early rocket experimenters in the United States, Germany, and Russia, Korolev had some hairy brushes with catastrophe. One of his assistants at GIRD, V. A. Andreev, recalls a mishap that could well have killed him.

> During testing of the OR-2 engine at Nakhabino proving ground . . . fuel was fed into the engine which was on the test stand. There was a hissing noise, but the ignition would not work. And that happened several times. Evidently, more fuel than necessary had accumulated in the combustion chamber . . . on the next try there was a deafening explosion. The tarpaulin which served as an awning filled out and there was a ball of flame. The situation was made worse by the fact that there were tanks of liquid oxygen and two tanks with 50 liters of alcohol near the fire.
>
> Sergei Pavlovich was the first to climb on the roof of the shelter. I jumped after him, and together we began pulling off the burning tarp. Despite the flames the fitters Frolov, Avdonin and Vorobiev ran up to the tanks and tried to remove them. It was impossible to disconnect them, so they flattened the connector pipes with cutting pliers and dragged them to a safe place.[41]

The euphoria from their eventual successes notwithstanding, the GIRD rockets were far less advanced than those that had been launched in America during the previous two years—from 1930 to 1932—by Robert Goddard. Goddard had improved his technology considerably since his first successful liquid rocket launch—which was the world's first—more than seven years before Korolev's success. Launched on March 16, 1926, from his Aunt Effie's farm in Auburn, Massachusetts, that rocket was of a very simple design and weighed less than half the 09 rocket—only 16 pounds (7.27 kilograms), including 10 1/4 pounds (4.67 kilograms) of liquid oxygen and gasoline. However, Goddard's technical description of the launch, as communicated to his sponsor, Charles G. Abbott of the Smithsonian Institution, on May 5, sounds much like Korolev's:

> After about 20 sec the rocket rose without perceptible jar, with no smoke and with no apparent increase in the rather small flame, increased rapidly in speed, and after describing a semicircle, landed 184 ft from the starting point—the curved path being due to the fact that the nozzle had burned through unevenly, and one side was longer than the other. The average speed, from the time of the flight measured by a stopwatch, was 60 miles an hour. This test was very significant, as it was the first time that a rocket operated by liquid propellants traveled under its own power.[42]

The understatement of the final sentence contrasts with the ebullience of Korolev's. In fact, it would be Goddard's reticence to seek publicity that would seriously inhibit the attention given to his achievements. In 1930 he moved his work to Roswell, New Mexico, where, with the endorsement of Charles Lindbergh, he was financed by two annual grants of $25,000 each from the Guggenheim family. With that support the Clark University physicist brought the technology to a new level of sophistication. He built a workshop replete with machine tools, a proving stand, and a launch tower, and bought large metal vacuum chambers for storing the liquid oxygen on site. His rockets were ignited electrically and stabilized by a gyroscope that controlled four vanes, actuated by gas-pressurized pistons at the open end of the rocket nozzle. The Guggenheim grant was then reduced to $2,500 because of the Depression and Goddard returned to his professorship at Clark in Worcester, Massachusetts. Other grants from the Guggenheim family and from the Smithsonian Institution enabled Goddard to resume his work in Roswell but he was not able to interest the U.S. military in his work. It seems ironic, in fact, that on August 29, 1933, only twelve days after the Russians' first successful liquid rocket launch, Goddard received a letter from the Acting Secretary of the Navy, telling him that, "Because of the great expense that would be entailed in development of the rocket principle for ordnance and aircraft propulsion, which under present stringency of funds appears hardly warranted, the Department regrets it is not in a position to further such development."[43] An even more discouraging turndown came seven years later, in 1940, from the U.S. Army Air Corps. Goddard had proposed to develop rockets to assist the takeoff of airplanes at a cost of less than $100,000 per year.[44] The response, from one Brigadier General George H. Brett, three months later, was:

> While the Air Corps is deeply interested in the research work being carried on by your organization under the auspices of the Guggenheim Foundation, it does not, at this time, feel justified in obligating further funds for basic jet propulsion research and experimentation.
>
> At such time as your experiments have reached a point which indicates the probability of successful reduction to practice of a device, capable of being incorporated in or attached to an airplane to assist in accelerating takeoff and upon which an evaluation can be made in order to determine the feasibility and practicability of military application, the Air Corps will then entertain further proposals involving the actual construction, installation, and test of such a device. . . .[45]

Goddard was so deflated by this letter he wrote to a colleague that " . . . after trying to do a good piece of work over a period of years and actually getting flights before anyone else, it is discouraging to have the implication made that nothing of value has been accomplished."[46]

Goddard eventually did get support for the application of rocket-assisted takeoffs to aircraft from the U.S. Navy in 1941. However, from 1932 to 1941 he was able to spend little time on the development of his rocket engines. He wrote that "   most of the work during the past nine years has been concerned with guiding and control mechanisms for sounding rockets, and similar aerodynamic problems. . . . The rocket motor has been used constantly since 1930 without any improvements."[47]

Goddard's pioneering work firmly establishes his credentials for being the father of world rocketry. He went far beyond Tsiolkovsky in the sense that he actually did extensive experimentation to demonstrate his ideas, which Tsiolkovsky did not, and he took out some 200 patents on rocket designs and components.[48] Still, the astigmatism of the U.S. military and his own reclusiveness and abhorrence of publicity kept him largely in eclipse for many years.[49] Consequently, the development of his rockets for weaponry and space exploration—with the exception of his wartime work in Annapolis on the use of rockets for jet-assisted takeoff of aircraft—was never properly followed up in the United States.

Not so with the Germans and the Russians, although both started out years behind Goddard. Johannes Winkler launched the first liquid-fuel rocket in Dessau, Germany, in March 1931. It was a tiny engine, generating only 7 kilograms (14 pounds) of thrust, using methane and liquid oxygen as propellants.[50] The next year Wernher von Braun began work on liquid-fuel rockets, supported by German Army Ordnance, and five years later he would be technical director of hundreds of engineers and technicians at Peenemunde. Development of the V-2 during World War II would follow. It would not be long before the Soviet military, too, showed interest in rockets.

Korolev wrote a bullish article on the future prospects for winged rockets and jet propulsion in the newspaper *Vechernaya Moskva,* on August 25, 1933, eight days after the Nakhabino launch. Under the title "Towards the Rocketplane" and with the byline "Engineer Korolev," he wrote that, "Jet flight vehicles can develop flight speeds of 3600 km/hr . . . and [can attain] immense altitudes [but that] practical resolution of this huge problem requires many years and persistent work." He also sounded a competitive alarm by writing, inaccurately, that "Professor Oberth [Germany] and Professor Goddard [America], after having achieved results with small experimental rockets, were accepted as Colonels of the corresponding armies, and their works are especially classified."[51] He then attributed to Oberth the development of a rocket capable of achieving several kilometers altitude and claimed that "Germany has allegedly built a rocket of such size that it will be able to fly a man." Korolev seems to have been misled, himself, by the exaggerations of sensationalist journalists.

The Russian military man who deserves the credit for first realizing the potential of the work being done at Moscow GIRD had already had his first exposure to rocket development in St. Petersburg—Marshal Mikhail Tukhachevsky (1893–1937). Tukhachevsky had visited GDL as Chief of Armaments of the Red Army in 1930, and had the perception to document the laboratory's "significant and important results," which, he wrote, "creates unlimited opportunities for firing shells of any size and range" and would "eventually lead to the solution of the problems of high-velocity stratospheric flight."[52]

Sometime later he had learned of the work being done at Moscow GIRD and was so impressed he wrote to the Army-Navy Commission supporting the group's experiments. "These efforts [those of Korolev and his coworkers]," he wrote, "in connection with the inventions of K. E. Tsiolkovsky in the areas of rockets and interplanetary travel, have great significance for the military and the USSR as a whole."[53]

Tukhachevsky, already funding GDL, began providing the GIRD group with modest financial support in August 1932.[54] Then, in a 1933 report[55] Korolev wrote that he would require at least 30,000 rubles to follow up the 09 rocket success with a series of launches aimed at achieving flight speeds of 800 to 1,000 meters per second and thousands of kilometers in range. Finding a mechanism for providing that kind of support by absorbing Moscow GIRD formally into the military was a bit more difficult for Tukhachevsky than with Leningrad GDL, which had such roots for some time. He decided that the best course was to combine the two organizations in what would become a new unified center of rocket development. The idea was appealing to both the Muscovites and the Leningraders and on October 31, 1933, the Council of Labor and Defense passed a resolution endorsing the action of the Revolutionary Military Council of the USSR in creating a Reaction Propulsion Institute (RNII, for Reaktivni Nauchno-Isledovatelski Institut).

Ivan Kleimenov (1898–1938), a military engineer from the Leningrad GDL, was named the first director. A graduate of the Zhukovsky Military Academy, Kleimenov had fought on the western front in 1918. The deputy chief engineer, from Moscow GIRD, would be Korolev. The Leningraders moved into the Moscow GIRD building on Sadovo-Spasskaya Street, not far from the present U. S. Embassy, and rocket development work was underway in earnest. The building still stands, marked by a plaque commemorating the GIRD work of six decades earlier.

While it was the prospect for using rockets in military weapons that provided the funding for RNII, a passion for space remained high in the minds of the rocket experimenters. In 1934 several of the group, including Tikhonravov and Korolev, went to Leningrad to participate in the All-Union Conference on the Study of the Stratosphere, convened under the auspices of the USSR Academy of Sciences. It extended over several days and, while it covered a gamut of subjects including aerology, acoustics, optics, atmospheric electricity, geomagnetism, the aurora borealis, cosmic rays, and biological problems, there were substantive papers, as well, on rockets. Some of the works presented seem remarkably perceptive, as reviewed some forty years later by Golovanov.[56] Alexei Likhachev, for instance, pointed out that in the future, "It will be quite advisable to use centrifuges to study the effects of the loads at different accelerations . . . up to 10g in experiments with human beings." He also foresaw the need for building chambers for simulating various altitudes, atmospheric pressures, humidity, and gas composition that space travelers would experience.[57] The participants were surprisingly knowledgeable of the work of Goddard and Oberth. Korolev reviewed in some detail the possibilities and limitations of solid rockets, air-breathing jet engines and liquid-propellant rockets, pointing out the need to develop sufficient power to carry "one, two or even three people . . . as well as instruments and devices for maintaining the crew's vital conditions. . . ."[58]

The space exploration diversion, however, was just that. The primary mission of RNII was the development of military weapons, no doubt in response to the encouragement of Tukhachevsky and his staff. "Work on ballistic missiles in the 1930's was primarily experimental, aimed at accumulating experience in designing and solving certain technical problems including the choice of the optimal configuration, thrust-to-weight ratio, the initial-to-final mass ratio, etc. The effort was unique at the time in its range and scope," wrote Viktor Sokolsky.[59]

Was it, in fact, "unique"? Hardly. In the late 1930s there was already ballistic missile development going on in the United States at the Jet Propulsion Laboratory. The Germans were deeply into work on what would become the A-4, or V-2, having received approval, in 1936, for plans to construct the Peenemunde facility.[60]

In 1934 work began at RNII on yet another approach to long-range weaponry—the winged rocket. About five times the range of ballistic rockets could be achieved by the installation of wings. Winged rockets were also easier to control with the help of autopilots, and work on them helped to accumulate experience for manned flight. However, the achievement of flight stability with these missiles was another matter, as "all efforts to achieve calculated flight trajectories failed" on the first winged rocket, the 06—designed by Korolev and Yevgeni Shchetinkov—in May 1934.[61]

In the midst of such intense development there was not much time for family life. But at the end of the summer of 1934, Sergei Pavlovich, now chief engineer of RNII, and his young wife Xenia, herself busily occupied with medical work, took a vacation in the Crimea with some friends. Xenia may have already been pregnant at the time since, on April 10, 1935, a daughter, Natasha, was born.

"My father wanted to have a daughter and he was sure that I would be a girl even before I was born," Natasha, a prominent lung surgeon in Moscow, told me in a 1991 interview. "He loved Lev Tolstoy and named me for Natasha Rostov from *War and Peace*. He had even told his friends before I was born that he would name me Natasha."[62] Like many young Russian families, Sergei, Xenia, and baby Natasha were not, at first, able to have their own apartment, and they lived with Sergei's mother, Maria Nikolaevna, and stepfather Grigory Balanin, on Oktyabrskaya Street. Then, in 1936, they got their own apartment at 28 Konyushkovskaya Street. Xenia "was always occupied professionally," as Natasha put it, and her father, of course, was putting in long days at RNII on his rocket work.

But the development of rocket engines was not the only technological priority. There were other difficult problems to solve if an operating missile system was to be perfected. Prominent among them was the development of reliable and accurate guidance and control systems. That task continued to occupy the RNII team for years. When young Boris Rauschenbakh came to work for Korolev in 1937, the first thing he was told was, "Well, all you young people necessarily want to build rockets or rocket motors, and believe that they constitute the full matter. . . . We already know how to build rockets and motors, but guidance of the flight and stability of movement have become a 'bottleneck.' "[63] Rauschenbakh, who became one of Korolev's most competent designers of guidance systems, recalled that, " . . . it soon became apparent to all that without the installation of rather complex devices on them, for example of the autopilot type, it would be impossible to expect to obtain any kind of serious results from the launches."[64]

The RNII team eventually developed automatic gyro stabilizers with two and three degrees of freedom—GPS-2 and GPS-3. These devices made possible automatic launching from a trolley and achieved stable flight along the climb trajectory.[65] The guidance and control design team at RNII was headed by Sergei Pivovarov. "However," wrote Rauschenbakh, "this group was engaged purely in design engineering, and Sergei Pavlovich understood very well that the automatic instruments which it developed would only function successfully if they were properly 'adjusted,' and their characteristics coordinated with the aerodynamic characteristics of the rocket and its proposed flight trajectory." In aviation, explained Rauschenbakh, the pilot can override the autopilot, but for a rocket "Everything had to be computed and planned prior to the flight . . . which was the only one."[66]

Rauschenbakh, now an Academician, currently chairs a group in the Academy of Sciences whose formidable name suggests the respect and attention given to the history of science and technology in Russia—the Commission on the Cultivation of the Scientific Heritage of the Pioneers of the Development of Cosmic Space. I interviewed him several times coincident with the annual history sessions of the International Astronautical Federation. On another occasion I spoke with him in his neighbor's dacha in the beautiful village of Abramtsevo, about an hour and a half north of Moscow. It was one of those cold, snowy days on which Russia seems most beautiful. Rauschenbakh is a very dignified man, given to wearing homburgs in formal situations, but on this day he was seated in front of a pleasantly crackling fireplace in his dacha garb—plaid wool sport shirt and jeans. Balding but healthy looking, he has blue eyes and the fair skin that reflects his German genes. "I was one of four engineers who worked every day with Korolev beginning from 1937," he told me. "The others were Shchetinkov, Dryazgov and Kasyatov. Sometimes we worked twenty-four hours straight, but usually we would start at 8 and keep going until 9 or 10 P.M. We couldn't go home as long as he was there. It was exciting, like a sport, the technical challenge was stimulating."[67]

It could also be intimidating under the severely demanding Korolev. Rauschenbakh recalls:

> I was assigned to work on the 212 winged rocket with automatic stabilization. One of the steps in preparing it for flight testing was the blowing of a freely suspended, fully assembled rocket with operating automatic stabilizer in the wind tunnel of TsAGI. This method was supposed to in some measure simulate the determined adjustments of the stabilization instrument with the inertial and aerodynamic characteristics of the rocket. . . . It is not surprising that both I and the team of mechanics sent with me to TsAGI encountered great difficulties in conducting one of the most complex experiments . . . filming the rocket as it oscillated in the air flow. . . . Sergei Pavlovich . . . who always considered it necessary to personally monitor the course of "key assembly" experiments, immediately evaluated the situation and [realized] that if everything continued in the same vein, the experiment would probably never be implemented. . . . Sergei Pavlovich did not reprimand me, but found a few very precise and impressive words, from which I clearly determined my inadequacy and subconsciously felt that if the filming is not performed within the next twenty-four hours, there would be some elemental disaster in Moscow such as an earthquake.[68]

Rauschenbakh solved the problem, to his own great relief and to Korolev's satisfaction, and cites the anecdote as an illustration of Korolev's style of disciplinary management.

In 1936, four years after the failure to fly a BICH-11 glider with a rocket, the RNII team had gone back to work on the design of another rocket-powered glider. This would be the first collaborative effort between Korolev and Glushko. Glushko's ORM-65 rocket engine, fueled by nitric acid and a hydrocarbon, developed up to 175 kilograms (385 pounds) of thrust. It was put on the SK-9 glider and underwent some three years of testing involving several hundred flight experiments.[69]

All in all, the work turned out by RNII from 1933 to 1938 under Kleimenov and his deputy Korolev would prove to be fundamentally important to the future capabilities of the Soviet Union.

It seems incredible that most of the achievements of this group would languish for the next seven years as a result of the Stalin purges that impacted brutally on many of the best of the Soviet Union's scientists, engineers, and military leaders, including Tukhachevsky, Kleimenov, Glushko, and Korolev.

4 THE GULAG AND SHARAGA YEARS

IN JUNE of 1938, Robert H. Goddard, the acknowledged father of American rocketry, received a letter from Harry F. Guggenheim telling him that he would get another $20,000, in support of his experiments in Roswell, New Mexico, for the next year.[1]

During the same month another rocket experimenter, Korolev, had a quite different experience. In the early hours of the morning of June 27, four men—two from the NKVD and two "witnesses"—entered his apartment at 28 Konyushkovskaya Street, near the American Embassy. In a matter of minutes he was taken away. His wife, Xenia Vincentini, recalled years later that she wasn't even allowed to give him a change of underwear, nor was he permitted to say goodbye to his sleeping three-year-old daughter Natasha.

At this point in his career, although only thirty-one years old (more than twenty-five years younger than Goddard), Korolev's work on liquid-propellant rockets was on a par not only with Goddard's but with that being done during the same period by Wernher von Braun and his colleagues in Germany. He was soon to learn, however, that he had been accused of "subversion in a new field of technology."[2]

The testimony that led to his arrest was given by three of his own fellow experimenters: Ivan Terentievich Kleimenov, Gyorgi Erikovich Langemak, and Valentin Glushko. They, too, had been arrested earlier on allegations that they had collaborated with an anti-Soviet organization in Germany. It is fascinating to learn from KGB archives that the Soviets had spies witnessing the rocket tests of von Braun and his colleagues as early as 1937, with regular reports being made by an informer through 1938.[3] Some of these reports went to the very highest echelons, with "delivery indexes marked" for Molotov, Timoshenko, Beria, and Stalin himself.

Was this a classic case of shooting the messengers of bad news? Virtually the entire command staff from German foreign intelligence was arrested, and some were, indeed, shot. Kleimenov had been a member of a Soviet trade delegation to Germany before moving into rocketry. He was arrested on November 2, 1937, pleaded guilty and "admitted that, during his work at RNII he had established a criminal relationship" with Langemak.[4] Both Kleimenov and Langemak were executed. Glushko, who was later to emerge as the leading rocket engine designer in the Soviet Union, and a rival of Korolev's, had been arrested three months before Korolev, on March 23, 1938, and sentenced to eight years in prison.

"That was a peculiar period," I was told by Arvid Pallo, who was working at RNII at the time. "When two guys would say a third guy was bad, that was enough. We lost the gift of speech in those days."

Pallo was eighty-one when I interviewed him. Tall, very lean, with weathered skin, his career in rockets dates back to 1936 and includes some major space milestones. On his fifty-fourth birthday, in 1966, Luna 9, which he helped design, soft landed on the Moon. He recalled Korolev's arrest vividly:

> He had been demoted about a year before his arrest, from chief of a group to a senior engineer. Kleimenov in that time wanted to focus on the development of solid rockets. Korolev wanted to work on a winged vehicle powered by liquid rockets. Some workers wrote to the NKVD that Korolev was holding back progress. . . .
>
> When Korolev didn't show up for work one day the other workers knew that he had been arrested. I had been on holiday at the time. When I got back to work two of the other engineers told me about the arrest. We were all very sad.[5]

Like most of the others imprisoned during the Stalin purges of the late 1930s, Korolev was given no trial. He was beaten and forced to confess to the trumped-up charges of his colleagues. It is not known for certain where his interrogation took place but the confession was extracted on July 27, 1938, at a court commission held under the chairmanship of one Vassily Ulrich.

Korolev's mother, Maria Balanina, could not believe that Stalin even knew what the secret police were doing. She had written directly to the Generalissimo for clemency towards her son a few weeks after the arrest. Her plea, in a telegram, was, "For the sake of my sole son, a young, talented rocket expert and pilot, I beg you to take the necessary steps to resume the investigation."[6] Alas, she was not successful. On September 27, 1938, Korolev was sentenced to ten years imprisonment with "disenfranchisement" for five years and his property was confiscated.

Korolev was then moved from one prison to another, although during this period there is some uncertainty about prisons and dates. It is known that in February 1939, he was in a transient prison in Novocherkassk, near Rostov, and that in that year he appealed personally for clemency to Stalin and to the Supreme Procurator, but to no avail.

It is also known that by October 1939, he was in one of the most dreaded of all prisons, a camp in the Kolyma area of far eastern Siberia, part of the notorious gulag system (gulag is the acronym for Main Directorate for Corrective Labor Camps), made famous in the West by the publication of Aleksandr Solzhenitsyn's *The Gulag Archipelago*. The pathological Stalin was eventually to execute, or send to dozens of gulag camps spread across the Soviet Union, millions of his real or imagined enemies, far outdoing the campaign of paranoia begun by the tsars. Estimates on the numbers vary greatly, and accurate figures will not be available until more archives are opened. The late Dmitri Volkogonov, biographer of Stalin, who was a Colonel General in the Red Army, gives an estimate of 4.5 to 5.5 million arrests in the 1937–1938 period alone, of which 8 to 9 hundred thousand resulted in death sentences.[7] Fellow Communist Party officials, scientists, engineers, industry managers, military officers, ordinary citizens—most of them innocent—were shot or exiled to inhuman prisons throughout Russia. The main reason for the purge was evidently to evoke terror, and therefore docile obedience, in the populace. So the arrest, quick conviction, and execution or condemnation to the gulag of innocent people was part of the strategy.

Conditions in the gulag camps were so bad—inadequate food, shelter, and clothing, brutal discipline and work assignments—that about 10 percent of the prisoners died annually from malnutrition, tuberculosis, or execution, according to Russian historian Roy Medvedev.[8]

One of the women doctors who worked in Kolyma told Korolev's daughter many years later that several thousands of the prisoners died each month. Robert Conquest, in his monumental book on the purges, says that Kolyma "had a death rate of up to 30 per cent [per year]. If we take its average population as 500,000, possibly an underestimate, this camp area alone probably accounted for over 2 million deaths up to 1950."[9] While Korolev survived Kolyma, the privations he suffered caused permanent damage to his health, including the loss of all of his teeth. He apparently also suffered a heart condition and a broken jaw, which would prove to be contributing factors to his relatively early death. But the details on his particular experiences during the imprisonment are not known because throughout his career he spoke to virtually no one about that period. However, just a few days before his death on January 14, 1966, at the peak of his career, in the middle of the competition to beat America to the Moon, as if to get it finally on the record he gave a bizarre account of his arrest and eventual liberation from the gulag to two of his favorite cosmonauts, Yuri Gagarin and Alexei Leonov. Leonov was the first man to walk in space, in 1965, and in 1975 again made history as one of the two Soviet cosmonauts who rendezvoused with three American astronauts in the Apollo-Soyuz mission. His respect and admiration for Korolev are very evident as he recounts this tale:[10]

Yuri Gagarin and I visited Korolev's home just a few days before he died to celebrate his birthday, which wasn't until the following week. [Korolev's fifty-ninth birthday was on January 12. He died on January 14.] At the party besides his wife Nina were Academicians Barmin, Ryazansky, Ishlinsky, Boris Petrov, Gherman Semyonov [the factory manager], a few others. When the party was over and the other guests had gone at about midnight, Gagarin and I were putting on our raincoats. Sergei Pavlovich said,"Don't go. I want to talk."

So Nina Ivanovna [the second Mrs. Korolev] got some more food and for the next four hours, over a bottle of cognac, we talked . . . until 4 A.M. Korolev opened up about his arrest and his experiences in Magadan, in Kolyma, in the gulag. It was deeply psychological talk, full of sensitive nuances.

He said that one night they came to his apartment, told him to collect a few things and took him away. Only recently he had been given a high position in the RNII and he was blamed for spending too much money on it. He was taken to Lefortovo prison, interrogated, beaten. He remembered asking for a glass of water from one guy who handed the glass to him and then hit him in the head with the water jug. He was called an enemy of the people.

"Today is your trial," he was told, and he was led down a long corridor into a room. The door opened and in came the interrogators, led by Voroshilov. [Kliment Voroshilov, one of Stalin's closest associates. Some historians say that it was highly unlikely that he would have been in on the interrogation.]

"Give me the list of accusations," he was ordered.

Korolev said, "I don't have it."

"Look in your hand!" Sure enough a paper was in his hand. He thought, well Voroshilov is the Peoples Commissar, he'll straighten this out.

But Voroshilov said, "Do you agree with the accusations?"

Korolev said, "I didn't commit any crime."

Voroshilov shouted, "None of you swine [*svolochi* in Russian] have committed a crime. Ten years hard labor. Go! Next!"

Those who had admitted crimes were executed. In Korolev's case many of the people associated with Marshal Tukhachevsky were arrested. Tukhachevsky, who had authorized founding of the RNII and provided its funds, was shot, as were a number of his staff.

Korolev was shipped off to Kolyma where there was a surface gold mine. He spent five months there, during winter, cutting trees, digging, pushing wheelbarrows. It was beastly work. Then a document came calling him back to Moscow for reinvestigation of his case. When he was leaving the camp, which was about 150 kilometers from Magadan, his fellow inmates collected clothing for him to take with him, to get him to Magadan. He remembers looking back at the black mass of his fellow prisoners, then turning and, noticing the sun come out, walking toward it. He stopped a truck and the driver said he'd take him to Magadan but he'd have to hand over his sweater. But when he got to Magadan he learned that the last ship for the season had already left via the Okhost Sea. It turned out that a few days later there was a heavy storm and everyone on that ship was lost.

So he was going to have to stay for the rest of the winter in Magadan. He looked for a place to sleep and found an Army barracks. It was 45 or 50°C below zero, there was a full moon, and about two meters of snow on the ground. But he was discovered and kicked out. He found another barracks but was made to leave that one, too. In his flimsy clothes he was freezing and he hadn't eaten for two days. Then, a kind of miracle happened. He was walking along a path in the snow when he came upon a loaf of bread, a *warm* loaf. He ate until he got hiccups, walked back to the first barracks and went to sleep under a bed, where he would be hidden. When he woke up his clothes were frozen to the ground.

The people in the barracks were under police control. Korolev said:

"Who played this joke, giving me bread?"

"Are you crazy? I was asked. All my life I've wondered where the bread came from."

He worked as a laborer, a shoe repairman, and on other jobs until Spring. Then he headed back to the mainland. On the way back he got scurvy and was taken off the train in Khaborovsk, near death. His body was swollen, his teeth bleeding and falling out. An old man appeared from somewhere and said,

"I'll make you better, Sonny." It was May and Spring weather was setting in. The old man put him under a tree facing the sun, and went away.

"When I opened my eyes," Korolev told us, "I could see something fluttering. It was a butterfly, something on this Earth still alive and beautiful. I was *alive!*"

The old man came back and massaged his gums with some herbs. It was *chirimsha,* or *kolba* they call it in Siberia. It grows like rice, it's close to garlic. Blood had been coming out of his gums but the massaging with chirimsha healed them. The next day he felt better, and in a week he was well enough to get back on the train to Moscow.

The story seems hallucinatory and is doubted by some Russian historians. But Leonov, a devoted admirer of Korolev, was quite obviously sincere in saying that it came directly from the Chief Designer. Leonov is another who speculates that Korolev's arrest was linked to his connection with Marshal Mikhail Tukhachevsky, although this theory, too, is questioned. Tukhachevsky had been responsible for organizing and funding RNII, and he and some of his fellow officers had been arrested and shot the previous year on charges of having collaborated with the Germans.[11] Tukhachevsky's treatment, in any case, is an egregious example of the incredible stupidity, not to speak of callous cruelty, of the Stalin purges. Whether or not it relates directly to Korolev's fate, it characterizes the environment of psychological terror that pervaded the Soviet Union during Korolev's time. One wonders what might have become of the Soviet Union if the likes of Tukhachevsky and literally millions of competent and dedicated professional countrymen had not been murdered or imprisoned.

Tukhachevsky was one of the most gifted, progressive, courageous, and intelligent officers in the Soviet military. As a young lieutenant in the tsar's army in 1915 he had been captured by the Germans and interned. He helped a young French officer escape the camp by dressing in the Frenchman's greatcoat at roll call. Both the Frenchman and a British officer were smuggled out of the camp while the ruse was being carried out.[12] Tukhachevsky himself escaped on his fifth try, became a Red Army commander at age twenty-four, after the Revolution, and fought many heroic battles for the Bolshevik cause. After one of his victories (the brutal assault on the Kronstadt naval fortress, which had rebelled in 1921 under the slogan "Soviet Rule Without the Communists"), Lenin personally praised him ("I am pleased, quite pleased, my dear fellow.[13]) and placed him in charge of putting down the rebellion of the kulaks, the wealthy farmers who were slaughtered by the thousands in those years. In 1927 Tukhachevsky wrote to Stalin that the Red Army should develop modern planes, armor, and automatic weapons.[14] It was in 1932 that he showed astounding vision by backing the early rocket experimenters and formally organizing and funding RNII from two zealous but, until then, moneyless groups of experimenters, the one led by Valentin Glushko in Leningrad and the other by Korolev in Moscow.

A quintessential example of the cultured Russian gentleman-soldier, Tukhachevsky is said to have spoken several languages, played Beethoven and Mendelssohn beautifully, and written some fifteen books. He and several other military commanders had been arrested on June 11, 1937, and shot. Also killed were his mother, one sister, and two brothers. He was rehabilitated at the 22nd Party Congress in 1956. Nikita Khrushchev said that Hitler's staff had taken advantage of

Stalin's paranoia by faking a document to the effect that Tukhachevsky and others had sent secrets to the German generals.

These incredibly stupid arrests and executions are directly responsible for the Russian failure to stay apace with the Nazis in weaponry, in the opinion of Gregory A. Tokaty-Tokaev, the Soviet rocket expert who defected to England. "I think," he wrote, "that the [now officially acknowledged] political arrests and murders of the 1935–40 period, known as Yezhovschina [after Nikolai Yezhov, then the NKVD chief[15]], caused much greater damage than was realized abroad. Far too many scientists, technologists and managers were destroyed, humiliated or disheartened. And rocket experts were no exception. Then . . . Marshal Tukhachevsky was the top government spiritual leader of military rocketry. But, thanks to Nazi provocation, he was shot in 1937 as a 'German spy,' and this sparked off a whole chain of disasters. Almost all who worked on a project discussed with and authorized by him, or who were in contact with him—as all leading rocket specialists were—had now to face the danger of being an 'accomplice of a spy.' "[16]

What is difficult to comprehend is how someone like Korolev, and many others, remained faithful to their government in later years in spite of the arbitrary infliction of such terrible acts on themselves, their colleagues, their friends and families. Loyalty to the old cause persisted even with many of the people who were themselves jailed.

"My own father was in prison from 1937 to 1940," says Leonov.[17] "But I never heard any accusatory words from him. He was a believer in Bolshevism." Not so, however, for many of the modern generation, including Cosmonaut Leonov, especially after glasnost gave them an opportunity to learn truths that had been hidden from them for decades. "It's a pity that our whole generation was disinformed," commented Leonov. "What we believed in wasn't really true. Everything went to rot."

Indeed, one wonders how the system worked at all with so many accomplished and distinguished people having been discredited by conviction of crimes, whether justified or not. As the American scholar of Soviet history, Adam Ulam, put it, "By 1938, if one believed the official line, one had to grant that beginning with the Revolution, Soviet society had been in the main organized and run by traitors and foreign agents. Not even the most inveterate enemy of Communism could have imagined or invented the picture of Soviet reality that was trumpeted daily in official propaganda."[18]

How did this propaganda affect the technical community? Mostly, it was so crudely presented that it wasn't believed. But open derision of it was not to be chanced either. And often it was necessary to go along with it to keep position, maintain funding, protect one's own family. The scientists and engineers themselves had to cover up the truth sometimes, giving themselves cover titles and organizations, for example, when they visited foreign countries.

Josef Gitelson, whose work with Korolev on the development of life support systems was mentioned earlier, says sadly:

> We developed a kind of voluntary schizophrenia in those days. You separate your soul into two independent parts. You can adapt yourself to the practice of lying in one part of your soul while you keep the truth in the other part. Both halves of the soul can be in absolute contradiction to each other, each remaining absolutely logical within itself. [We] listened to or uttered the ritual political wording which we could not believe, or which we would doubt or which we would not care to think about and beyond these limits people remained absolutely honest in science, business, in personal relationships. The truth part you shared with your family and a few very close friends. That's why it was so important to have real friends, so much more important than in the Western world. . . . Voluntary schizophrenia was characteristic of all or most thinking people, except those few who made their choice to be dissidents. I think that Sakharov was like this before openly confronting the system.[19]

General Volkogonov, himself, "admits to having lived two mental lives, pursuing a successful career in military education, while gathering material on the evils of the past as assiduously as any persecuted dissident."[20] But think of the damage to progress on any and all fronts—political, economic, scientific, or technological—when so many people have had their reputations destroyed by accusations of crime or espionage. As Ulam writes, "How did one approve a construction project when one of the architects had confessed to being a wrecker whose blueprints caused the collapse of several edifices?"

In some cases there was mass sentencing of entire ethnic groups. Boris Rauschenbakh was one of thousands of Soviet citizens of German ancestry shipped to a gulag camp in the Ural mountains at the end of 1941. He was from one of the families of Germans who had been invited to take land in the Volga region of Russia in 1766 by Catherine II. Some 300,000 of these ethnic Germans, called Volga Germans, continuing to speak their native language, lived together in that area for almost 200 years. Rauschenbakh spent four years in prison, returning in 1946 to resume his technical career.[21]

Incredibly, Stalin himself escaped popular blame for these horrendous violations of human rights until Khrushchev's time. In his 1988 book Fedor Burlatsky, a former Khrushchev speech writer, says, "We should remember the majority of people destroyed by Stalin continued to believe in Stalinism. Many of them before they were shot cried out: 'Long live Comrade Stalin.' That was the cry of Yagoda [G. G.], Stalin's executioner, who was destroyed by his own ruthless machine."[22]

Korolev's case was reinvestigated in 1939 when Yezhov was replaced by Beria as Minister of Internal Affairs after Yezhov himself had been convicted of violation of criminal prosecution laws. "After the investigation," says Gyorgi Vetrov, "his [Korolev's] sentence was reduced from 10 to 8 years. . . . He was supposed to be sent back to Kolyma. But many people interceded on his behalf . . . including Grizodubova [Valentina], the woman pilot who was a deputy of the Supreme Soviet, and also the famous pilot, Mikhail Gromov."[23]

As referred to earlier by Leonov and Tokaty-Tokaev, it was perhaps through the intercession of Andrei Tupolev, who had been Korolev's teacher at Moscow Higher Technical Institute, that he was allowed to leave Kolyma, although there is no documented proof of this. Tupolev himself had been arrested on October 21, 1937, and sent to another uniquely Soviet penal institution, albeit not nearly as harsh as the gulag camps. This was the *sharaga,* also known by its diminutive, *sharashka.* The sharashkas were under the supervision of the 4th Special Section of the Ministry of the Interior, reports Volkogonov. "In this area Stalin adopted a purely pragmatic approach, the world outlook or political views of those serving sentences being no longer of any consequence. Quick results are what mattered. . . . Beria's agency kept Stalin constantly informed of the work of the scientists in the prisons and camps."[24] Solzhenitsyn provides a major source of information on how these prisons operated in his novel, *The First Circle.*[25] Solzhenitsyn's translator, Thomas P. Whitney, explains that the word *sharashka* "derives from a Soviet slang expression meaning 'a sinister enterprise based on bluff or deceit.'"[26]

Whether or not it was Tupolev who got him out of the gulag, the fact is that Korolev was moved, in September 1940, to Tupolev's sharashka. Most of the sharashka occupants were intellectuals, scientists, engineers whose alleged misdeeds had been reported by fellow professionals, party officials, neighbors, friends. "Thus," writes Burlatsky, "we can see that in Stalin's times people's lives and feelings were not at all straightforward: some spent their time putting people in prison, others spent their time in prison; some played the role of the hammer, others that of the anvil; some were informers, others were victims of denunciation."[27]

It is very difficult for a Westerner to fathom the motivation that drove Tupolev, Korolev, and many other competent engineers and scientists to devote themselves as fully as they did in the sharagas to the projects assigned by political leaders who were violating their most basic rights, preventing them from seeing their families, allowing no mail or packages, hounding them with security guards. But they not only worked, they worked creatively and diligently. "It is easy to forget," writes Volkogonov, "when we think of the horrors of Stalinism, that there was also the amazing socialist phenomenon of labour zeal and achievement. True, people were often motivated by fear, but they nevertheless gave material reality to plans which today would be regarded as fantasies."[28]

For example, the Tu-2 light bomber and the Ilyushin-2 attack aircraft, which had notable records in World War II, were designed in the Tupolev sharashka—otherwise known as Central Design Bureau No. 29 (KB-29), actually attached to the NKVD. The building, on Radio Street in Moscow, is still today part of the Tupolev design bureau, and is marked by a plaque commemorating its wartime use.

Andrei Sakharov, in his memoirs, wrote, "Legend has it that one professor confined there was pleading his innocence of any wrongdoing when Beria interrupted him to say, 'My dear man, I know you're not guilty of anything. Get the plane in the air and you'll go free.'"[29]

KB-29 was, in effect, a special design bureau for prisoners, but without cells. There was a large dormitory, or sleeping room. Another area was for work. The place was equipped with special fences. It was on the top floor of the building. Prisoners worked side by side with nonprisoners, with guards watching over them. They were also permitted to meet with family members, but at Butirskaya, another prison in Moscow, rather than KB-29.

Lev Kerber, one of the aeronautical engineers who worked with Korolev in the Tupolev sharaga, was asked to recount his experiences of more than five decades earlier.[30] At age ninety, Kerber, a tiny fellow with a scruffy white beard, was living with his even tinier wife in a dingy apartment on the upper floor of a typically rundown Moscow high-rise when I interviewed him. The walls of his study are decorated with memorabilia from his career, including several photos of Tupolev, but none of Korolev. He wore a small skullcap, a thick woolen sweater, slippers, and both he and Mrs. Kerber coughed throughout the interview.

> I was put in charge of the air navigation training of the pilots Gromov and Chkalov who were planning a flight to the U. S. in a Tupolev TB-3. This brought me in close contact with Tupolev. It was a very

pleasant and productive time. Then, on October 21, 1937, Tupolev was arrested. I was actually in the army at this time, and a few months after Tupolev's arrest, I was arrested, too. I was sentenced to ten years in prison and accused of counter-revolutionary terrorism and spying, although there was no basis for these accusations.

At one o'clock in the morning three guys from the NKVD came to my house. They confiscated my six-year-old son's rifle, which had a broken barrel. They ordered me to dress and took me to Lyubyanka prison. I spent seventeen days there. I was thirty-five years old at the time. I was then transported to Butirskaya prison, where I was beaten severely and not permitted to sleep. I signed all the confession papers, which said that I had sabotaged the radio equipment on the Gromov/Chkalov TB-3. The TB-3 radio equipment was very large. It took three cars to carry it.

I spent one year at Butirskaya, then I went through a jury process in which they showed a piece of equipment with holes in it that was evidently supposed to prove my sabotage. I had never seen it before. I was sent to the Kuloilag, a gulag camp in Archangelsk. I spent sixteen months there. My quota was to cut down sixteen trees a day. We got 800 grams of bread and two plates of soup a day. I got scurvy after a while.

Tupolev was told to make a list of people who should be brought to his sharaga in Moscow who had the qualifications to help design new aircraft. He made a list of 200 people and Korolev, who came from Kolyma, and I were both on it. We were sent to Moscow to the sharaga, which was named Central Design Bureau 29. Korolev was put to work in the department in charge of wing design for the 103 airplane, which was later designated the Tu-2 light bomber. There were about 800 of these planes built during the war. Our beds were near each other's and so we talked a lot, always with low voices. We would have breakfast at 7 A.M. of kasha, butter, white bread, coffee, chocolate, or tea. We got good food and got fat. We would work from 9 to 1. Our lunch would be bread, butter, meat or fish, dessert, juice. We would have a one hour rest and then work from 2 to 6. Then we had free time. We could go for a walk in an area we called "The Monkey Place," because it had big metal bars around it. It was small, about the size of three of our tiny rooms. At 9 P.M. we had dinner of meat or fish and tea. We went to bed at 11. They counted us at night. There were three dormitories, with about sixty guys in each. We worked six days a week, were allowed no visitors or letters at first, but after a year we were allowed visitors.

I spent March 1940 to August 1941 in that sharaga. Then we were moved to Omsk (in Siberia, about 1,400 miles from Moscow) because of the approach of the Germans to Moscow. In the Spring of 1942 I was released as a prisoner, but we stayed in Omsk because of the importance of the aircraft work we were doing. During this period Korolev, myself, and Sergei Mikhailovich Yeger, a long-time deputy of Korolev's, became friends. Korolev, though, was often occupied with his own thoughts. He was sometimes difficult to talk to. He seemed very pessimistic. He was healthy, working conditions were good, but he was pessimistic because he thought that since prisoners were designing these secret airplanes we would all be shot in order to preserve the secrets. When I went back to Moscow in 1943 Korolev was still in prison, in Kazan. I looked up Xenia Vincentini, Korolev's wife, and his daughter Natasha and gave them a letter from Korolev. This made Xenia very happy.

Korolev was moved, with Yuri Rumer and others, to a new rocket sharaga in Krasnapolyana in the Caucasus. It was the former hunting castle of the last Russian emperor, Nicholas II. This was now 1944. Eventually the rocket sharaga was transported to Moscow to a big building named for Lenin that you can see over there from this window. Now they make electrical equipment there but they started working on rocket engines back in 1944.

I continued my friendship with Korolev but after he came back from Germany, in 1946, he was given better housing. The rest of us from the sharaga got very poor housing. He told us a lot of stories about Germany and respected the German scientists for what they had accomplished. But he was still pessimistic and still worried about being shot. I think that he was badly damaged, psychologically and mentally, during his imprisonment. He had neuroses. I am more of an optimist. In any case we remained good friends although he had few close friends like Yeger, Rumer, and me.[31]

Sakharov wrote, "I don't think he [Korolev] ever forgot his time in the camps." As Kerber acknowledges, he did not forget his fellow prisoners either. Sakharov agrees, recalling that, "He once tried to organize support for Yuri Rumer, a theoretical physicist who had also worked in Tupolev's sharashka, when Rumer became a candidate for corresponding membership in the academy. Korolev failed to get Rumer elected, though in other instances his efforts were more successful."[32]

Korolev's daughter, Natasha, remembers visiting her father during his Moscow sharashka imprisonment:

I remember when I was five or six, in September 1940, all of the prisoners were transferred from the Tupolev sharashka on Radio Street to Butirskaya prison for one day and we were allowed to visit for a half hour. My mother had told me that father was a pilot, and I asked him how he could land in such a small yard. We visited him several times in

> Butirskaya. Then in 1941, he was transferred to Omsk in Siberia and at the end of 1942 he was moved to Kazan, about 400 miles west of Moscow, but he was still a *zek* [Russian for prisoner]. He designed the liquid-fueled rocket boosters for the Petlyakov 2 [Pe-2] dive bomber. My mother and grandmother would tell me that my father was on business trips. He was not released until 1944 [on August 10]. But he didn't return from Kazan until the spring of 1945. Then he went to Germany in September 1945.[33]

Korolev had kept up his letter campaign for release from prison, directing a stream of petitions to the prosecutor on June 2 and 10, July 5 and 23, and on September 13, 1940. In the July 23 letter, he termed "absolutely wrong and deeply unfair" the charges of anti-Soviet Trotskyist activity leveled against him and pointed out that he had been involved in " . . . exceptionally important, and essential to the USSR, defense business—the creation of rocket planes which significantly improved the flight and technical performances of the best modern propeller-driven types. This business is downplayed, contemptuously ignored."[34] He protested that " . . . arrested enemies of the people led the NKVD to an error by their base slander . . . " and asked that his case be reinvestigated and that several of his professional colleagues be interrogated on his behalf. He got no answer to any of the letters and apparently gave up asking for clemency until after the war.[35]

There has been speculation about why Korolev's wife, Xenia Vincentini, is not on record as having protested her husband's arrest. Vetrov offers one theory:

> In 1940 or 1941 Xenia Maximiliana Vincentini defended her thesis for her candidate of science degree. This is while her husband was in prison and was categorized as an "Enemy of the People." Whereas most relatives of these prisoners were oppressed—lost apartments, jobs, educational opportunities—Vincentini didn't. One possible explanation is that she might have renounced her husband—just as others did. There is no evidence that she renounced him, however. This is just speculation. In fact, Xenia probably saved Korolev's mother's life by keeping her in her apartment. Another reason to speculate about a possible alienation between Xenia and S.P. is that in 1944, when both Glushko and Korolev were released as prisoners, while continuing their work on rockets in Kazan, Glushko's wife went to Kazan to join her husband while Xenia didn't.
>
> One more inconsistency: Xenia said that she got messages from Korolev from his train trip to prison but that she could not keep these messages. It was too dangerous. But Korolev had been given the right of communication at the time of his sentencing. His sentence was ten

years in prison, the loss of his rights for an additional five years, and confiscation of his property. But he did have the right of communication.[36]

In spite of his isolation from his family, and the oppression he must have felt as an innocent victim of the purges, Korolev's work in the sharashkas was exemplary and showed his passion for developing the rocket for substantive use in weapons and aircraft. A document dated August 6, 1941, when he was in Omsk, deals with various applications of a rocket, one of them using nitric acid and kerosene as propellants for an aerial torpedo.[37]

When, on November 19, 1942, Korolev was moved to the sharashka in Kazan, his host organization was again the NKVD, the place having been designated a "Special Design Bureau of the 4th Special Department of NKVD of the USSR, attached to the aircraft plant No. 16." There he joined another group of distinguished engineer-prisoners including the aircraft engine experts D. D. Sevruk, A. A. Kolesnikov, A. D. Charomsky, B. S. Stechkin, and his former accuser Glushko. On January 8, 1943, Korolev became chief designer of Group No. 5, which worked on jet engines for aircraft based on Glushko's designs. There ensued a period of creativity that would surely have impressed Goddard, then working on rockets at Annapolis, Maryland, and von Braun in Peenemunde. Among their designs: the four-chamber RD-1 liquid rocket engine with turbopump that generated 1,200 kilograms of thrust, for application to the Pe-2 dive bomber. Some fifty-eight sheets of calculations and four pages of drawings, dated December 16, 1942, were made for this design, which, it was felt, provided a solid base for designing future rocket interceptors. Work continued through 1943 and in October of that year the first startup of the engine in flight took place. The engine operated for two minutes, increasing the aircraft's speed from 340 to 420 kilometers per hour.

In the summer of 1944 the entire experimental design bureau was released from custody. A protocol dated June 27, 1944, from the presidium of the Supreme Soviet states that Korolev was discharged and the prior convictions expunged.[38]

It was at this time that the NKVD handed over its responsibility for the experimental design bureau to the People's Commissariat of Aviation Industry. Glushko was made chief designer, with Korolev his deputy. During 1944 numerous other designs emerged from the group, including the application of auxiliary rocket engines to the Lavochkin 5 fighter plane. Three years later, on August 3, 1947, this application would be given a public flight demonstration. Also under development were a long-range unguided solid-propellant ballistic rocket, the D-1, and a winged guided missile, the D-2.[39]

Letters from Korolev to the deputy People's Commissar for the Aviation Industry, P. V. Dementiev, dated October 14, 1944, and June 30, 1945, detail a D-1 rocket with an 1,100-kilogram launch weight, including a 30-kilogram warhead, with a range of 12 to 30 kilometers, and a D-2 with a launch weight of 1,200 kilograms, carrying also a 30-kilogram warhead for 20 to 70 kilometers. One variation of the D-2 is even designed for manned flight. These letters offer to create in Kazan a "special collective for developing long range missiles" and go so far as to specify the fellow Kazan prisoners that he would need to carry out his work, mentioning B. V. Rauschenbakh, A. I. Polyarny, and M. P. Dryazgov.[40]

One marvels at the realization that Korolev, after some seven years as a prisoner, apparently stayed on in Kazan—albeit as a free man, or as free as any Soviet citizen at the time could be—and continued to work on rockets for about a year.

Then, in the late summer of 1945, he was commissioned a colonel in the Red Army. He flew to Germany on September 8 to join other Soviet colleagues gathering information on a weapon—the V-2—which had been developed by a team led by the same Wernher von Braun whose career would so eerily parallel his own.

5

The German V-2: Bedrock Technology

For years Soviet space leaders put down the contribution that captured Germans and their V-2 technology made to the Soviet ballistic missile and space programs. "Not significant," they would say, "we got mostly the technicians. The Americans got von Braun and his top team. We sent our Germans back after a few years."

That explanation is no longer the Party line. In fact, it is now acknowledged that German rocket technology was bedrock to the USSR, just as it was to the United States. One of the Russian rocket pioneers who tells the story candidly is Boris Chertok, still active as a consultant to RSC Energia, the successor organization to the Korolev design bureau. Chertok was sent to Germany in 1945, along with a small number of Soviet rocket specialists, to gather V-2 technology, documentation, and specialists.

But that wasn't the first Soviet foray into German rocketry. In an interview in *Izvestia* in 1992, Chertok recalled a clandestine visit in 1944, almost a year before the war ended, instigated by none other than the British Prime Minister:

> A personal and strictly secret message from Winston Churchill to Marshal Josef Stalin on July 13, 1944, in essence marked the beginning of a broad interest by the USSR in German rocket technology.
>
> Churchill reported that evidently the Germans had at their disposal a new rocket weapon which introduced a serious threat for London, and asked permission for English specialists to go into Poland for investigation of the test range, which was located in the region of Soviet attack forces. Stalin answered that he understood the anxiety of the Prime Minister and promised to take the matter under personal control. Personal control for Stalin in that time meant a categorical command to carry out instructions at any price. To Poland immediately went a group of specialists from NII-1.[1]

The Russians had already known about the German V-l, the jet-powered buzz bomb, and judged that it wasn't powerful enough to produce a massive bombardment. But the V-2 was a different weapon. Honoring Churchill's request, the British were invited to Poland, and came with espionage agents and experts who had mapped missile impacts and launch sites carefully in advance.[2] The Soviet expert team, which included Viktor Bolkhovitinov and the rocket specialists Mikhail Tikhonravov and Yuri Pobedonostsev, arrived in Poland on August 5,

1944, according to Gyorgi Vetrov, "even before the front lines had moved past that point. They found fragments of the A-4 despite careful destruction by the Germans. Remnants were then shipped to Bolkhovitinov's institute [NII-1]. Vassily Mishin, who would later become Korolev's first deputy, studied these remnants in a group that also included Chertok, Alexei Isaev, and Nikolai Pilyugin."[3]

For some days only "high authorities" were able to examine the hardware. But eventually, says Chertok,

> sense took over and engineers were also allowed in. So I come into this hall. Several hours before me our engine man, Alexei Mikhailovich Isaev—one of the future stars of our rocket technology—was let in. I see the lower part of his body and his legs sticking out of the rocket engine nozzle, while his head is somewhere inside. . . . I approach Bolkhovitinov . . .
>
> "What is this?"
>
> 'This is what cannot be,' he replies. . . . Understand, one of our most talented aircraft designers simply did not believe that in wartime conditions it would be possible to develop such a huge and powerful rocket engine. We had at the time liquid engines for our experimental rocket planes with thrusts of hundreds of kilograms. One and one half tons was the limit of our dreams. Yet here we quickly calculated, based on the nozzle dimensions, that the engine thrust was at least 20 tons.[4]

The Russians then deduced that the rocket could lift vertically some 12 to 14 tons. "We were shocked," said Chertok, to note that the rocket used neither nitric acid nor kerosene, the Russians' customary fuels, but rather alcohol and liquid oxygen.

In fact the V-2 engine had undergone a quite remarkable development, under the supervision of Walter Thiel, between 1936 and 1941. Thiel had scaled up a 1.5-ton engine to 25 tons, while reducing engine size very substantially, by solving difficult problems of fuel injection and burn-through of the walls of the combustion chamber.[5, 6]

The V-2 had a gross weight of 12.52 tons, confirming the Russian estimate. It could carry a 1-metric-ton payload 200 kilometers. It was 14.3 meters long and had a diameter of 1.65 meters. It was the first missile to incorporate a turbopump, powered by an 80 percent hydrogen peroxide steam generator, and the first to use a guidance system with a three-axis platform for stabilization, although the latter was developed too late for incorporation in the production V-2.[7]

On September 8, 1944, the first attacks on Paris and London were launched from mobile V-2 installations in Belgium. Eventually, 3,225 V-2s were fired successfully, mostly from Holland and Germany.

Estimates are that 12,685 people were killed and 33,700 homes and buildings destroyed by the weapon.[8]

Those numbers notwithstanding, it was a fantasy to believe, as Hitler apparently did, that the weapon might be decisive in the war. The fact is that the total explosive power of all of the V-2s fired was less than that of one RAF bomber raid. Nonetheless, the V-2 was a remarkable technological achievement, especially when one realizes that the 3,225 firings were made over a period of only two and a half years from the time of the first experimental launch on June 13, 1942. Even more important to history, updated V-2 designs formed the basis of the first ballistic missiles and space launchers of not only the United States and the Soviet Union but eventually of France, Britain, and China. It is not very well known, for example, that some eighty Germans came to France in 1946 to help design the Viking engine used on the Ariane launch vehicle.[9] Ariane has become a very successful launch vehicle, carrying, by the 1990s, more than 60 percent of the world's commercial communications satellites into orbit. In 1947 German V-2 experts came to Westcott in England, eventually helping the British to design the Black Knight missile, later abandoned in an economy move. In the 1950s, when the Russians and Chinese were collaborating intensively, a cadre of Chinese engineers were taught rocket design based on V-2s in Moscow.

But far and away the countries to benefit most from the V-2 technology were the USA and USSR. The Americans did, indeed, get the cream of the V-2 crop. Wernher von Braun kept most of his best engineers together, escaping the Russians by fleeing Peenemunde, which was in the Red Army zone, and giving himself and his team up to the Americans near Reutte, Austria, in May 1945. Eventually, 127 of the von Braun Germans came to the United States. After a period at Fort Bliss in Texas they settled in Huntsville, Alabama, forming the technical strength of the Army Ballistic Missile Agency. There they escalated the V-2 technology into the design of the army's Redstone missile and its first intermediate range ballistic missile, the Jupiter. Later would come substantial contributions to the development of America's first satellite and eventually to the Apollo lunar landing program.

But not generally known is the fact that the Russians, even without von Braun and his top team, recruited a very substantial group of Germans, albeit not just from the V-2 program, and certainly exploited the V-2 technology to a degree that was at least equal in resourcefulness to that of America, and they carried out their work in a style even more Machiavellian than their counterparts from the USA.

The 1944 visit to Poland by the Russian rocket experts was followed by a much more concerted effort, beginning in April 1945, to round up V-2 hardware, launch facilities, blueprints, and as many engineers and technicians as could be found. Since all of the launch and production facilities were in the Soviet zone, which became East Germany, the potential haul was huge. Although most of the von Braun team was gone, and the Americans had already spirited off the blueprints, assembly jigs and fixtures, and as many of the intact V-2s as they were able to grab, there was still plenty of materiel, albeit in pieces, to salvage.

Chertok, a big hulk of a man—bald, craggy-faced, still full of conviction although in his early eighties—recounted the story to me decades later at the dining room table of Korolev's daughter, Natasha, after a tasty Sunday night supper.[10]

"I arrived in Germany," he said, sipping modestly from a tiny vodka glass, "on April 23, 1945. Isaev, the engine designer, and I were put in charge of an organization which we called the Institut Rabe [Raketenbau und Entwicklung—rocket manufacturing and development] in Bleicherode, near Nordhausen." Although unofficial ("We never got Moscow permission for this"), it was the first Russian-German venture after the war. Chertok, who had been promoted from soldier to major in the Red Army, declared himself chief of Rabe, and began to recruit specialists. He and Isaev, ironically, were advised to take over the house in which von Braun had lived.

"[We were] dirty and dusty," he told his *Izvestia* interviewer, "we looked for a place to rest. . . . [We] discovered four toilets and three great bathrooms. A bedroom was on the second floor. Isaev threw aside a snow white feather blanket and plopped on the bed. There was a mirror in the ceiling. Alexei Mikhailovich smoked a Belomor [Russian cigarette]. 'You know, it's not bad at all, this Fascistic beast's pit.' "

"Eventually," said Chertok, "the institute comprised about 1,000 people, half of them Russian, the other half German workers, plus about 50 to 60 Peenemunde veterans, including mostly technicians and a few senior technical people—among them Helmut Grottrup. There were also Kurt Magnus, a gyro specialist, and a fellow named Hoch, and there were some ballistics specialists from Krupp and Siemens.[11] Our main job was to restore all of the V-2 documentation that von Braun had taken to the U.S."

Work at Rabe was started in a three-story building that had been an electric power station. The lure for the German engineers was more the accessibility of food than money. The American occupation zone was only a few miles away, and the Americans weren't watching the border carefully, so the Russians started an enticement campaign, offering good jobs and lavish rations to those who would come over. "Both the Russians and the Americans were conducting intelligence operations," said Chertok. "We were trying to understand what the Americans were up to. They had 'Operation Paperclip' underway [code name for the U.S. operation that gathered up the von Braun team as well as the V-2 documentation and hardware]. We had agents in the U.S. zone to grab von Braun but he was guarded too well. Grottrup came to us for idealogical reasons, but with a great deal of propaganda on our part. I promised the German scientists that they could go back if they didn't like it. Professor Schuler did go back eventually. He was a navy navigation expert and he didn't want to work for either the US or USSR."

I later learned from Magnus, now a professor at the Technical University of Munich, that Grottrup's decision for the Russians was not for idealogical reasons. He refused the American offer because he did not wish to be separated from his family. Schuler, Magnus told me, visited Bleicherode for only two or three days. He never intended to join the team.[12]

Although they may not have been vigilant about border crossings, American intelligence specialists were conscientiously watching these operations, as evidenced by a secret report that was eventually released by the Central Intelligence Agency some fifteen years later, in 1960.[13] The report accurately identified Grottrup, who had been a valuable deputy for guidance and control to Ernst Steinhoff at Peenemunde (Steinhoff joined the von Braun team in the USA), as the man designated by the Russians to lead a group that eventually numbered some 5,000 Germans. Their assignment, using the words of the CIA report, was to "recreate the design data, test equipment, drawings, manufacturing jigs and tools, ground support equipment, and operating instructions" for the A-4 (A-4 was the Peenemunde group's designation of the V-2. A-4 was renamed V-2, V for *Vergeltung*—meaning retaliation—by the Nazi Propaganda Ministry).[14] Further, the report goes on, they were to set up an assembly line to build A-4s from surviving components, test the engines, build two complete railroad trains for transporting the support equipment and staff for launches, instruct "one Soviet and one German launching and testing crew" and develop "proposals for further improvement of the A-4."

The Russians were awed by the underground Mittelwerk plant near Nordhausen to which manufacture of the V-2s had been moved after Allied planes had bombed the plant in Peenemunde. Built into the Kohnstein Mountain, it had two 3.5-kilometer galleries that opened on terrain level to the outside. Railroad tracks permitted an entire train to enter the plant. Production capacity was designed for thirty to thirty-five missiles per day.

"You won't find a plant on either American or on our territory, that would produce as many surface-to-surface missiles today," said Chertok. The fact that the plant had employed thousands of slave laborers brought in from the concentration camps under SS General Hans Kammler, many of whom died under the cruel working conditions imposed on them, could have been the reason that the Russians didn't like the idea of working in the "gloomy underground" plant. They decided, instead, to use a nearby V-2 repair facility called Klein Bodungen to assemble and complete the missiles.

Grottrup soon found that his decision to stay in Germany was not in accord with the plans of his captors. Only a year after going to work with the Russians, he, along with thousands of other engineers, scientists, and technicians would be shipped, literally, by trains, freight cars, and trucks, with their families to the Soviet Union.

Before that would happen, however, he would direct the blueprinting and dismantling of Mittelwerk, as well as the original missile development center at Peenemunde, and the engine development facilities and test stands in the Frankenwald Mountains at Lehesten. The latter were in excellent condition, and under the direction of Valentin Glushko, the first firing of a V-2 engine took place on September 6, 1945. Eventually the Lehesten facilities were shipped in toto to the USSR.

For his work, Grottrup was treated generously. He, his wife, and two children were given the home near Bleicherode of a rich merchant who had been summarily ejected. He was paid a good salary and provided with excellent food rations.

Eventually Institute Rabe was reinforced with Soviet specialists from the army and from industry. In August 1945, Mishin, Pilyugin, Bolkhovitinov, and such other future missile luminaries as Boris Frolov, Viktor Kuznetsov, and Mikhail Ryazansky appeared, having been given military ranks and uniforms. They were confronted with some stiff challenges, including the lack of technical documentation, which had been gathered up by the Americans. Reconstructing it was very difficult. Mishin went all the way to Prague for some of the archives. Some valuable pieces of hardware were discovered, such as an especially crucial gyro-stabilized platform,[15] which had been under development for application to a two-stage missile.

Not so impressed by what the Soviet rocket specialists found, however, were the bureaucrats back home. "We examined the V-2 test sites and tried to reconstruct the documentation," remembered Mishin. "We sent a report to Shakhurin [Alexei] and Pyotr Vasilievich Dementiev at the Ministry of Aviation, but they said it wasn't their business, it was Ministry of Weapons business. The report was ignored."[16]

In spite of that disinterest, the Institute Rabe team worked diligently. "We solved puzzles every day," said Chertok, first understanding a device's functioning, then reconstructing the documentation. Pilyugin would usually disassemble a device "down to the last screw" personally, then direct the Germans to make detailed drawings. Then he would build the device from the drawings.

Russian exploitation of V-2 technology took place in many sites throughout their zone of Germany. Besides the locations mentioned above there were a missile controls plant in Berlin; a design office in Sommerda, near Erfurt; an engine assembly plant called Montania near Nordhausen; and an electrical equipment factory in Sonderhausen. The last must have been especially appreciated because the Russians were not well versed in modern electrical technology. As one veteran Soviet electrical expert, Nikolai Sheremetyevsky was to say later, "All the electrical equipment that was later mounted on Korolev's rockets we initially took from the Germans and then improved."[17]

Korolev himself did not arrive in Germany until September 8, 1945, still not officially rehabilitated, fresh from Kazan. Resplendent in his Red Army colonel's uniform, however, he had apparently wasted no time getting his perks, as Chertok recalled.

"My first meeting with Korolev," said Chertok, "was in his lavishly decorated study, which had been the home of a German electrical contractor [in Bleicherode]. I had been given notice that he would be arriving and I was told to be polite to him and show him everything. Korolev drove up by himself in an Opel. He asked me why I, a Soviet official, had such a pretty German secretary. I said she could type and do stenography and speak both German and Russian. We were allowed to hire Germans, even if they had been Nazis, but we couldn't hire Russian prisoners of war—a paradox. Korolev looked very healthy, even though he had been in prison. He had a prominent forehead, lively black eyes. He looked directly at you as though x-raying you. His uniform looked good on him. He was full of energy. It was a brief meeting. I explained the work of the institute to him but he turned down a tour. He said he was in a hurry to go to Berlin and asked me for some gasoline, saying, 'I'm sure I'll see you later.'"[18]

Soon the brass from the Ministry of Armaments arrived from Moscow—Vassily Ryabikov, Deputy Minister, and then the Minister himself, Colonel-General Dimitri Ustinov. The latter would be a particularly influential figure in the development of the Soviet missile and space technology capability from the 1950s to 1970s.

The decision was made to create a more substantial organization, the Nordhausen Institute, which would incorporate Rabe as the guidance system component. Chertok continued as head of the latter, with Pilyugin his chief engineer, but reported henceforth to the overall institute director, General Lev Mikhailovich Gaidukov. "Gaidukov was a high-ranking Party member," said Mishin. "We called him the 'Parquet General' which is like 'Desk General.' We were part of a group which included representatives of the army, air force and navy."[19] Gaidukov's deputy and chief engineer was Sergei Korolev.

The job of reconstructing all of the documentation was assigned to a joint Soviet-German design bureau in Sommerda. In command at first was Vassily Budnik, who was succeeded later by Mishin.

Glushko, already in charge of engine testing at Lehesten, got the task of reconstructing engine production technology at Montania. An institute for studying air defense missiles was created in Berlin under Vladimir Barmin. Reconstruction of ground equipment was assigned to Barmin's deputy, Viktor Rudnitsky.

Korolev, says Chertok, "was a wonderful organizer . . . able to unify people around himself. . . . He inspired us with the idea that this work was not just a reconstruction of German technology but the source of a major new direction."

Still, he was "outwardly reserved and uncommunicative," wrote Gyorgi Tyulin in a 1989 article.[20] "Once I asked him: Sergei Pavlovich, what do you think about all this? He did not answer at once. . . . Much later I realized that the black period of repressions, prison, Kolyma, Magadan had made him cautious and he avoided opinions on acute matters as if he feared that everything might be repeated again."

He was especially rueful that the groundwork that he and his GIRD colleagues had done in the 1930s, work probably equal in significance to that of the youthful von Braun and his colleagues during the same period in Germany, had been aborted by the imprisonment of Glushko and himself, and the execution of Kleimenov and Langemak.

"We had accomplished much good groundwork back in the late 1930s," he told Tyulin, who wrote that, "We returned to this subject often. Korolev would present his views and pay due respect to 'von Braun's team' and the 'German breadth' but even then he believed that the design solution of the A-4 [V-2], the most sophisticated rocket of that time, could not be used for future developments."[21]

Aware that the Americans had made the biggest V-2 coup, the Russians must have been especially delighted when they were invited, in October 1945, to a series of V-2 demonstration launches organized by the British near Cuxhaven on the North Sea just west of the mouth of the Elbe River.[22]

Authorization for the firings, known as Operation Backfire, had been planned by the British Air Defense Division as early as May 1, when it was realized that Nordhausen, where the V-2 was manufactured, would soon be in the Russian zone. Over the next few months, some twenty-five V-2 specialists were transferred to Cuxhaven from the Peenemunde group at Garmisch Partenkirchen, although eventually 1,000 German nationals, including 274 prisoners of war, and 2,500 British military personnel were involved in the operation. Search parties scoured Germany, France, Belgium, and Holland to find the weapons and their launch apparatus. They turned up enough materiel to fill 200 trucks and 400 freight cars, although much of it was in such bad shape that they could only assemble eight complete V-2s.

There were three demonstration launchings, on October 3, 4, and 15, with the Russians present only for the third. Most of the British and American witnesses were military officers with no special expertise in rocketry. Exceptions among the Americans were Theodore von Karman, founder of the Jet Propulsion Laboratory at California Institute of Technology; William H. Pickering, who would become JPL Director in 1954; Lieutenant Commander Grayson Merrill; and Howard S. Seifert, also of Cal Tech. Pickering, who eventually led the team that developed the payload for America's first satellite, Explorer I, and went on to send scientific spacecraft to the Moon and every planet except Pluto, doesn't recall that there were any Russians present for the one firing that he witnessed, but perhaps they kept a low profile.[23] Tyulin reports that the Russian representatives included, besides himself, General A. I. Sokolov, and two first-rate rocket engineers from NII-1, Pobedonostsev and Glushko, and also Korolev. Another report, however, says that Korolev had arrived unannounced from Nordhausen with a colleague and was kept stewing and fuming outside the fenced area.[24]

Pictures taken during this period show a handsome, ramrod straight, if somewhat short and hefty Sergei Pavlovich. Only thirty-nine years old, bull-necked, stern of face, he could have been portrayed by a young Rod Steiger in a movie. For a time he had his wife and daughter with him. "My mother and I joined him in Germany in May 1946, and stayed there until August," Natasha Koroleva told me. "I had a lot of fun with him in Germany. We visited a lot of interesting places. We even took an automobile tour around the country in an Opel."[25]

Korolev's interaction with the Germans was substantive, and dealt mainly with technical questions. Magnus, in a very interesting book, describes his first meeting with Korolev, a meeting that revealed, even at the time, a deep interest in space flight.

> We were particularly impressed by a medium-sized, somewhat stocky man in the uniform of a colonel, who, by his looks and his gestures, revealed an alert intelligence. His precise questions fortified this impression on our part. What may be his name? His colleagues called him Sergei Pavlovich, but he was introduced to us as "Gospodin [Mister] Korolev." Well, who among us, ignorant of Russian habits, would have known that the first name and the name of the father are used among friends, but the family name is used more formally? So, Sergei Pavlovich Korolev was our partner in our conversations. It was only much later that it dawned on us that we were meeting here with the most important and the most successful development engineer of rocket technology in the Soviet Union. He was to remain close to us for the following several years.
>
> In our discussions during that evening, Korolev's interest focused mainly on ascent trajectories of rockets, characterized by the complicated interplay between propulsion and guidance, which finally determines the range of a rocket. "And if we increase the range more and more," he remarked, "then we will finally be able to build artificial satellites which continue to orbit the Earth."
>
> Such a statement, although absolutely correct on the basis of the well-known laws of planetary motion, appeared to us as a utopian project whose technical realization would be a matter of the far distant future. However, Korolev, unimpressed by our restraint and hesitation, continued enthusiastically: "And if we increased the cut-off velocity further by about 40 per cent, then we could visit the Moon. Let's all work together to achieve this."[26]

Anticipating the day when the Soviets would move back to Russia, the idea developed, says Chertok, to build a special laboratory train that could be used to measure, test, and prove out missile elements. Thus, specialists could carry out their work in "any deserted place in the Soviet Union" under tolerable living conditions. Tyulin gives a different version of why the trains were built. Korolev, he says, wanted to conduct test launches of the V-2 in Germany, just as the British did in Cuxhaven.

"Many of our comrades supported the idea. By that time we had everything required for a test launch. . . . L. A. Voskresensky [later to become Korolev's deputy for testing] was appointed chief of the flight test group. We queried Moscow. We were refused on the grounds that it was advisable to conduct this kind of work in the Union. It cannot be said that we were disheartened by such a decision; to a certain extent we expected it and so the task arose immediately of building a special train that would consist of everything from cars with missiles and ground equipment, laboratories for testing gyroscopes and radio gear, and shops housing general service rooms."[27]

A German railway firm built the train, then a second one, under contract to the Soviets. They were such a success, and so readily built by the Germans, that the military wanted one too, and they proved to be very valuable back in the ravaged Soviet Union.

The Russians were remarkably prescient in their use of the German know-how. They assigned one officer to the job of supervising the V-2 specialists in the compilation not only of detailed memoirs on their Peenemunde experiences but also on future projects that they had hoped to develop.

Astoundingly, a "Russian version of the book was later destroyed by our censorship. However, what we learned shook our imagination," said Chertok. "Many modern ideas were developed during the war years in Germany. One of the most audacious was the A-9/A-10 two-stage missile, intended, prospectively, for bombardment of America." The first stage A-10 was to have been a more powerful version of the A-4. The second stage A-9, which was winged, would have ignited at 200-kilometer altitude. Launch was envisioned from the coast of Portugal. Radio signals from the missile would be sent back to a control center, which would reconcile its position to a map. At the right time the controller would signal the missile to descend to the target. The concept, remarks Chertok, except for modern guidance techniques, is not greatly different from the terrain-following cruise missiles used during the Gulf War.

Chertok's speculation shows that he is one more victim of a misunderstanding about A-9/A-10, according to Ernst Stuhlinger, a veteran of Peenemunde who later made very significant contributions to the U.S. space effort as a member of the von Braun team. "The designation of the A-9/A-10 rocket as a missile intended for the bombardment of America is a horrendous misstatement perpetrated by those who try to prove that von Braun was a war criminal."[28] Adolf Thiel, who wrote his Ph.D. thesis on this project, and later became a vice president of TRW in California, explained to me that this two-stage rocket was studied in connection with the problem of bringing a spacecraft back from a satellite orbit, using a long glide path through the atmosphere.[29] It was one of the advanced studies commissioned by von Braun, said Stuhlinger, whose objective was "to build two- and three-stage rockets that could reach orbit with a decent satellite payload" and "a 'glider rocket' that could return to Earth from orbit by using the aerobraking capability of the atmosphere, rather than the thrust of a counter-rocket [in other words, early studies of a space shuttle!]."[30]

In any case, it was a concept, according to Thiel, that was beyond Peenemunde's technological grasp. As historian Michael Neufeld has pointed out, "the guidance requirements were too extreme, the aerodynamics were unknown, and the materials did not exist to prevent the upper stage from burning up during reentry into the atmosphere."[31] Nonetheless, the concept must have been intriguing for the Soviets when they began to develop plans for their own long-range missiles.

There were other very exotic ideas catalogued by the Germans. One even called for a manned version of the A-9/A-10. Another was to develop atomic reactors for rocket propulsion.[32] Still another was to reflect sunlight to Earth with an orbiting mirror. The latter is an idea still being fancied by advanced thinkers to supply energy to Earth. The Russians, in fact, tested a miniature version of the concept by deploying a 21-meter solar mirror from their Mir space station in 1993 to show that solar energy could gradually accelerate a "solar sail" to interstellar velocities.[33]

Another concept from the V-2 Germans that presaged actual developments by not only the USSR but the United States, Britain, France, and modern Germany: the launch of spacecraft from airplanes. There was even rumor of a "kamikaze" plan. Some twenty-five volunteers, according to Chertok, were to have been recruited to be launched in specially modified V-1 plane-missiles from long-range bombers. Targets were even designated: the industrial complexes at Kuibyshev, Cheliabinsk, Magnitogorsk, and other cities.

This rumor, says Stuhlinger, might have been sparked by a quite different idea, for a "one-way fighter plane, loaded with an explosive charge. The pilot would maneuver his plane into a collision course with an enemy bomber, and shortly before impact would push his ejection button, causing his entire cockpit to be jettisoned and parachuted to Earth. This project was not seriously pursued, as far as I know."[34]

By the fall of 1946, the Germans under Grottrup had done what their Soviet captors had directed them to do. So what happened next must have been a big shock. Perhaps as many as 5,000 skilled Germans—not only the Grottrup group of rocket specialists, but also other engineers, scientists, and technicians experienced in weapons systems, submarines, and aircraft—were literally kidnapped and shipped, with their families, by trains, freight cars, and trucks, to workplaces outside of Moscow. None of them was given a choice in the matter. On the contrary, the transport of the people was as sudden and peremptory as during the Stalin purges, even if their destination was much more salubrious. They were, in fact, treated quite deferentially.

A first-hand account of the mass movement was given in a television program, aired in the United States in 1993, by Magnus and Grottrup's widow, Irmgaard.[35] From the TV transcript:

> *Narrator:* In October, 1946, shortly before the first atomic reactor in the Soviet Union started working, a Soviet delegation came to see the rocket plant in Germany. The delegation, led by General Gaidukov, had plans for the German rocket experts.
>
> *Magnus (through interpreter):* General Gaidukov was present and listened to all the proceedings. Then he said, with a large gesture, "Well, you've been very industrious, I'd like to invite you to a dinner tonight." A huge table had been set up in a big hall there. The whole thing was lit in a festive manner, and we were offered a meal, the like of which we, at the time—remember, it was autumn 1946, and there was nothing to eat in Germany—it was a meal the likes of which we'd never seen before. Fruit in absolute abundance, which was unheard of at the time, and, of course, vodka, vodka, vodka, nothing else to drink. When the party was over, shortly before midnight, each of us individually was taken home in a car by Soviet officers. And three hours later, they got us out of our beds.
>
> *Mrs. Grottrup (through interpreter):* . . . I went to sleep, and around 3:00 A.M. I was wakened by the telephone. Someone, I can't remember who, said, "The Russians are at the front door. We're going to be taken away," and I thought it was a joke. . . . We went off if in a terrible rush that night—it all seemed chaotic to us—but the train left with all of us aboard.[36]

Magnus tried, on the night of the dinner, to have Korolev intercede for him. He told Korolev that he was about to get married but didn't yet have the marriage documents.

"'You can marry without any problem in the Soviet Union!' he told me."

Magnus said no to that idea and Korolev then urged him to come along, that he would be forced to anyway, and promised that he could return to Bleicherode in a few weeks.

Magnus was skeptical. "But do you really believe that we can travel back to Bleicherode? After what we have experienced today? I doubt it.

"Korolev then said, 'I give you my word of honor as a Soviet officer,' " and Magnus decided that he was not in a position to argue any longer. "It would not be advisable," he wrote, "to look for new arguments if the other side brought up such an emotional and dignified matter as a word of honor. . . ."[37]

Chertok says that a special directive had been issued to treat the Germans gently during the mass movement, and allow them to take "any woman" they wished, "even if she is not a wife," and any possession they wanted. Mrs. Grottrup tested the latter directive to an extreme. The Russians had given the Grottrups use of a farm, and having been warned of famine in Russia, she said she would not move without the cows and their hay.

"What to do?" says Chertok. "Loaded [cows and all]."[38]

Sergei Kryukov, later one of Korolev's key deputies in the development of ballistic missiles, recounted his personal experience as one of the Russians assigned to escort the Germans back to Moscow. "We had one passenger car with five to six German families plus several cars containing their personal effects. One of the Germans, who had no family with him, abandoned the train somewhere in Germany or Poland, leaving his belongings. This caused a problem with officials, resulting in a delay, and then we lost three days because the Poles stole our locomotive and we had to get a new one."[39]

Despite their forced transport, the Germans were given excellent treatment on their arrival. The top people got much more commodious living quarters in the Moscow environs, two or three times the salaries, and much better food than their Soviet counterparts, who were still suffering wartime privations. It was embarrassing at times, remembered Mrs. Grottrup, to be faring so well while the Russian neighbors were so obviously on hard times. Not so well provided for were the German underlings, although they too were adequately fed and housed, first in modest wood-framed houses in various villages near Podlipki, about a forty-minute drive northeast of Moscow, along the Moscow-Yaroslavl highway.

Most of the rocket specialists reported to a newly created institute called NII-88 (Scientific Research Institute-88), which absorbed the former NII-1, in Podlipki. Some of the NII-1 rocket specialists, including Vassily Mishin, Boris Chertok, and others, were transferred to NII-88, and NII-1 would eventually become the separate Keldysh Research Center.[40] The institute's establishment by decree #1017-419ss ("Issues of jet armament") of the Party and the Council of Ministers, was a clear indication of the escalation in importance of missile development in the postwar USSR. It had been signed on May 13, 1946, some five months before the Germans came to Russia, by Josef Stalin himself. It was, in effect, the "birthday of the rocket industry of the Soviet Union," as characterized by the Russian space historian, Maxim Tarasenko.[41]

The decree gave responsibility for ballistic missile development to Minister of Armaments Dimitri Ustinov, who would reign over the USSR's rocket and space flight programs throughout Sergei Korolev's career, sometimes stormily, and often in conflict with the Chief Designer.

It is interesting that it was the Ministry of Armaments, representing the artillery industry, which got this job rather than the Ministry of Aviation Industries. As Yuri Mozzhorin, one of the veterans of that era, recalled, the aviation people—already loaded with aircraft developments, and probably fearful that it would be their necks if they failed in this new and risky challenge—ducked the assignment.[42] One major consequence was that, unlike in the United States where the aviation and rocket companies essentially merged into the aerospace industry in the 1950s and 1960s, they remained divided in the USSR. Even today in Russia, with a few exceptions, companies are either in missiles and space or in aviation.

NII-88 was headed by Lev Gonor, former head of a large artillery plant, and had three branches—one was an experimental plant; another was dedicated to specialized disciplines such as materials, aerodynamics, engines, fuels, controls, telemetry; the third was a Special Design Bureau, SKB (Spetsialnoye Konstruktorskoye Byuro), with various missile design departments.[43] Named by Ustinov to head Department 3 —devoted to long range ballistic missile design—was Sergei Pavlovich. Freshly returned from Germany, Korolev was still technically under conviction for his alleged counterrevolutionary activity prior to his imprisonment.

It is ironic that the NII-88 plant had been built by a German firm, Rhein-Metall Borsig, as a machinery factory in 1926.[44] A further irony is that virtually all of the technical leaders appointed during this period to head the Soviet missile and rocket effort had become acquainted with each other in Germany, beginning with director Gonor. The chief engineer of NII-88 was Yuri Pobedonostsev, who had been at Nordhausen. His deputy was Chertok, who had headed the Institute Rabe in Bleicherode. Other German veterans were Valentin Glushko, who was made chief designer of the OKB-456 rocket engine plant in Khimki, also near Moscow; Mikhail Ryazansky, who was made chief designer of NII-885, responsible for guidance systems; and his deputy, Nikolai Pilyugin.

Korolev was made chief designer of NII-88's SKB-1, responsible for long-range ballistic missiles. It is mind-boggling that such a dispersal of responsibility among facilities, some reporting to different ministries, could produce a working system. But it did work, and it was largely because of Korolev. Each of the principals had become used to accepting the authority of Korolev from the days in Germany, when an unofficial council of chief designers came into being with Sergei Pavlovich its chairman. As Chertok put it, "They perfectly well understood that without it [Korolev's authority], and with departmental compartmentalization, with dispersal among different ministries, we could not succeed in anything."[45]

This highly competent group of engineer-managers not only functioned but achieved remarkable results in what could have been a paralytic government structure by following their own pragmatic rules under inspired leadership. It may be one more reason why the Soviets owe a debt to the V-2 Germans, since the working relationship had been started in Deutschland.

Those V-2 Germans not assigned to NII-88 went to work for either Glushko or Ryazansky. Some 150 Germans worked at NII-88, including 13 professors, 32 doctor-engineers, 95 diploma-engineers, and 21 engineer-practitioners. Only 17 of these men had actually worked at Peenemunde, according to one of this group. With their families they numbered nearly 500 when they were moved to the island of Gorodomliya, where they operated a kind of workshop as a branch of NII-88. Chertok says that, although it was surrounded by barbed wire and guarded by women gunners—the Germans, who were used to being guarded by the Gestapo, joked about that—it was actually a luxurious place. They received, as Chertok puts it, "quite tolerable feeding . . . and on weekends were periodically transported to Moscow, to theater, museums."[46] That would seem to be a much exaggerated recollection, according to Magnus, who remarked that, "I suppose that no one of our group will agree with this strange opinion of Mr. Chertok. He has a flourishing fantasy. . . . 'Periodically' means really perhaps 10 times for small groups (four to eight people in seven years)! Most of us could enjoy this never at all."[47]

The Germans were organized into work sectors, each supervised by a Russian and a German leader. In general, the propulsion, controls, and construction sections were at NII-88 itself, while the ballistics, aerodynamics, design, physics, chemistry, shops, and static test sections were at what was called Branch 1 at Gorodomliya. In the beginning there was much moving back and forth, but after May of 1948 all of the Germans were at Branch 1.

Their first jobs were to help install the A-4 assembly line brought from Nordhausen in a large building at NII-88 and to restore thirty German A-4s. Half of the latter were completely assembled, and the other fifteen were in the form of major components.

The exploitation of the Germans' skills and experience was very carefully organized to minimize contact with Soviet peers. Magnus recalled the strategy used by the Russians.

> Most of the Soviet engineers who were to take care of us did not yet have any professional experience. Some of them were still students . . . undergoing practical training. We often asked ourselves why—perhaps intentionally—we were not brought in contact with experienced specialists. . . . Only much later did we realize what tricky means the Soviets applied so that we would not learn what they were really doing: the Russian top experts, organized in work teams like us, worked in the same factory, but carefully separated from us. They worked on the same subjects as we did, and they studied our reports with greatest care. The contact between them and us was maintained by young engineers.
>
> But how could they debate with us about the technical questions, some of them really sophisticated, which the experts had asked them to clarify? As a consequence of the frequent delays caused by the need to report back to the experts, and to ask for new instructions, our anticipations turned into certainty: there was a group of experts parallel to ours, but anonymous to us! Its members were first class specialists, they followed and controlled our work, but remained invisible to us. . . . There were indications that Korolev was the chief of that phantom group. He, who had talked with us almost daily and very openly in Bleicherode, contacted us only rarely in the Podlipki works. And when he did come, he was always very curt, asked very precise questions, and then disappeared again quickly.[48]

One of the young Russian engineers who was assigned to work with the Germans in early 1947 was Oleg Ivanovsky, later be to a key designer on the Korolev team. "The Germans were working with us cheek by jowl," he remembered. "They showed us how the V-2 was put together. There were about ten or fifteen of them where I worked, not hundreds. They lived in a rest home, then later on an island [Gorodomliya]. They were delivered to the office by bus every morning. They got twice the money we got and they were allowed to visit Moscow. I learned practical skills from them. We assembled several V-2s, and SAMs [surface-to-air missiles], from the hardware brought from Germany."[49]

In June of 1947 the Germans were asked to design a new long-range ballistic missile, which was given the designation G-1. Grottrup headed the project and the Russians were asked to give them support. "However," says Chertok, "starting in 1948" there was an "active search for Russian authors of all inventions, discoveries and the newest scientific theories."[50] This put the kibosh on any German-originated development. Korolev himself, furthermore, was not enamored personally of the idea of supporting the German concepts.

As Tyulin put it, "He, one of the first pioneers of rocket technology in our country, had to drink a full cup of humiliation as a prisoner and to learn, after his release in 1944, that some of his own ideas had already been realized, and that German rocket men had gone farther than his most ambitious plans."[51] It was galling to him to have been made chief designer and then asked to test, not his own concepts but the R-1, which was to be an exact replica of the V-2.

It was already recognized that the V-2 was obsolete, but it was also recognized that mastering its production, learning how to launch it, and then instructing the military in its operation were of vital importance. In fact, Stalin himself, according to Tyulin, was the one who insisted that a thorough understanding and construction of the V-2 was needed before going on to indigenous rockets. Korolev, said Tyulin, "was convinced that a group of our leading designers was capable of creating a more reliable rocket . . . with longer range. Stalin listened to those words from Korolev in a private briefing. His response: "First we must complete work on the R-1."[52]

A description of the first of the A-4 test launches was given on the TV show by Magnus and Mrs. Grottrup:

"In the summer of 1947," said Magnus, "six or eight colleagues from our group suddenly disappeared. We weren't told where they'd been taken. We were very worried that they'd been abducted again, but it soon transpired that they'd been taken to this rocket testing site."[53]

The site was Kapustin Yar, in the Astrakhan area of southern Russia, which became the main test area for ballistic missiles over the next decade, until Tyuratam, known better today as Baikonur, took over in the late 1950s. It was in Kapustin Yar that the first Soviet A-4, rebuilt under the tutelage of the Germans, was fired successfully. Mrs. Grottrup was told that, "After that launch, they were so happy and enthusiastic. They jumped up and down like little children. High ranking ministers or not, they were just like little children. Then they grabbed their vodka bottles and got drunk."[54]

Chertok recalls that in September 1947, a contingent arrived at what he calls the State Central Range at Kapustin Yar for the first test launch. "Officers were somehow accommodated in a small town. Soldiers lived in tents and dugouts."[55] Tyulin describes the place as a "Bare, lifeless steppe with dry sagebrush gray from dust, camel's thorn, and sparse little islands of spurge. There was essentially no water. The hot wind chased the swirling dust and balls of tumbleweed."[56]

The test stand, says Chertok, was "very close to our special train. An airfield, where planes landed on a ground strip, was also in the vicinity." Assembly and test were in a wooden barracks. "It was cold and the wind blew through it.

"At last, the missile was rolled out to the firing stand. However, we couldn't ignite the engine for a long time. The igniters . . . kept failing . . . one or another relay switch failed every time." These failures were "hotly debated at sessions of the State Commission" chaired by Marshal of Artillery N. D. Yakovlev. On the commission were "not only such high chiefs as D. F. Ustinov but the deputy of Beria, Serov [Ivan], as well . . . a 'Damocles sword' of reprisal hung above everybody.

"The first launch was performed on 18 October, 1947, at 10:47 A.M. The second on 20 October. Early in the powered flight a strong deviation of the missile to the left was recorded. Reports of the expected impact did not come, and on site observers commented, not without humor: 'It went towards Saratov.'

"Serov ticked off to us: 'Can you imagine what will happen if the missile landed in Saratov? I won't even tell you, you can guess yourself, what will happen to all of you.' But we understood that Saratov was more than 270 kilometers away from where the missile was launched [the range of the A-4], and therefore we were not nervous."

It was learned later that the deviation was 180 kilometers. Ustinov decided that the Germans should be consulted.

"Before that, Dr. Magnus, a specialist in the area of gyroscopes, and Dr. Hoch, knowledgeable in the area of electronic controls and guidance, had sat on the range without a particular job. Ustinov said to them: 'this is your missile, your devices,' " and they were ordered to figure out what had gone wrong. They adjourned to the lab on the train, began analyzing the guidance data, and soon determined the cause of the deviation. [Tyulin's account says it was gyroscope error caused by vibration.] Changes were made and the effect was immmediate. The deviation on the next launch was small.[57]

Ustinov directed that all German specialists and their assistants were to get "huge" bonuses—15,000 rubles each, and a "canister of alcohol. . . . We amicably celebrated a successful launch."[58]

Quite a different version of the unsuccessful and successful launches is described in the book by Magnus. After the failure, he writes, "Ustinov . . . looking at us with a stern face . . . said: 'The day after tomorrow, the next rocket will be launched. Can . . . each of you guarantee that the launching will be successful as planned?' What could we answer? We had been asked the same question again and again during these past few days."[59]

Grottrup replied that the preparations were careful and there was no reason to expect failure. But Ustinov pressed relentlessly and there were "questions, blame, explanations, and justifications . . . for some time." Korolev kept the discussion "within bearable limits by translations, interpretations, explanations. . . . We could not help being drawn into a jungle of technical details." Eventually Ustinov said, "You know that the next test launchings are very important. Do your job as you are expected, then you can count on an award."

"Grottrup, quick-witted as always . . . replied . . . 'Sir, the best award would be your decision to allow the German specialists to return to their home country.' "[60]

The next launch was, indeed, successful, and, as Magnus describes it, there was "Relief, even jubilation everywhere. Korolev is beaming with joy, he thanks us with a handshake." Later, however, Korolev asked the Germans to stay on for a talk. "There was something in the air; we sensed that immediately. As soon as we were alone, Korolev began with a serious, even worried expression on his face which did not fit at all with his victorious smile of a few minutes ago: 'The Minister ordered me to investigate whether the failure of the first two test launchings could have been caused by sabotage of the German specialists. It makes me very suspicious, he said, because the error could be corrected in such a very short time.'

"We thought that we had not heard correctly. That was a blow below the belt which showed us quickly that we lived in a land of institutionalized distrust. . . . 'How could we have committed sabotage, since we did not participate in the preparation of the tests at all?' I asked." Korolev persisted by asking if such failures had occurred in Peenemunde. Of course, said Grottrup, and they were always analyzed carefully. Korolev asked for a report, including guidelines for component tests in the future, and said that he personally did not suspect malicious intent. He was apparently under orders to carry out the grilling.[61]

Eleven of the German-built, Russian refurbished, V-2s were launched, and five of them reached their targets, the same percentage of effectiveness as the Germans had during the war. Five of the eleven had been assembled at Nordhausen, six at NII-88. "Both proved equally unreliable," was Chertok's comment.

After this the Russians set to the job of building their own missiles, a task "unbelievably difficult for our country. . . . We had to develop a lot of technologies from scratch. Furthermore, where to get the materials . . . absolutely new materials . . . never produced by our industry before."[62] There were "only" thirty-five research institutes and design bureaus and sixteen principal plants involved in the R-1 program.

Back to Kapustin Yar went the Russians in the special train in the fall of 1948, this time with twelve NII-88 built R-1s. Nine were launched and seven reached their targets, a better score than the "trophy" missiles, but still not very reliable." The first R-1 launch, on September 7, 1948, had a control system failure. The first successful flight was on October 10.[63] Chertok directs a barb at contemporary manufacturing practices: "Just try now, in our great shops," he says, "with wonderful equipment, to produce a new missile in a year. Give the personnel good rations and bonuses and you still won't be able to do it. Yet we did it. We made a first series, a second, a third. In 1950 the R-1 was commissioned for service. In parallel, work proceeded on other missiles and in ten years we followed the path from the R-1 to the R-7—the first ICBM—it is still performing for cosmonautics."[64]

The path to the R-7, however, had first to go through the R-2, R-3, and R-5. It began with the R-2 development starting in June 1947, paralleling Grottrup's work on G-l, which had the same design objectives—600-kilometer range, more than twice that of the V-2/R-1, and with a capability of warhead separation, propelling only that payload rather than the entire vehicle to its target. It is not clear, says Chertok, whether it was Korolev or Grottrup who first thought of warhead separation.[65]

In September 1947, Grottrup submitted the G-1 design to the NII-88 science and technical council, proudly pointing to the fact that G-1 was not only capable of 600-kilometer range, but was no larger than the V-2 and promised ten times the accuracy. Furthermore, he said, it should provide a basis for development of much longer-range missiles. Korolev's similar R-2 project, he recommended, should be carried on independently and in parallel right up to pilot production.

How to respond to that recommendation was the quandary of one Sergei Vetoshkin, Ustinov's principal assistant for rocket technology in the Ministry of Armaments. " . . . he asked me directly," says Chertok, "'Boris Yevsevevich, you started all this activity in Germany . . . and know what they are capable of better than I. Now they design a new rocket . . . the main question . . . for which Dmitri Fedorovich has tortured me, is what to do with this project? The Germans can't build it on the island with their own capabilities.' "

"The question wasn't simple," Chertok goes on. "There couldn't even be talk of Korolev working under Grottrup. What about Grottrup under Korolev? That's also unrealistic because Korolev would immediately declare that 'We'll do it ourselves.' "[66]

The Soviets couldn't afford two parallel design bureaus, each with its network of subcontractors. Chertok's recommendation was to use the Germans' ideas and experience as resourcefully as possible but to gradually send them home. He went to see Grottrup in Gorodomliya in the winter of 1948. Grottrup made a long speech to the effect that they had not been able to implement any of the experiments fundamental to the G-1 design. They had not been permitted to carry out aerodynamic studies, use the engine test stands. How can we prove, he complained, that it is possible to drive a turbine with gases bled directly from the combustion chamber? Calculations won't provide the answer. Experiments are necessary. Radio communications must be tested using aircraft and test ranges. He asked what the future of the Germans was.

"I don't have the right to kill his hope," Chertok thought, and advised Grottrup to continue his work. The G-1 project was upgraded and by the end of December, the draft design showed substantial improvements in prospective performance, including a range of 810, rather than 600 kilometers. But there simply was not enough capability at NII-88 to keep the project going in parallel with Korolev's, and so in 1950 the Ministry of Armaments cancelled further works on long-range missiles in the German collective. For some months the Germans worked on secondary tasks.

Magnus describes some of these jobs in his book: "a barometric altitude meter for airplanes . . . chemists studied novel rocket propellant combinations. . . . Electricians and electronic experts worked on a quartz clock [that] achieved an accuracy that approached that of the best mechanical clocks known at that time. . . . One task that had particular importance for the island was the design of a 'hermaphroditic' vehicle, a crossbreeding of car, boat, and sled . . . to guarantee contact between island and mainland at times when ice formation excluded traffic by boat, but when ice cover was [not] thick enough to support a car or sled." Magnus himself worked on such devices as a "position gyroscope [a positional reference system] with a special double rotor . . . to improve the . . . accuracy of existing instruments with non-symmetrical rotor profile."[67]

The 1960 CIA report, which is so detailed that it must have been compiled by a particularly competent spy or even by one of the Germans working at Gorodomliya, says that most of the work performed up to December 1949 was on designs aimed at improving the A-4, but some "presented new and radical concepts."

There were three categories of design projects—the preliminary, or Vor-Projekt; the sketch, or Skizzen Projekt; and the technical, or Technisches Projekt. The Vor-Projekt was just a ten- to twenty-page concept. The Skizzen Projekt was a detailed design including not only the chief project engineer's concept but comments and proposals by engineers responsible for major components. It could consist of fifteen or twenty reports numbering up to five hundred pages and drawings. Most advanced was the Techniches Projekt, which offered new designs complete with all calculations, and parts drawn in enough detail to permit production drawings.

Some eight design proposals emerged. One was the R-10, which, based on the data in the CIA report, seems to be the G-1, whose ill-fated end was described by Chertok earlier. It would have achieved its increased performance over the R-1 by increasing the engine combustion chamber pressure; by eliminating the hydrogen peroxide gas generator system and bleeding combustion gases directly into a turbine that would power the fuel and oxidizer pumps (this was the concept referred to earlier, causing Grottrup to complain that they weren't allowed to do the experiments that would have enabled it to develop); and by reducing the vehicle's weight through the use of integral fuel tanks.

The German version of R-11 never got beyond the sketch stage. It would have increased the R-1's range by increasing the length of the vehicle by 2 feet, enabling it to carry more fuel. It also called for a redesigned engine that would have upped the thrust to 35 tons. The Russians did build an R-11 as their first tactical missile, however, with an engine designed by Isaev, using nitric acid and a hydrocarbon fuel. Range was 270 kilometers and the first launch was on April 18, 1953.[68]

The R-12[69] would have had three engines, two of them to be jettisoned after initial boost, to increase range to as much as 2,500 kilometers. This project was ended between the Vor-Projekt and Skizzen stages.

R-13 would have had a 1,000-kilometer range, using a single engine of 120 tons of thrust. An alternative design would have had a cluster of four 35 ton engines. This project never reached the Skizzen stage.

Minister Ustinov, in the spring of 1949, says the CIA report, asked the Germans to produce a design for a ballistic missile that could carry a 3-ton warhead 2,500 kilometers. They came up with R-14[70] and R-15. R-14 would have been a cone-shaped vehicle, about 25 meters long, 3.7 meters maximum diameter, calling for engine thrust of 100 tons. "Range control," says Magnus, "would have been by fuel cutoff using either inertial or radio control. Guidance would have been by special gyros and by radio-beam."[71] It was carried through most of the final, or Technisches Projekt, stage. R-15 was a two-stage concept. First stage, an improved R-1, would boost the second stage onto a horizontal flight path at an altitude of about 140 to 200 kilometers. Second stage was a cruise missile powered by a winged ramjet, 12 meters long, with a 5-meter wingspan. This concept reached only the Skizzen stage.

Both the R-14 and R-15 concepts were submitted to the NII-88 Scientific Technical Council. But unlike the case of the R-10/G-1, Grottrup was not given a chance to participate in the council deliberations, although numerous requests were made for more details. In fact, in 1952 two items that seemed to be related to the R-15 were actually ordered from the Germans.

In the summer of 1949, while they were working on the R-14, one of the senior NII-88 administrators, General Spiridonev, asked the designers to consider possible ways to counter a ballistic missile attack with an anti-missile. The Germans said they were too busy on the R-14. "Your work on the rocket R-14, with the exception of some details, will now be terminated. From now on you will work on a new project," the Germans were told by Ustinov himself, according to Magnus.

> Our surprise could not have been greater. Ustinov went on to explain that an anti-rocket rocket should now be developed. Its purpose should be to encounter and destroy approaching rockets at such high altitudes that their explosive charges would not do any damage on the ground. At first we thought that we had misunderstood what he had said. . . . Grottrup was the first to grasp the situation . . . he remained admirably cool and, quite factually, asked the precise question: "Are there Soviet instruments that allow us to locate approaching rockets precisely and at an early time? And how much time do we have between having located the oncoming rocket and its destruction?"
>
> Embarrassed silence. The Soviets looked at each other in surprise and whispered to each other. There was no translation. Finally, they said: "Details will be clarified after work on the project has been started."[72]

A few of the Germans at Gorodomliya were put to work gathering data on the two surface-to-air missiles being developed during the war—the Schmetterling (Butterfly) and the Wasserfall (Waterfall). Neither of these missiles had been used operationally, not having matured sufficiently, but some work was done on their development while the Russians were in Berlin in 1946.

Schmetterling (HS-117) was a midwinged monoplane about 4 meters long, with a 2-meter wingspan. It carried a 25-kilogram explosive charge about 15 miles against targets below 10-kilometer altitude. The Russians tested the control system of one of the two Schmetterling working models that had been reconstructed under Russian direction in Berlin. One of these had actually been launched from an aircraft to test the control system. In Russia the work continued, and at one point the control system was tested in one of the wind tunnels at TsAGI. TsAGI is still today one of the premier aerodynamic testing centers in the world, with such an enormous range of low-speed to hypersonic-range wind tunnels that U.S. aerospace firms are contracting to carry out research there.

Wasserfall was an 8-meter-long surface-to-air ballistic missile with a 1-meter-diameter body stabilized by four stub wings. Steering was by four vane assemblies in the rear. It weighed 3.5 tons and could carry about 90 kilograms of explosive to a 20-kilometer altitude. Thrust was from an engine using nitric acid as an oxidizer and aniline as a fuel. A small group at NII-88 worked on reconstituting the missile but the effort was low-key. It seemed that the Russian interest in both Schmetterling and Wasserfall was primarily in understanding the weapons technically rather than in producing them.

Much higher interest was manifested by the Russians in the R-113, a more advanced surface-to-air missile that would incorporate some of the Wasserfall features. A group at Gorodomliya was asked in September 1950 to do a design study of R-113, using a Wasserfall power plant modified to give higher performance. Slant range was to be about 50 kilometers and the missile was to be effective in the range of 5 to 30 kilometers. The Germans worked six months on the proposal and handed it in in April 1951.[73]

During the German study, speculates the CIA report, the Russians themselves gave indications that they were doing surface-to-air missile development. " . . . most definite indicator was the Soviet preference for the extensive use of wood and metal construction in the missile design. When the Germans raised questions about the difficulty of bonding the two together, the Soviets provided excellent glues and practical solutions. The timing of the requirement, which came towards the end of the Germans' exploitation, and the interest of numerous Soviet visitors to the island, who were not part of the NII-88 staff, tend to support the conclusion that the Soviets wanted solutions to problems their own designers had encountered."

One of the interesting conclusions of the CIA report was that perhaps the Germans' "most valuable contribution" to the Soviets overall was in the field of instrumentation and test equipment. Most of the German work in the USSR, both in rebuilding German missiles and in researching new designs, was "the repair, refurbishment, and calibration of scientific and technical equipment without which the missile research and development program could not have proceeded. In certain cases the Germans created new pieces of test equipment, such as a flight simulator which would simulate the flight conditions of various missile designs by means of electrical circuits and save many hours of practical testing."

Another statement in the report signals a policy that has served the Russians in a major way throughout the development of their space program, namely, the introduction of recent graduates of universities, and even those who were still students, into practical on-the-job engineering tasks. " . . . when the Germans were removed from their work on classified material," says the report, "the Russians assigned different groups of young, inexperienced engineers to Branch 1 on Gorodomliya Island so that they might learn from the German personnel." Throughout the 1950s and 1960s it was Korolev's consistent policy to integrate the university education of young engineers from such schools as Moscow State University (MGU), Moscow Aviation Institute (MAI), Bauman Moscow Higher Technical School (MVTU), Moscow Physical Technical Institute (MPTI), Moscow Engineering and Physical Institute (MIPI), and Leningrad Mechanical Institute with stints at NII-88 and later OKB-1, the Korolev design bureau.

This practice, with some exceptions, is not typical of the U.S. aerospace industry. Generally American undergraduates studying aerospace engineering have little or no exposure to the industry until they are hired by one of the companies. There are summer internships at NASA, and at most companies, but the Soviets follow a practice of systematically bringing students into the design bureaus as a regular part of their education. This probably accounts for the practicality of the Russian engineers' training, in contrast to the largely theoretical training of undergraduate American aerospace engineers.

It was in October 1950, says Chertok, that all of the Germans' classified work was cancelled, although the Wasserfall modification described above must have been an exception. According to Gyorgi Vetrov, who visited Gorodomliya in those days, "they did essentially nothing, playing tennis, planting flowers and waiting for repatriation."

Magnus says that, "We read, did some personal work, or simply loafed around. Well, one cannot just loaf around for months. Therefore, we began, mostly in groups of two or three, to beef up our professional knowledge. For example, one of these groups worked through a test book of theoretical physics, chapter by chapter, discussed the content and deepened everyone's understanding by solving the problems. . . . For me, several months were filled with translation of a newly published monograph about the theory of dynamic stability."[74]

Back to Germany went the whole team. The first group departed on January 20, 1952; the second on June 10 and 13, 1952. On November 16 and 20, 1953, the last sizeable group of some twenty-four families went home, including Grottrup and family. In 1954 a small final contingent, including Magnus, returned to the Fatherland.[75]

By then the Russians had substantially milked their ideas and, led by Sergei Pavlovich, were well into their own long-range rocket developments, with the intercontinental ballistic missile as top priority.

6

The World's First ICBM: Aimed at the USA

It is amazing to realize how quickly the Soviets, led by Korolev, developed an intercontinental ballistic rocket out of a postwar military-industrial capability that was pitifully weak. Beginning virtually from scratch in 1946, the Russians, albeit with the help of very effective turncoats from the West, had exploded their first atomic bomb in August 1949, only four years after the first U.S. test demonstration at Alamogordo. Then, a team that included Andrei Sakharov, working at an intense pace, detonated a hydrogen bomb on August 12, 1953, only nine months after the United States.

But it would be the bomb's delivery system, Korolev's R-7 "Semyorka" (Little Seven, idiomatically), that would become the world's first ICBM. On April 19, 1957, Korolev had finally been notified of his rehabilitation by the Khrushchev government. It was now official that he had been unjustly imprisoned. Four months later, on August 21, 1957, as if in gratitude, his R-7 propelled a dummy warhead an intercontinental distance of about 6,400 kilometers, some 4,000 miles, from Baikonur to the Kamchatka peninsula.[1] That feat would not be matched by the USA until fifteen months later, on November 28, 1958, when Atlas carried a simulated H-bomb payload from Cape Canaveral more than 10,000 kilometers into the deep South Atlantic Ocean.

For his accomplishment, Sergei Pavlovich deserved national recognition in a country whose paranoia about alleged American aggression was at a peak. But he got only the personal satisfaction of the achievement, since his very identity was a state secret.

Only obliquely was he referred to by the Soviet leaders. In a July 9, 1958, speech in Bitterfeld, East Germany, Khrushchev eulogized his rocket specialists extravagantly while stating the rationale for preserving their anonymity:

> We shall erect an obelisk to the people who have created the rocket and the artificial earth satellites. We shall inscribe their glorious names in gold to make them known to posterity for centuries. In due time the photographs and names of those glorious people will be widely known by the people. We highly value these people, we prize them and protect their safety from hostile agents who might be sent in to destroy these outstanding people, our valuable cadres. But now, in view of the safety of the country and of the lives of those scientists, engineers, technicians

and other specialists, we may not as yet make public their names and their photographs.[2]

While the creators of the rockets remained a secret, the fact of the test launch was trumpeted around the world. It was particularly satisfying to brag to Americans. In an interview with publisher William Randolph Hearst Jr., on November 22, 1957, Khrushchev said, "The Soviet Union possesses intercontinental ballistic missiles. It has missiles of different systems for different purposes. All our missiles can be fitted with atomic and hydrogen warheads. Thus, we have proved our superiority in this area."[3]

And so went the crowing for the next several years. On September 7, 1958, Khrushchev scoffed at U.S. naval power in a letter to President Eisenhower, "In the age of nuclear and rocket weapons of unprecedented power and rapid action, these once formidable warships are fit, in fact, for nothing but courtesy visits and gun salutes, and can serve as targets for the right type of rockets."[4]

Although this was mostly bluster, and it was some time before the Soviets had enough of an H-bomb arsenal to be a strategic threat to America, the perceived threat was palpable, and it was Korolev's R-7 that backed up the posturing.

The development of the R-7 is a remarkable story not only of technical inventiveness, but of resourcefulness against tough obstacles. It took the United States seven years to develop the Atlas, although the first four years, 1951 to 1954, were spent convincing the government that it would work. A sizeable number of distinguished engineers had been convinced that the cruise missile, like the Navaho being developed at the time at North American Aviation in California, was the more logical way to carry a warhead over great distances, and that the ballistic missile was not feasible.

Eventually, however, the arguments of ballistic missile pioneers like Karel Bossart of Convair, considered the father of the Atlas, won out. As described by one of Bossart's colleagues, Richard Martin, the "ballistic missile was simpler, cheaper, not the huge technology leap many thought it to be, and particularly it was less vulnerable to defensive measures."[5] Similar arguments were going on at the same time in the USSR, where Korolev himself, as well as other designers, were working on cruise missile concepts. And there the answer came out the same: the ballistic missile got the nod. The argument won by 1954, but it took four more years, until 1958, to develop the Atlas. The R-7 had made it from authorization in early 1954 to first successful launch in 1957. The "rocket packet," or staging principle on which R-7 was based, however, had been drawn up by Mikhail Tikhonravov in 1947 and went through progressive evolutionary changes on paper through 1953.[6]

But the start-up difficulties, beginning in 1946 at NII-88, where the R-7 would eventually be developed in the small town of Podlipki, outside Moscow,[7] were monumental. Following the loss of more than 20 million of its people in the Great Patriotic War, and perhaps tens of millions more, including some of its best citizens, in the pre-war purges, the Soviet Union was in desperate condition. The area around the NII-88 plant, which had been dedicated to the manufacture of artillery pieces for the war, was in an awful state.

"Hardly anybody suspected that the plant was destined to become a production base for such complicated and demanding technology as rockets and space vehicles for traveling to other planets," wrote Gyorgi Vetrov.[8]

NII-88 director Lev Gonor, a Red Army general who had managed an artillery factory during the war, found that he had inherited a plant that was in very bad shape. There were no working tables for the designers, so equipment boxes had to be used. The production buildings were in disrepair. The roofs leaked so that puddles appeared on the factory floors when it rained. "The heating didn't work," wrote Vetrov, "so it was colder inside the shops than outside."[9]

The food supply, as in all of the Soviet Union in the postwar period, was dismally inadequate. "People who had a cow," said Vetrov, "or more than three goats," were not allowed to have bread. Most of the workers had to depend on their own "kitchen gardens," which were commandeered around the nearby countryside. "We went across swamps and wherever we found a piece of land for five to ten people we took it." That sometimes caused trouble with the local authorities. "Last year [from a March 1947 account of a local Communist Party meeting] we were brought before the Prosecutor. Five hundred slots were taken away, while there are 2,000 more people who have applied for kitchen gardens. We don't know what to do. . . . "[10]

Disease was rampant among the workers, expanded hospital facilities were desperately needed, housing problems were acute. Workers had to be moved from tent camps, degenerating barracks, and basements and reaccommodated in decent structures. During 1946, reported Vetrov, 1,832 people quit, even though the tasks assigned required expansion of some specialties by factors of two or three.

Trying to keep his staff content during these times was perhaps one of Korolev's most difficult challenges. He spent almost as much time dealing with the welfare of his people as he did in developing the technology. His technical secretary, Antonina Zlotnikova, says, for example, that he followed faithfully a practice that would have been unusual even for the personnel departments of American aerospace companies:

> People would come to him with all kinds of requests, and he would see everyone. And he always checked to make sure that his orders had been carried out. Every Thursday was reception day. Sergei Pavlovich would begin seeing visitors in the morning, and stayed until he had seen his last appointment. This was on average 30–35 people. It was a difficult, post-war time. And Sergei Pavlovich always helped people in trouble. He got them medicine and interceded about their housing.[11]

The tribulations of trying to develop new rocket technology in such an unstable enterprise were hardly diminished by events in Korolev's personal life. His marriage with Xenia was coming apart. Partly, perhaps, it had to do with Xenia's deep preoccupation with her own medical work. But another reason was the emergence of a rival, a pretty divorcee named Nina Ivanovna Kotenkova, who was thirteen years younger than Sergei Pavlovich. Nina Ivanovna was an English interpreter in the Podlipki office, and Sergei Pavlovich wanted to keep up with the American rocket journals. He asked her out and she fell in love with him on the first date. At this time Korolev was living in an apartment in Podlipki, near his work. Nina Ivanovna lived in the same apartment. Xenia was infuriated by the affair.[12]

Daughter Natasha, who was only twelve years old at the time of the breakup, and who knew about the infidelity, recalled the sad period in a 1991 interview, only a month after the death of her mother Xenia. The divorce, she told me, took place in 1948.

"It was a tragedy for our family because my mother loved my father. After this I met my father not so often, usually only at his mother's house, or in her dacha in Barvikha, about 7 kilometers west of the Moscow beltway."[13]

In the year before the divorce, Korolev had already begun to share the problems of his daunting job with Nina Ivanovna, although their marriage did not take place until 1949. For example, in a letter to the young interpreter he described the terrible conditions prevailing at the launch centers where the missiles were tested. One of Korolev's Soviet biographers, P. T. Astashenkov, quotes from letters penned during the earliest launches of the R-1 at Kapustin Yar.[14]

> [On October 8, 1947:] "The dust is horrible, it is hot by day and cold by night. There is a shortage of water. And this dismal saline steppe all around. Our mobile living quarters are simply an oasis. . . . I am convinced that we will return with good, great accomplishments." On October 12 he told Nina Ivanovna of his typical day's work, "My day

consists approximately of the following: I arise at 0430 hours, Moscow time, quickly breakfast and stroll a bit in the field. We return sometimes by day and sometimes in the evening, but then as a rule comes an infinite succession of all possible questions until 1 or 2 o'clock in the morning; it is rare that we can go to bed earlier. . . . Our work abounds with difficulties, with which we can as yet not cope. It is comforting that our ground crew turned out to be so friendly and unified. For under these conditions it would have certainly been impossible to work otherwise. The mood of the people is cheerful, as the decisive days approach. . . . [15]

Nonetheless, Korolev's nasty moods often erupted. "He makes a lot of noise, and drives everyone out," Atashenkov quotes one of his colleagues. "And this is all because failure in all technical questions he takes as a personal one. He could not distinguish the physical from the pathological. If things went badly he could not live peacefully. He said, 'I can never forget, going home, that something is wrong with the technique.' "[16]

Going to work for him during this time, however, were some of the brightest young engineers in the Soviet Union. Stimulated by Korolev and the challenge of pushing technological frontiers and developing the world's first intercontinental ballistic missile, they worked with great zeal. One of the new engineers, Oleg Ivanovsky, who was a key designer of a number of the bureau's missiles and spacecraft, remembered his initial exposure to the Chief Designer:

It was in 1947 that I saw Korolev for the first time. My partner and I came out of the building one day. We were standing near the gates when they opened and a powerful German trophy car came in and passed us at high speed. Inside was a guy in a brown jacket who looked very intense. Who is this crazy guy? "That is the King," we were told, "he can't drive slower."[17]

Another engineer who joined the design bureau in this period, Sergei Kryukov, told me about his entry and personal growth at the Korolev design bureau. It reveals a mode of training and discipline that seems to have characterized the whole working force that eventually produced so many innovative rockets and spacecraft.

About seventy-five years old at the time of the interview, Kryukov was only twenty-eight when he first went to work at NII-88 in 1946. In apparent good health today, he is tall and lean, bald, well-tanned, and well-tailored. He wore a gray suit for the interview in my hotel room, a neatly pressed tan shirt, and he took his jacket off to display natty colorful suspenders and a silk tie. His shoes, though, were characteristically Russian—scruffy, black, with thick rubber soles to protect against Moscow's omnipresent mud puddles.

"I graduated from school in 1936 and entered Stalingrad Mechanical Institute. The war began in 1941. I had already begun to work in an artillery plant in Stalingrad, continuing my studies as well. Then we were let out of the Institute, without diplomas, in 1942, and evacuated to Siberia."[18]

Kryukov was sent to Germany in July 1946, to study surface-to-air missiles, and was one of the Russians assigned to escort German specialists back to the USSR that October. He continued:

"I joined the artillery department of Moscow High Tech [Bauman Institute], got married, had a daughter, and needed a job. I asked someone in the Ministry of Defense Industry, then headed by Ustinov, for a job and I was sent to Podlipki. There were several artillery plants converting from military production at that time and there were few jobs around. I was assigned to a design research department at Podlipki, under Bushuyev [Konstantin, later the Soviet leader of the USA–USSR Apollo–Soyuz rendezvous in space in July 1975], my first teacher. He was very smart and chose the right directions for our work."

Bushuyev was one of Korolev's three deputies. First among them was Vassily Mishin, who was to take over the entire design bureau when Korolev died in 1966, including the responsibility for the ill-fated manned lunar landing program. The other was V. S. Budnik, who, like Mishin, had worked with Korolev in Germany, when Department 3 was being formed. Budnik was in charge of construction, while Bushuyev ran a design department. Mishin, a strong-minded and very competent engineer, was Korolev's right-hand man, however, and was in charge whenever Korolev was absent, as he was in August 1946, on a mission to Germany. From the beginning Korolev and Mishin were intent on staffing up with dedicated people. Those who could not tolerate the severe working conditions and disciplined management soon departed. The size of the early staff was 140, including 60 engineers, 55 technicians, and 25 "practicians" (usually equal to engineers in knowledge and experience but without formal diplomas) but by the end of 1947 it had grown to 310.[19] Adding good people was not easy, though. Not everyone in the Ministry of Defense Industry realized how important was the work of the department. But Ustinov did, and it was his energy and persistence, as well as Korolev's, that forced the bureaucratic apparatus to work actively to supply the needed personnel.

Of even more importance than the staffing challenge was the generation of political and funding support from the top levels of government. Ustinov was a vital ally, but it was Stalin, himself, who was making the big decisions on how to allocate defense rubles.

On April 14, 1947, Korolev was summoned by Stalin to a key meeting of a special commission that called for the development of a long-range plan for rocket technology development.[20] Korolev was ready with a plan whose basic premise was that the German V-2 technology, which was being reproduced exactly at Podlipki—on Stalin's orders—was clearly past its time, even though mastering it had been fundamentally important. New approaches were needed for systematic development of an ICBM.

Korolev's ballistic missile ideas had been germinating as far back as the early 1940s when he was still a prisoner in Kazan, although at that time he thought that solid-fuel rockets were more likely to produce the tremendous thrusts needed than liquid rockets.[21] It was the German experience with the V-2 that persuaded him differently. But new technology—larger rocket engines with more advanced fuels, stronger fuel pumps, more accurate guidance systems, more effective launch techniques—would be required.

An early product of the plan would be the R-3, a 75-ton missile, which would have put all of Europe within its 3,000-kilometer range. However, the development of the R-1, R-1A, and R-2 would continue simultaneously and serve as an experimental base for testing out promising construction designs, and manufacturing these missiles would provide a base of production experience for the R-3. R-1 was simply a Russian copy of the V-2. R-1A was the same missile, but it tested out the concept of a detachable warhead, which would be used on the next development, the R-2.

R-2 called for a true technological jump. The R-1/V-2 frame was much heavier than it had to be. The entire vehicle had been designed to take the high dynamic loads experienced during launch, whereas the loads in flight, and on atmospheric reentry, were much less, and could be handled by a lighter structure. The weight saved could be used for fuel and payload. The solution was to separate the payload from the launch portion of the rocket. Korolev had first mentioned this idea to his staff in March 1947.[22]

The R-2, including the separable warhead, was a 20-ton vehicle designed for a 600-kilometer range. It incorporated some of the ideas conceived by Grottrup's German team, which had been working in Gorodomliya on a similar project, called R-10 or G-1. One was the use of the missile's own outer skin, rather than separate internal tanks, to hold the fuel. This saved on weight. Another was to incorporate a more powerful rocket engine, the Glushko RD-101, which had a thrust of 35 metric tons compared to the 28-ton thrust of the R-1A's RD-100 engine.

R-2 was launched from Kapustin Yar October 26, 1950, and flew 600 kilometers. It went into service in the Red Army in 1951.[23]

Korolev was very hard on his staff during these years, but he also made sure that they received special amenities over which he had influence. It was possible, for example, for high performers in industry to get what were called *Profsoyuzny* or Trade Union Doctorates for job performance, without having to defend a thesis at a university. Korolev, who got one of these doctorates himself, also saw to it that some of his top deputies got them. When, in 1953, Korolev was elected a Corresponding Member, or *Chlen Kor,* of the Soviet Academy of Sciences, he again stepped in to help staffers receive the same honor. But as ready and willing as he was to help his key staff, he was just as quick to cut them off if he was crossed. He often threatened to fire people in his anger but, Vetrov writes:

> . . . [I] can't recall that he ever did. What he didn't tolerate, though, was betrayal. His attitude was "I am the Chief Constructor. I have the need to get all of the information first."
>
> An example of how disturbed he would be if he didn't get the information: he had greatly valued one employee—Nikolai Nikolaevich Zhukov, who was in charge of telemetry measurements. Every time Korolev went to the test range he took Zhukov. One day, during an R-2 test at Kapustin Yar in late 1950 or early 1951, a warhead blew up on reentry. But Korolev was informed of this failure by the military representatives, not by Zhukov, even though Zhukov had it first. After that he lost his respect for Zhukov, stopped talking to him and did not interfere when he was reassigned to Yuzhnoye.[24]

While the Korolev plan for ballistic missile development was clearly bold and innovative, the NII-88 collective of designers from the other departments was skeptical that it could be carried out. What's more, some found Korolev's tough, demanding style and his independent character objectionable. And the fact that he was an ex-prisoner didn't add to their confidence.

It is interesting that one of the main forums for airing company problems in those days was at meetings of the local Communist Party group. At one such meeting Korolev got a dressing-down from the Party Secretary, a fellow named Pokrovsky, who pointed out that Korolev had hired as one of his deputies a *zek,* a prisoner, from Kazan days, K. I. Trunov.[25] Trunov was subsequently fired, and Korolev had to deal with this kind of interference for several years to come.

He was also accused of juggling personnel irresponsibly. One of his critics, named Yakovlev, whose wartime experience commanding an artillery battalion at the front presumably gave him some extra credibility, told a party meeting that, "You won't find an engineer in our department who hasn't changed jobs two or three times . . . it must be said frankly that for three years complete anarchy and chaos exist in the disposition of cadres in our department."[26]

NII-88 Director Gonor even implied that Korolev was not doing right by the Communist Party, "the Party organization fights poorly for the brave promotion of Communists to scientific jobs . . . we still have non-Party members as heads of departments. . . . "[27] The target of the remark was obvious: Korolev was the only such department head. He continued to suffer attacks on his management style at party meetings, which were usually held in the plant itself, including implications that he was not allowing criticism and self-criticism to be aired sufficiently.

Nonetheless, as Vetrov reports, "intensive creative work continued in Department 3 which laid the basics for development of domestic rocket technology for many years."[28] Even the Party meetings began to reflect satisfaction at the progress. A June 8, 1949, meeting of the department notes that "Flight testing of the R-1A rocket successfully performed, integration of the first R-2E machine [an experimental rocket] completed, draft design of the R-3 is being examined." Then a historic decision was adopted, "The Partcom considers it necessary to proceed to the preparation of a general plan of development of the Institute in the direction of solving the tasks of creating missiles with super-long range." To carry out the plan it was recommended that the Institute establish a "highly qualified commission" by July 1, 1949.

Working on the design of the R-3 was Kryukov's first job at Podlipki.

> I was to do a research study on the characteristics of the so-called R-3 rocket, which was designed for a 3,000-kilometer range. This took two years, from 1947 to 1949. It was the first time we studied the limits of a one-stage rocket, and then the boundaries between the first stage and second stage—the longitudinal and cross-sectional parameters, the optimization of the number of stages, fuel capacity, propellant components, ballistics and gliding approaches. These studies are still valid. We also did studies of the ballistics, flight dynamics. There were about 100 to 200 people in Bushuyev's sector, part of Korolev's Department 3, engaged in this research. It's difficult to say how many people in other departments were involved in this project. Chertok and Voskresensky were involved, too. The Korolev enterprise was already quite large at this time.
>
> The main result of our research was the determination that a one-stage missile was unfeasible beyond 1,500 to 2,000 kilometers.[29]

In July 1949, one month before the explosion of the first Soviet atomic bomb, a meeting was held to discuss the prospects for a rocket that could deliver the bomb once it was operationally ready. Attending the meeting were Korolev; Igor Kurchatov, leader of the Soviet atomic weapons effort; Minister Ustinov; Marshal Mitrofan Nedelin; and others. The R-3 was offered by Korolev as the logical launch vehicle, and six months later he was ready with a detailed design for the rocket, which would be powered by a new engine conceived by Glushko, the RD-110, with 120 tons of thrust. In April of 1950, when he was named head of his own NII-88 design bureau, designated OKB-l, Korolev forecast that the R-3 would be ready for flight test by 1953. In defending the project, Korolev wrote Ustinov that it was so important it should involve scientists and engineers around the whole country, as well as an immense amount of resources.

In the meantime, Korolev transferred the job of producing the R-1 and R-2 to a newly created factory, formerly devoted to automobile production, in Dnepropetrovsk, Ukraine, but he kept in continual touch with the plant's work. Vladimir Utkin, now head of the Central Scientific Research Institute for Machine Building (TsNIIMash), remembered the transition from the days when he was a young engineer in Dnepropetrovsk:

> In the 1950s we began to produce the R-1 and R-2 missiles in Dnepropetrovsk. Korolev sometimes visited us because of the problems of getting started. I was the supervisor of production. Our group of designers was involved in changes made for serial production, such as substitutions of materials. We would have to defend our proposals for making changes and he would listen carefully, and he would assign deputies to monitor the changes and send them to see us on a strict schedule. They would come with specialists and documentation, we would explain our suggested changes. Example: let's change the thickness of the tubing so it doesn't touch another structural member. Korolev would sometimes be there and would listen patiently to the problem, and the proposed solution—he probably knew about it in advance—and he would make a decision. He never said "We'll think about it."
>
> Sometimes he disagreed, though, and would offer short, well-thought-out words on why.[30]

While R-2 was essentially a derivative of V-2 technology, with very limited potential as a weapon, Korolev had very high hopes for the R-3, which would be able to reach U.S. bases in England. But the missile's demanding technological advances caused insurmountable problems. Glushko was unable to increase the specific impulse (the main yardstick of performance of a rocket engine, it is the thrust per unit flow of propellant, usually measured in seconds) of the RD-110 to the 285 seconds required. This would have been more than 35 percent higher than that of the RD-101, which powered the R-2. But Glushko simply did not have the ability to make a combustion chamber that could withstand the high heat and vibration that the tremendous thrust would generate.

The R-3 project was terminated in 1952. Nonetheless, the team working on it in OKB-1 made significant contributions to advanced rocket technology.[31] Kryukov is quite proud of them: "We made recommendations on how to upgrade single stage missiles by carrying fuel tanks. We were able to increase the range of the 270 km R-1 to the 1,200-kilometer R-5. We also made recommendations on further design improvements leading to an ICBM."[32]

R-5 (SS-3 Shyster) was, in fact, the first truly Soviet-designed strategic missile. With a 30-ton mass, it required a new Glushko engine, the RD-103, with 44 tons of thrust. The project wasn't even begun until 1952, but R-5 flew its full 1,200-kilometer range from Kapustin Yar on March 15, 1953.[33]

One can appreciate the level of intensity of OKB-1 during the early 1950s by the realization that this relatively small enterprise was simultaneously developing a whole family of tactical and strategic ballistic rockets of varying ranges.

Testing of the R-5 was not without difficult problems. Vsyevolod Avduyevsky, now an Academician in charge of promoting conversion of military and space technology for commercial applications, recounted an example based on an experience that he had in 1952 with Korolev and the R-5.[34]

> While I was testing our missile [Vladimir Chelomei's cruise missile] at Kapustin Yar, Korolev was also there, testing the R-5 ballistic missile 5 or 10 kilometers away. We lived in the same town: "Number 2." Once my friends and I decided to get a close-up look at the R-5. We needed Korolev's permission.
>
> "Where is Korolev?"
>
> "In the canteen."
>
> At lunch we repeated, "Where is Korolev?"
>
> We were told, "When he comes in you'll see him."
>
> Indeed, when he came in the whole atmosphere changed. Everybody stopped talking. When Korolev walked over to get his food, he stopped three or four times to shoot words at people. He never stopped working. He seemed intense, constrained. I waited until he finished his meal, approached him, and asked for permission to see a test. He looked at me hard. I was only about thirty or thirty-one. Then we started talking

rockets and his closedness disappeared. He told us about some of the problems he was having with the rocket launches. When the engines started there was a kind of burst shock.

He gave me permission to see a launch but with this condition:

"Don't just be a silent witness. Tell me what you see."

He was afraid that Glushko hadn't told him the whole truth about what was happening during the alcohol-oxygen combustion. A couple of days later [there was a launch every few days] I was there for a test launch, watched some of the complex procedures, and saw the rocket rise vertically.

I was impressed, but it seemed to me that there was a shock noise. On the next test, I was there and again I heard the engine shock. But Glushko rejected my opinion.

"It seems to me it was smooth," he said to his deputy, Dominic Sevruk. Sevruk supported Glushko. Korolev asked us over. Nobody knew us.

"What do you think?" I hesitated.

"It was not quite smooth," I said. "I heard a shock."

Korolev said, "OK, I appreciate your opinion. You need to be an engineer not a diplomat. If you try to be a diplomat we stop our acquaintance as quickly as we made it."

R-5 was an achievement to be proud of, but what Korolev's country needed was a much-longer-range weapon than even the aborted R-3. The need became especially apparent when news reached the USSR of the work underway at that time in the United States on a long-range supersonic cruise missile. By 1950, at North American Aviation in California, a vigorous team of engineers headed by William Bollay and including Dale Myers (later to be an important contributor to the Apollo program), was handed a difficult set of requirements for a cruise missile: 5,500-nautical-mile range, 0.25-mile accuracy, Mach 2.75 speed, 15,000-pound payload. It became the Navaho.[35]

As a consequence, in 1952 Korolev began work on a similar cruise missile concept. His first effort was to produce a test vehicle that could carry a half-ton payload over a 900- to 1,300-kilometer range. But what happened next leads one to speculate that Soviet espionage must have gleaned excellent information on the progress being made at the time on another California project—the development of the Atlas ICBM at Convair-San Diego. A directive was issued to Korolev: forget the cruise missile and get back to long-range ballistic rockets.[36] Korolev's concepts for cruise missiles were reassigned to two aircraft design bureaus. Now the Chief Designer would be competing not only with the United States but with his own cruise missile concepts.

One of Korolev's cruise missile concepts—called Sorokovka, which had the cruise vehicle with warhead mounted over a first-stage rocket booster—was adapted by an aircraft design bureau headed by Vladimir Myasishchev. Myasishchev, however, chose to use four rockets as the booster rather than Korolev's single rocket. A ramjet, or perhaps more than one, designed by the Mikhail Bondaryuk design bureau, would have powered the cruise vehicle. Glushko took on the rocket engine.[37]

The other Korolev cruise missile design, called Burya, was assigned to Semyon Lavochkin. Lavochkin's Burya concept, however, was much more advanced in technology than Sorokovka, and would have carried a 150 percent larger payload. It looked like a large aircraft, with a delta wing and a cruciform, swept-back tail. The cruise vehicle, also powered by a Bondaryuk ramjet, had two booster units, each with four 20-ton-thrust rocket engines designed by Alexei Isaev. It was one of the first vehicles to use lightweight, high-strength titanium, capable of withstanding the high temperatures expected at Mach 3+ speeds. It had an inertial navigation system, corrected by a celestial system, that was sufficiently effective to have guided the vehicle in its first test over a 1,500-kilometer distance and back.[38]

In later tests, I was told by Mikhail Rozhdostvensky, who had been chief theorist at the Lavochkin design bureau, Burya covered "the full range of some 9,000 kilometers from Kapustin Yar to Kamchatka."[39]

Back in the United States, Navaho, after some difficulties, flew over a range of 1,075 miles on November 6, 1956. Boosted by two 120,000-pound-thrust kerosene/oxygen rockets, sustained by two 48-inch-diameter ramjets, it covered its flight path at altitudes up to 65,000 feet.[40]

The competition between cruise missiles and ICBMs lasted for some five years, but eventually the ICBM won in both countries. Just as Atlas prevailed over Navaho, so did Korolev's ballistic missile candidate, R-7, win out over Sorokovka and Burya. The cancellations of Navaho and Sorokovka occurred, in fact, in the same year, 1957. Burya lasted three years longer, but was cancelled after reaching its 9,000-kilometer design objective in 1960.

"It was a great mistake for the government to cancel Burya," Avduyevsky told me.[41] A Burya veteran, Yakov Natenzon, imputes dirty politics, writing in 1993 that the "decision was made to cease flight tests due to hidden intrigues. This is a typical example of the voluntaristic decisions which were made without any substantiation by the leaders of the Soviet Union, even in the area of rocket technology where money and manpower were unlimited."[42] The accusation itself seems a throwback to the cavalier attitude towards funding that characterized the military-industrial complex in the 1950s. The likelihood is rather that the cancellation was based on pragmatic technical and strategic reasons. The cruise missiles of the 1950s had made important contributions to rocket engine development, aerodynamics, materials, and navigation, which in many cases added to the technology base for the ballistic missiles. However, they were considered too slow and vulnerable to anti-missiles. ICBMs, on the other hand, were making good advances on virtually all technological fronts.

Kryukov recalled, for example, that "In parallel with our work on R-1, R-2, and R-5, we had been conducting studies on the use of low-boiling-point propellants, hypergolics, and the design of winged warhead systems as preliminary studies leading to the R-7. I worked under Bushuyev as deputy for rockets. Other deputies were Viktor Makeev and Mikhail Reshetnev."[43]

Vetrov, who was himself a designer at NII-88 during this period, reviewed some of the studies that show the astounding range of thinking that Korolev stimulated. There was the N-1 study,[44] an experimental rocket based on the R-3; N-2, possible propellant components; N-3, various studies of missiles for ranges up to 10,000 kilometers; T-1, a study of different ICBM designs, including one with a 3-ton payload; T-2, a study of winged ICBMs. All of these works preceded the development of the R-7.[45] It was T-1, however, whose design for carrying a 3-ton payload had been authorized in February 1953, which became the technological forebear of R-7. Like R-7, it was a "packet" design, that is, the upper stage took off from the spent lower stage. However, the T-1's 3-ton payload proved inadequate for the bulky thermonuclear bomb, which was tested in October 1953. A 5-ton payload would have to be carried.

Kryukov recalls that, "In late 1953 Malyshev [Vyacheslav Malyshev, head of the strategic missile and bomber program] came to the factory and directed us to redesign the vehicle to take a 5-ton nuclear warhead. Initially the liftoff mass was 170 tons. We had to rework the whole design to scale up the liftoff mass to nearly 300 tons."[46] The result was the R-7, which could carry up to 5.4 tons.[47] Authorization for its development took place by decree of the Council of Ministers shortly after the successful bomb test.

"I was young and inexperienced during these times," continued Kryukov, "but I was not hesitant to raise questions, even of the chief designers. 'If you continue to question I don't see how we can work together anymore,' said Bushuyev. Korolev, hearing of this, invited me to talk it over. I don't remember the details, but I had a good relationship with Korolev at the time, even though he kept all of his workers at a distance. Only Voskresensky, of his deputies, could use the familiar 'ti' [you] with him."

"At the time," said Kryukov, "low-boiling-point propellants were considered preferable for an ICBM, but we began to work with high boiling point fuels in preparation for the design of R-7, which would have an 8,000-kilometer range. . . . We faced major difficulties in developing the system. It called for big advances in pumps, we had to deal with hydrodynamic shocks, we had many failures. The strapons fell apart. We had large heat fluxes, overheating. But we overcame these difficulties, gained experience, became more sophisticated."[48]

Accumulating that experience was hard gained, but Korolev throughout this period maintained a conscientiousness about grooming young engineers to enter the mainstream of the design bureau's work.

Gyorgi Grechko, the cosmonaut who flew four times into space, started out at OKB-1 as an engineer. He remembered:

> The first time I met Korolev was in 1955, one evening at the end of the working day. He asked many questions.
>
> Where was I educated?
>
> What kind of lectures did I have in my Institute?
>
> Why did I ask to work at OKB-l?
>
> What kind of books and theater did I like?
>
> We talked for about one half hour. He treated me like an equal. There was no sign in his manner that he was a giant and I was a pigmy. I was nothing, only a student. I had no experience! . . . He was very famous, not with the public but with specialists. He asked not scientific questions but about life, my salary, my dormitory.[49]

Throughout the problems Korolev faced with staff and technology development, he was careful to maintain the backing of his most important supporter, Josef Stalin, whose word was the definitive one in all major decisions. Khrushchev wrote in his memoirs that:

> . . . while Stalin was alive, he completely monopolized all decisions about our defenses, including—I'd even say *especially*—those involving nuclear weapons and delivery systems. We were sometimes present when such matters were discussed, but we weren't allowed to ask questions. Therefore, when Stalin died, we weren't really prepared to carry the burden which fell on our shoulders. . . . Not too long after Stalin's death, Korolev came to a Politburo meeting to report on his work. I don't want to exaggerate, but I'd say we gawked at what he showed us as if we were a bunch of sheep seeing a new gate for the first time. When he showed us one of his rockets, we thought it looked like nothing but a huge cigar-shaped tube, and we didn't believe it could fly. Korolev took us on a tour of a launching pad and tried to explain to us how the rocket worked. We were like peasants in a marketplace. We

walked around and around the rocket, touching it, tapping it to see if it was sturdy enough.[50]

Apparently, following Stalin's death (in March 1953), Korolev's competence won over his new leaders quickly. Khrushchev wrote, "We had absolute confidence in Comrade Korolev. We believed him when he told us that his rocket would not only fly, but that it would travel 7,000 kilometers. When he expounded or defended his ideas, you could see the passion burning in his eyes, and his reports were always models of clarity. He had unlimited energy and determination, and he was a brilliant organizer."[51]

Certainly one of Korolev's achievements was to build a solid working relationship with Mstislav Keldysh, then head of NII-1. Keldysh, considered the "Theoretician of Space," became a close ally of Korolev's in both technical and political initiatives over many years.

The technical work done at NII-1 was first-rate, including fundamental contributions to the development of the reentry cone for the R-7. Avduyevsky, one of the key engineers on that project, gave me an insightful account of how the materials for the cone were selected. American researchers were going through the same cut-and-try experiments at the time.

> I didn't want to continue to work at Chelomei's [after that organization was put under Sergei Beria, son of Lavrenti, after the death of Stalin] so I went to NII-1. This is the outfit that developed the Katyusha weapon. Glushko and Korolev worked there before being arrested.
>
> This was a good place to work. Keldysh was in charge and we communicated easily without a lot of hierarchy. Keldysh asked me what I wanted to do. I said that I was working on propellant combustion and I wanted to continue. Keldysh drew a sketch on the table—a sharp cone. He said,
>
> "This is the warhead for an ICBM. It's necessary that this cone shape be a good shell for rocket reentry."
>
> I said, "This is engineering work. I'd prefer scientific work."
>
> He said, "Don't be in such a hurry. Immerse yourself more deeply into the problem. Go to a meeting at Korolev's bureau and you will hear more about it. Then we'll meet again in a week."
>
> This was probably around July 1953. At the meeting Korolev spoke, and so did his deputies. He told us about everything they went through in testing the cone-shape configuration on the R-5, using a special trajectory to simulate the correct reentry velocity. He stopped, reviewed all the details, then said to everyone present,
>
> "Your comments?"
>
> I said that I understood that silicon carbide, fiberglass and carbon had all been tested, that the carbon burned and so it seemed no good.

But I said even though it burned it won't melt. He acknowledged my opinion and announced,

"You are now responsible for the carbon experiment."

Keldysh then organized a special department in the Institute of Thermal Processes with Valentin Likhushchin in charge [twenty-five years later he became director]. I headed the division in this department called "Thermal Shielding for Warheads." We weren't using the ICBM designation in those days. Korolev gathered us together regularly. We kept pressing him to see the advantages of carbon and it was decided that 1) carbon and 2) carbon-silicon would be tried. Carbon competed poorly against carbon-silicon, but several launches were made with carbon. Konstantin Bushuyev from Korolev's bureau was in charge of the tests. Both carbon launches were unsuccessful. Korolev informed Keldysh and asked for his official response. I made the point that carbon would be better but that its problem was that it didn't have enough internal strength. We monitored the tests with kinotheodolites and could see the well-illuminated warhead breaking into pieces.

Korolev got us together and said,

"Carbon has a better possibility because we understand it better, and with carbon-carbon filaments we can increase the structural strength, but we are short on time. If we are to succeed by 1957 we must use the silicon-carbide."

A department was set up in TsNIIMash to fabricate the cone experimentally, allowing for about 100 millimeters of ablation. The technology was later transferred to a factory. The development of this cone-shaped warhead was a crucial milestone in my career. It lashed me to Korolev, and gave me many opportunities to observe the relationship between Korolev and Keldysh. Korolev was always very strong, but he was delicate in his relations with Keldysh. Korolev often visited us in the evening from his bureau, usually accompanied by different people. We continued to develop the carbon-carbon material from a scientific point of view.[52]

Solving the nose cone reentry was only one of the hurdles faced by the R-7 designers, and not all were surmounted without failures. "Our early tests were unsuccessful," remembered Kryukov. One of the reasons for the first failure was axial vibration of the vehicle, called the "pogo" effect (after the axial motion of a child's pogo stick), a serious problem encountered by U.S. designers as well. It occurs when the dynamic characteristics of the propellant feed system resonate with the thrust oscillations in the engines. "The result," as described by Kryukov, "is an ever-increasing vibration, producing loose connections, failed tubes, etc. The problem was eventually solved by pogo suppressors which isolate the feed system from the vehicle."[53]

Vetrov, however, disagrees on the reason for the failure of the first test. It was not pogo effect, he says, but a fire in the tail section caused by a leak in the fuel line. But perhaps it was pogo effect that caused the fuel leak.

"At the insistence of Korolev," Kryukov went on, "we conducted intensive studies to cure these problems and eventually a complete family of launchers was developed for ICBM, interplanetary, manned spacecraft, and heavy reconnaissance satellite missions."

In contrast to the Atlas, which was powered by two booster engines and a sustainer engine providing about 200 metric tons of thrust, the propulsion system for the one-and-one-half-stage R-7 had, and still has today in its updated versions, twenty thrust chambers, providing more than 500 metric tons of thrust. There are four RD-107 Glushko engines, each with four thrust chambers, around the circumference of the vehicle, and one four-chamber Glushko RD-108 core engine in the center (the "half" stage). A clever design of Glushko's was to use only one turbopump for each set of four chambers. Korolev's bureau designed the vernier engines for thrust vector control—two on the RD-107 and four on the RD-108. The more advanced method of steering the engines, by gimballing the combustion chambers, was beyond the state of the art at the time. All of the engines use liquid oxygen and kerosene fuels.[54]

Why four chambers for one engine? Because at the time, Glushko couldn't make a single chamber with the needed thrust without severe vibration problems. He tried valiantly, even building one experimental engine with a cross pattern in the combustion chamber, but it didn't work. However, each chamber on the engines generated about the same thrust as the German V-2 engine. The R-7 engines were much more efficient and weighed only about a third as much as the V-2 engines. The four-chamber arrangement also enabled the lengths of the engines to be reduced.

The four groupings of four RD-107 chambers in the R-7 ICBM were designed to burn out in about 140 seconds and drop off, and the four RD-108 chambers continued to fire for a total of about 320 seconds, placing the vehicle into a ballistic trajectory to its destination, which could have been one of many major cities in the United States. "Launch of the missile from the territory of the USSR could level the response nuclear attack practically at any point of the possible enemy territory," reads the 1994 historical brochure of RSC Energia, descendent of the Korolev design bureau.[55]

One of the technological breakthroughs that made ballistic missiles feasible was very accurate guidance. Both the Soviets and the Americans went through a difficult development period during which the guidance system was perfected. Initially both were forced to use a mixed guidance approach.[56] Just after launch, radar would be used to track the missile, then the small vernier, or control rockets, would be sent radio signals that adjusted their thrust angle to correct the trajectory, after which an inertial system took over to target. The initial Atlas vehicles likewise used a radio-inertial guidance system. Small rate-integrating gyros, developed primarily at the Massachusetts Institute of Technology by a team in the Instrumentation Laboratory headed by C. Stark Draper, were used to maintain the vehicle's attitude reference. A set of radars at the launch site accurately tracked the vehicle's position and velocity through the powered phase, which took it out of the atmosphere. Radio signals to an onboard programmer adjusted the flight profile during the sustainer phase. Then a radio signal would cut off the small vernier engines in the final phase of powered flight, resulting in a velocity that assured that impact would be reasonably close to target—about one mile in those days—since warhead separation and reentry aerodynamics had insignicant effect. Radio-guided missiles, however, were difficult to protect from jamming. Consequently, as soon as more accurate gyros and accelerometers (instruments that sense changes in position and velocity) were developed it became possible to install onboard all-inertial systems.[57]

Khrushchev himself was sensitive to the USSR's guidance problems. "First and foremost," he said in his memoirs, "we had to develop an electronic guidance system. It always sounded good to say in public speeches that we could hit a fly at any distance with our missiles. . . . I remember that in the first days of our Semyorka program, while the missile itself had a range of 7,000 km, we could direct it to a target only by placing guidance systems every 500 km along the way."[58]

This is an exaggeration. While the R-7 did require one radio correction post down range, the inertial guidance took over for the rest of the way to the target. Then, with the R-7A, which started testing in 1960, and had a longer range than the R-7, the guidance was purely inertial all the way.[59] The validation of the guidance system was one of the important aspects of the earliest test launchings.

Boris Chertok recounted for me his recollection of the very first R-7 test launch: "We had our first test launch of the R-7 intercontinental ballistic missile at Tyuratam/Baikonur on May 15, 1957. Before that our testing of the R-1, R-2, R-5, and R-11[60] took place at Kapustin Yar [near Volgograd in Kazakhstan]. We were five months away from our families. Our work day was intensive. Korolev checked all aspects of the vehicle assembly and launch preparations himself. He had divided the staff into assembly and launch specialists. I was responsible for assembly and test. Leonid Voskresensky had charge of launch. We worked twelve- to fourteen-hour days, seven days a week, and Korolev himself was the hardest worker. The primary workers were soldiers. While we were there Korolev would look through the reports from OKB-1 in Moscow, like research proposals. He was scrupulous in examining new ideas, never interrupted whoever was offering them. If he approved an idea he would introduce it immediately. For example, the design of the R-7 launch system was changed several times. The final design didn't look like the first proposal. Korolev, for example, dropped the carbon vanes in favor of small steering rockets."[61]

That first test was a failure, as were the next four.[62] Three of those attempts, on June 9, 10, and 11, 1957, were aborted on the pad because of a balky nitrogen valve. The other two, the first test on May 15 and another on July 12, were failures that resulted in destruction of the vehicles after launch.[63] But on August 21, 1957, the vehicle flew over the full range on a ballistic trajectory to Kamchatka, although it disintegrated into thousands of pieces at an altitude of 10 kilometers.[64] As mentioned earlier, this was more than fifteen months before Atlas duplicated that feat, no doubt an achievement very satisfying to Korolev, who was constantly evoking the American competition.

"He always read the unclassified U.S. literature on the development of the Atlas ICBM," said Chertok, "but didn't follow that design. He found the open American journals interesting, but he used them mostly to motivate his staff. 'Look, those Americans are going to beat us to orbit,' he would say. He referred also to the Dornberger memoirs. [Walter Dornberger, who had been the boss of Peenemunde as a general, was then working for Bell Aircraft in Buffalo, and had written a book on his V-2 days.][65] 'All those German guys in charge of the U.S. program, we left those guys.' " Korolev was sensitive to the gossip that the Russian program was German-dominated.[66]

In fact, building a Russian expertise in rocket and space technology for the long term was one of Korolev's enduring passions. That was one of the motivations that drove him to a deep and continuing commitment to training young engineers to move into his missile and space projects. Even though his own work was virtually all-consuming, he often took the time to communicate with students who might one day be on his staff. It is very impressive, for example, that this leader of the Soviet ballistic missile program lectured a group of students during one of the busiest periods of development and test. Equally impressive is the realization that he was sharing very sensitive, highly classified information with them. He gave a series of five lectures in 1949 to students at his alma mater, MVTU. The lectures developed in some detail the rationale for the approaches being carried out by the Korolev design bureau for long-range ballistic missiles. It's as if Karel Bossart would have lectured Cal Tech engineering students on his approaches to the design of Convair's Atlas missile. It simply would not have happened. The complete text of the lectures, with engineering drawings and mathematical equations, was printed in 1980.[67]

It was during these early years that Korolev introduced the practice of hiring students, while they were still at university, to do their "diploma work" and sifting out the best of them for permanent jobs.

One such student was Arkady Ostashev. When I interviewed Ostashev in 1993 he seemed a fragile old man—white-haired, needing a cane to get around—although probably still only in his sixties. The interview took place at the Center of Flight Control in Kaliningrad, the Russian equivalent of NASA's control center in Houston. In front of us, on a huge display panel, we watched in real time while the youthful cosmonauts, Alexander Serebrov and Vassily Tsblieyev, spent several hours outside their Mir space station adding a new solar power panel. But we talked of events that had happened more than four decades earlier, when Ostashev was a young space engineer himself.

> I remember going to the Podlipki station, seeing a large water tower, where Korolev's monument now stands, and going to the department of personnel. I was twenty-two years old. Two security people let me in and delivered me to Korolev. After listening to my presentation he said,
>
> "No, I won't take you."
>
> My hopes crashed, but then he said,
>
> "I won't take you alone, you misunderstood me. Select some good guys from your class and come next week and I'll take all of you from MAI [Moscow Aviation Institute]."
>
> I brought thirteen fellow students the following week. Korolev said that thirteen was a lucky number.
>
> He started our meeting with a two-hour speech about the prospects for developing rockets and carrying out space exploration. The program he described—placing a satellite in orbit, sending spacecraft to the Moon, Mars, Venus, putting man into space—he performed completely before his death. We were so excited we could hardly sit in our chairs. He finished the speech with,
>
> "We are close to the time when people will be issued vouchers by their trade unions to fly in space."
>
> He made suggestions for our diploma work and said,
>
> "All of you will work for me until December [1947], as senior technicians [not yet engineers]. Then from January to April you must do your diploma work, then I'll check for myself. If you're not working on your diploma I will fire you."

So, until December we got full salary as senior technicians, then from January to April we got about half salary. I got my diploma on May 13, 1947, on a Monday, at 1300 hours. Myasishchev [a famous aircraft designer] was chairman of the State Commission and he listened to our diploma presentations here in Podlipki.

We celebrated our transition from technician to engineer twice. The first was at a dacha with all twelve of my colleagues. Korolev came to it with Nina Ivanovna. Then and there we realized his sense of humor, and his penchant for using proverbs. His proverbs had educational meaning, like, "Don't be afraid of taking new initiatives. . . . Always try to work long and make a good thing rather than work fast and make a poor thing. . . . "

The second celebration involved about 300 graduates of MAI, about 200 from Moscow, others from Alma Ata. Some very distinguished people were there including Yakovlev, Myasishchev, Tupolev, Lavochkin, Antonov, Mikoyan, they all contributed so that 1,000 rubles went to every graduate. We were the first major graduating class after WW II. All of the design bureaus got their share of graduates.

I worked on the R-2, on which I had done my diploma work, on flight control, and the possible use of R-1 jet vanes on the R-2. I got this task because my diploma work was on stability. I worked under Gyorgi Stepanovich Vetrov, who reported to Moisheyev and then to Bushuyev. I proved that the R-1 vanes worked on the R-2 and my report resulted in their use.[68]

The R-7, designated the SS-6 Sapwood by the North Atlantic Treaty Organization, had a relatively short life as an ICBM. Although it was test launched before the Atlas, it did not go into the weapons inventory until 1959, the same year that the first Atlas squadrons were declared operational. Eventually some 129 Atlas ICBMs were deployed in the United States, mostly in the western states. R-7s and R-7As were deployed in only two launch pads at Baikonur and, eventually, four at Plesetsk, a launch center readied by 1959. Located at 63 degrees north latitude, about 300 kilometers south of Archangelsk, Plesetsk soon became the busiest of the USSR's three launch facilities, having responsibility for placing in orbit reconnaissance and other military satellites.

Atlas was a much more sophisticated missile than R-7, capitalizing on advanced materials and structures technology, as well as on the big American lead in electronics microminiaturization, to produce a vehicle much lighter than its Soviet counterpart. Atlas had a stainless steel skin so thin that it required pressurization to keep it from collapsing. The R-7 structure was so rugged it could be walked on. R-7's empty weight was 23 metric tons, compared to Atlas's 7.7 tons, including warhead.

Atlas had only two main engines and a sustainer engine. R-7 had five clustered engines, each with four thrust chambers. Atlas was originally designed to carry a separable entry vehicle with an H-bomb payload of 1.8 metric tons over a 10,000-kilometer distance. R-7 had to boost a bigger payload—5.4 tons—because the Soviet H-bomb was much clumsier. It was this lift capability that later shocked the West when it enabled Sputnik 1 and then the gigantic 1,327-kilogram Sputnik 3 payload to be orbited.

After the August 1957 success, and up to April 1958, some ten to fifteen tests of the R-7 were carried out, including some failures, but none of these flights went beyond about 6,000 kilometers. Then there was a pause in the testing until March 1959, leading to speculation by CIA officials—who had been keeping track of the tests systematically, using U-2 overflights as well as a large radar installation in Turkey—that problems had arisen.[69] The problems were apparently caused by the plaguing pogo effect. In March, tests resumed at a rate of about four per month and in July 1959, Secretary of Defense Neil McElroy predicted that "a few" Soviet ICBMs would be ready for launch by the end of that year.[70] In fact, as the U.S. spy satellite program would show, only a few R-7s were ever deployed.

Even these few installations, however, were very costly. Sergei Khrushchev said that his father was given estimates of about a half billion rubles per R-7 site. "What will we do, we'll be without our pants?," he complained to Korolev.[71]

Successors to R-7 would be even more expensive, but the reality was that Semyorka itself would not get a large investment as a ballistic missile. It simply was not a weapon for the long run because it lacked the quick readiness capability required of a practical ICBM, depending as it did on volatile propellants that boiled off quickly, and therefore could not be fueled in advance. Including checkout, it took some twenty hours to prepare an R-7 for launch, much too long to provide rapid response to an enemy threat. Atlas' storable propellant RP-1 (essentially kerosene) was always loaded, but it still required fifteen minutes to pump in the liquid oxygen prior to launch.

By 1958 storable propellants began to be introduced in the USSR as well. Their use had appeal to the military because they enabled missiles to stand fully fueled for as long as two days, as opposed to the limit of a few hours for cryogenic propellants such as liquid oxygen. But Korolev was against these substances—like unsymmetrical dimethylhydrazine (UDMH) and red fuming nitric acid—because they had a lower specific impulse, were expensive, and because of their toxicity were difficult to handle.

But Korolev's stubborn opposition to storables inevitably brought him into competition with a new cast of developers on the next generation ICBM. His concept for an advanced long-distance missile, the R-9 (NATO designation SS-8 Sasin), called for a launcher with an 81-ton initial mass, and a nose cone of 1.7 to 2.2 tons that could be propelled 12,500 kilometers.[72]

His long-time supplier of rocket engines, Valentin Glushko, was already a convert to storables, and wanted to put them into the R-9. Korolev stood his ground, however, and eventually Glushko acquiesced, perhaps unwillingly. But he soon got into trouble building the new liquid-oxygen and kerosene design that Korolev wanted. He was bedeviled by such problems as high-frequency vibration in trying to perfect an open-cycle 100-plus-ton thrust engine with those propellants, and found himself seriously behind schedule.

Korolev, impatient with Glushko's work, went directly to Brezhnev, then Secretary of the Central Committee of the Party, responsible for the defense complex, to demand that the engine responsibility be switched to Nikolai Kuznetsov, whose background was primarily in aircraft engines. Dimitri Ustinov, Minister of Defense Industry at the time, supported keeping Glushko on the job, over Korolev's objections, and Brezhnev paid heed to Ustinov's opinion. Glushko was eventually able to suppress the vibration by switching to a closed-cycle, staged combustion design with higher combustion chamber pressures.[73]

This was just the beginning of the troubles that Korolev would have with Glushko throughout his career. "It was a big blow to Glushko when Korolev was put in charge of the development of long-range ballistic missiles after the war," I was told by Vassily Mishin.[74] "I think that Stalin made the decision." At any rate, Glushko was continually in Korolev's shadow, despite having being made chief designer of his own bureau—OKB 456—in 1946. Mishin told me sardonically that a pair of albums in the Korolev museum, depicting Glushko's engines in different eras, are symbolic of the deterioration in the relations between the two. The earlier album, he said, "is inscribed, 'I am giving this volume to a genius designer as a sign of respect.' A couple of years later another edition of the book came out and this time it's addressed to Korolev without 'Dear' and says 'I am presenting an album of horses you are riding and will ride for a long time.' "[75]

General Gyorgi Tyulin remarked on their differences. "It began with little things. [Glushko] said that the most important thing in rocket design was the engine. 'Even a stick will fly if you tie an engine to it.' "[76]

While Korolev was sometimes bull-headed, so was Glushko. "His [Glushko's] character, formed in his youth, led him to become monopolistic in the development of rocket engines," said Gyorgi Vetrov. "He had too much influence on their development."[77]

For example, the resourceful Glushko was not only successful in keeping the engine responsibility for Korolev's R-9 missile, he then did not hesitate to go to work for Korolev's competitor as well. He had begun four years earlier—in 1954—to develop engines with storable propellants for a new intermediate-range ballistic missile—the R-12 (SS-4 Sandal), designed by Mikhail Yangel.

Yangel's rise to a high leadership position in Soviet missilery is one more story involving humble beginnings, political imprisonment—in this case close relatives rather than Yangel himself—a hard-gained but high-quality education in aircraft engineering, and then a rapid rise to the top under political sponsorship.

The son of a Ukrainian political prisoner who had been sent to a desolate, remote Siberian village, young Mikhail managed, nonetheless, to have a good early education. He was sent to Moscow where he attended Moscow Aviation Institute, graduating with honors in 1937. Before the war he worked with the famous aircraft designer, Nikolai Polycarpov. In 1938, at age twenty-seven, he was assigned to tour aircraft firms in Europe as well as in the United States—including Sikorsky Aircraft in Connecticut and Douglas and Consolidated Vultee in California.

In 1939 one of his brothers was arrested for alleged disloyalty, causing Mikhail to lose his job. This was even though two of his other brothers were senior military officials, one of them a general. However, he got another job working under the aircraft designer Artem Mikoyan (brother of longtime politico Anastas) until in 1950 Dimitri Ustinov entered the picture.[78]

Ustinov, always looking for specialists to recruit into defense technology, steered the promising young engineer to NII-88. He was put in charge of missile guidance work, replacing Boris Chertok, who became his subordinate.[79] According to Yaroslav Golovanov, Chertok, a Jew, was demoted as part of a campaign of anti-Semitism launched by Stalin after World War II. It also resulted in the deposing of another Jew, Lev Gonor, who had been NII-88's director, and who was replaced by Konstantin Rudnev.[80]

When Rudnev retired, Ustinov made Yangel the director, putting him over Korolev. "He had a more attractive personality than Korolev, and Korolev had the record of being a political prisoner," Mishin told me.[81]

This was one of the first of Ustinov's numerous moves to counter the growing power of Korolev, who was so offended he didn't even show up at the first meeting of NII-88 department heads called by Yangel.

Korolev was to learn not only that Yangel was in charge, but that his younger colleague was very astute both technically and politically. Yangel knew how to work effectively with the Communist Party, having become a member in 1931 at age twenty, and was always careful to get his decisions blessed at the Party meetings. Eventually, however, Korolev was to regain the upper hand. Yangel was lowered to chief engineer, in charge of science only. "He reacted negatively to the loss of design responsibility," I was told by Yangel's widow, Irina Strazheva, and a feeling of enmity between the two simmered until the summer of 1954, "when the ultimate solution to Korolev-Yangel differences was reached. Ustinov, who was always friendly to Yangel, asked him to head the design bureau at Dnepropetrovsk." Having an alternate source of ballistic missiles was a way for Ustinov to dilute Korolev's power. It was also, however, insurance against a U.S. assault on Korolev's bureau in Kaliningrad, the only ballistic missile source at the time.

At Dnepropetrovsk Yangel's design bureau, NPO Yuzhnoye, became expert in storable rocket propellants, which were attractive for military missiles. So Yangel's missiles became the military standard. In those days, Strazheva said, there was an expression in the military:

"Korolev works for TASS, Yangel for us."[82]

R-12, one of those military missiles, would make headlines day after day in October 1962, when it would be emplaced in Cuba, triggering the historic Kennedy–Khrushchev confrontation. With a range of about 2,000 kilometers, it was meant to be aimed at U.S. and British bomber bases in Europe and Asia, but its range was just right to threaten major cities in the United States—if emplaced in Fidel Castro's country.

One of the engineers who worked on R-12 was a bright young designer named Vyacheslav Kovtunenko, later to become Chief Designer of the Lavochkin organization. Kovtunenko had an unusual career path that involved him over the years with the Korolev, Yangel, and Lavochkin bureaus. I interviewed him in 1993 at Lavochkin, in Khimki, just outside of Moscow, not far from Sheremetyevo airport.[83]

Kovtunenko's office looked classically pre-glasnost, with a big Lenin photo on the wall. The whole establishment had been supersecret until a few years earlier. Lavochkin, which built thousands of fighter aircraft during World War II, inherited from Korolev in 1965 the responsibility for designing and building lunar and planetary spacecraft. They also build early warning satellites that monitor Western ballistic missiles from their highly elliptical "Molniya" orbits. A balding, somewhat portly man in his early seventies, Kovtunenko was one of the few engineers allowed to attend international space conferences before glasnost. I had met him several times at those meetings. He greeted me warmly, smiling to show a lot of gold in his crooked teeth. After telling me in some detail about his own background, back to his graduation from Leningrad State University following World War II service as a soldier, he told me about how he got involved in the R-12 development, and then the R-16, the ICBM that would win out over Korolev's R-9.

> In 1951 it was decided to transfer some of the NII-88 people who worked on missiles to Yuzhnoye. A group of young specialists were assigned there under Vassily Sergeevich Budnik, who had worked for Korolev. Glushko sent one of his people to Yuzhnoye, too. I was in the group that Budnik wanted, but Korolev crossed my name off the list. He said that an aeronautical engineer should have nothing to do with missiles.
>
> But in March of 1953 my name was again on the list for Yuzhnoye. At that time Yangel was Korolev's deputy in Podlipki. I talked to Yangel, Korolev was out of town. He said, "It's too early for you to go there," but I decided to go. When Korolev came back from his trip he summoned me. He didn't like someone to leave him on his own. He asked me why I decided to go. I said that I'd like to have independent work. He liked my answer. "OK, go," he said, "but you're welcome back here."
>
> I worked on R-2 and R-5 aerodynamics at Yuzhnoye. There's not much interesting aerodynamics in ballistic missiles. But it was unusual to be able to work on one's own missile design. We were given the job to develop the R-12, which was to have a 2,000-kilometer range. We started to compete with Korolev. He could have blocked our attempts because Yuzhnoye was supposed to build his missiles, but he didn't. We usually went to Podlipki to work together and then the design would be approved by the military. In 1954 Yangel was made Chief Designer of Yuzhnoye and Budnik became his Deputy. It was then that the Design Bureau was separated from the plant. The R-12 went to the military and we started on a 4,000-kilometer-range missile [R-14] and then an ICBM [R-16]. I was chief of the design department, as a deputy to Yangel. We started to compete with Korolev seriously, but Korolev had a noble attitude about this.[84]

Korolev did what he could to modernize the R-9 to meet the R-16 competition. He felt that with new high-speed fuel pumps he could make the missile's response time acceptable to the generals. Mishin did a lot of work on how to keep topping the oxygen. "All prelaunch operations were fully automated," says the RSC Energia historical brochure.[85] Operations may have been automated, but testing was still primitive. One of the weapon's designers, Arkady Ostashov, told me this anecdote:

In 1961 we had the first test of the R-9 ICBM. The test pad had two parts—a movable part attached to the missile, and a fixed part on the pad itself. The two parts were mated on the pad. When the engine was ready to start we detected a small leak of liquid oxygen between the two parts. You could see a small cloud of oxygen vapor. Voskresensky said,

"Let's go to the rocket."

He and I and another guy approached to make sure that the leak was small. Fortunately nobody was watching. Voskresensky pissed on the leaky joint, the liquid froze, and the joint held until ignition.[86]

R-9's rival, R-16, having storable propellants, could be fueled and fired in thirty minutes, and therefore could be siloed. But Korolev and his deputy, Vassily Mishin, put up a vigorous fight against it. Kovtunenko recalled that Mishin said flat out that, "It's a dead end to make ICBM's with storable propellants."[87]

Mishin prepared a paper backing up his claim, and it led to a request for an investigation, which was signed by all of the members of the Council of Chief Designers.

A commission, headed by Keldysh, was established, working in the Kremlin, to arbitrate between R-9 and R-16. It included such experts as Gyorgi Petrov and Boris Petrov (unrelated, although both were Academicians).[88]

Another Yuzhnoye designer who remembers the controversy, Boris Gubanov, told me that, "Korolev put up a lot of arguments against the storable propellant UDMH proposed for R-16, calling it 'this deadly substance.' " It was not a simple debate. Kovtunenko told me:

> In evaluating the use of storable propellants it is necessary to realize that the design depends on crucial parameters—like the thickness of the thermal insulation of the reentry body, the launch weight variations, the residual fuel in the tank. The missile must be designed much more carefully than the missiles that use oxygen, which have more flexibility.
>
> . . . The critical comments on our design were being received day after day. We'd deal with one and then face another. We thought that Korolev's people were the sources of most of these comments.
>
> Some critics said that we could not get a good enough accuracy with autonomous guidance at these ranges. Pilyugin was behind that criticism. But Academician Viktor Ivanovich Kuznetsov said that was not true. He had developed gyros that were capable of producing high accuracy. Alexander Ishlinsky supported Kuznetsov. The Commission decided for the Yuzhnoye design. But this was not the end. The final decision had to be made at the Khrushchev level. The decision documentation is in the classified archives.

It was decided to proceed with both R-9 and R-16, with the final decision postponed until both could be tested on the launch range.[89]

R-16's first test was disastrous. In mid-October 1960, Nikita Khrushchev returned to Moscow from a bombastic appearance at the United Nations General Assembly where he was quoted as bragging that his "missiles were being turned out like sausages from a machine." It was especially ironic, therefore, that only a few days later, on October 24, 1960, one of the few "sausages" that he actually had, R-16, exploded catastrophically as it was being prepared for firing at Baikonur. A large number of people were killed, including Marshal Mitrofan Nedelin, then head of Soviet strategic rocket forces, and many other high officials and members of the launch and design teams.

An eyewitness, S. Averkov, described the scene of horror thirty years later:

> . . . a flash of fire erupted from the second stage engine nozzle. The powerful jet immediately ruptured the oxidizer tank. Nitric acid gushed out onto the concrete. Both the rocket and the launch structures were engulfed in a firestorm. At that moment, the motion picture camera that was to photograph the launch was activated. The dispassionate film conveyed to us a frightful picture—people still alive becoming torches. . . . The rocket broke in half and fell on the launch pad, crushing those who were still alive. . . . Some people were devoured by fire; others, still running, were overcome by poison gases. . . . [90]

The event was carefully covered up, with Nedelin's death attributed to an airplane crash. But various versions of the accident, and speculations on its cause, emerged over the years and only in the glasnost era was the real story revealed. On the thirtieth anniversary of the explosion, in 1990, a Red Army newspaper revealed that 165 people died in the accident, many of them burned beyond identification. A monument at the site lists only fifty-four names of servicemen who were killed, most of them based in Baikonur. The bodies of Nedelin and other ranking officers were taken to Moscow.[91]

"Many were even buried in different parts of the cemetery so that no one would think that, God forbid, all those people were burned alive by the same hellish flame," was Averkov's comment.[92]

Yangel missed the accident by leaving the launch site to have a cigarette. Kovtunenko told me that he, too, had left the site, ten minutes before the blast. "Yangel didn't recover soon from the tragedy. He was seen weeping at both the cosmodrome and back home in Irkutsk Oblast," Averkov wrote.[93]

"It's very clear what caused the accident," Kovtunenko told me. He said that Leonid Brezhnev himself was directed by Khrushchev to conduct the investigation. It was learned that during the pre-launch procedures the operators had misconnected cables on the launch pad. When ignition failed to occur, a check of several systems was performed simultaneously on the fully fueled missile. The programmer on the second stage was returned to T minus zero, and the signal went through to ignite the second stage. It then burned through to the fuel tank of the first stage and set off the gigantic explosion.[94]

In spite of the disaster, the R-16 worked out its problems and became the Soviet Union's first successful ICBM, taking advantage of a new, much lighter nuclear warhead, and having been given an improved inertial guidance system. Deployment began in 1961, and by 1965 some 190 of them were installed in different places, some in silos and some in "coffin" or surface launchers.[95]

Putting a developed weapon system out of business, however, seems to have been just as difficult in the USSR as it sometimes was in the United States. Korolev's R-9 was also deployed operationally, although it was clearly not the weapon the Soviets counted on, with only nineteen complexes put in place by 1965.[96]

Korolev apparently took his setback as premier ICBM designer philosophically, preferring to be in the space exploration business anyway. Kovtunenko recalled a friendly reunion with his erstwhile boss.

> During 1961 we had a series of meetings to review the R-9 and R-16 tests. I had been given a Hero of Socialist Labor award for the R-16. With it goes a gold star. It's not common to wear the gold star, but you do wear them at the cosmodrome. Korolev saw mine and shook my hand. Korolev was a hard man but he respected the good work of others.
>
> At this time we already had a space program. For example the awards for the R-16 and for Gagarin were given at the same time.

Although Korolev's preoccupation with space exploration dominated his thinking, as well as OKB-1's agenda, the bureau continued to work in ballistic missiles, eventually developing two solid-propellant missiles.[98]

But the major responsibility for ballistic weapons was taken over by other firms. Korolev for some time had been focusing on Sputnik and manned flight.

7

SPUTNIK: NO BIG DEAL TO KHRUSHCHEV—AT FIRST

WHILE IT JOLTED the rest of the world, the successful launch of *Sputnik 1* on October 4, 1957, received casual treatment, at first, in Moscow. Korolev's former colleague, Academician Boris Rauschenbakh, told me, some thirty-five years later, "Look up the pages of *Pravda* for the first day after the launch. It got only a few paragraphs. Then look at the next day's issue, when the Kremlin realized what the world impact was."[1]

The article in the October 5 *Pravda* was, indeed, tersely phrased. Positioned modestly in a right-hand column partway down on the first page, it did not even mention the satellite in its head. Titled routinely, "Tass Report," it gave the facts of the launch clinically, and the editors apparently felt obliged from the very first sentence to educate the readers on what it was all about:

> In the course of the last years in the Soviet Union, scientific research and experimental construction work on the creation of artificial satellites of the Earth has been going on.
>
> As already reported in the press, the first launches of satellites in the USSR were planned for implementation in accordance with the program of scientific research for the International Geophysical Year.
>
> As the result of a large, dedicated effort by scientific-research institutes and construction bureaus, the world's first artificial satellite of the Earth has been created. On 4 October, 1957, in the USSR, the first successful satellite launch has been achieved. According to preliminary data, the rocket launcher carried the satellite to the necessary orbital speed of about 8,000 meters per second. At the present time the satellite is moving in an elliptical trajectory around the Earth and its flight can be observed in the rays of the eastern and western Sun with the help of simple optical instruments [binoculars, spyglasses, etc.].[2]

The article went on to give the basic information—size, weight, orbital inclination, radio frequency on which the beep could be heard—and it credited the great Tsiolkovsky with having established the feasibility of artificial Earth satellites decades earlier.[3]

The next day's *Pravda* was something else.[4] "World's First Artificial Satellite of Earth Created in Soviet Nation" stretched across the top of page one, which was devoted almost entirely to the achievement. But the lead story in the right-hand column did not recount the feat itself, with firsthand reports from the Soviet protagonists—their names were top secret, after all, so they could not even be contacted. Instead, the column was datelined New York, and it quoted in detail the congratulations of Russia's fiercest Cold War rival, the United States. The words, generous in their praise, were from Joseph Kaplan, chairman of the U.S. National Committee for the International Geophysical Year. Both the Soviet and American satellite programs were carried out under IGY auspices, with the results available to the world's scientists. Below the Kaplan item were congratulatory bulletins from A. C. B. Lovell, the British astronomer, and from a member of the USSR's political family, Pavel Novatsky of the Polish Academy of Sciences—the latter headed "Big Victory."

Big victory it certainly was. Poems lyricized the event, like "Leap into the Future" and "Scouting the Celestial Deep." An ephemeris, showing the times when the carrier rocket would be visible over cities in the USSR, as well as Detroit and Washington, was printed like a train timetable. It was the moment to cash in on the performance of a feat that Nikita Khrushchev could never have dreamed would have so powerful an effect. He, after all, had been apathetic about "just another Korolev rocket launch," as Rauschenbakh described the attitude of the Soviet premier and his claque on first hearing the news.

In the days to come, though, *Pravda* was delighted to print the praises of friends and enemies. Reactions from Peking and Shanghai (that friendship would dissolve only a few years hence), Warsaw, Paris, Vienna, Rome, London, and an especially long one from New York, ran under a big headline that ran across the page, "Russians Won the Competition."[5]

It was a competition that the Americans should have won hands down. The concept of putting up a satellite had been known to the world's space enthusiasts for many years. In America, serious proposals to launch a spacecraft into Earth orbit had been discussed since the mid-1940s. Robert P. Haviland recalls that when he was working in the Navy's Bureau of Aeronautics in 1945, motivated by a report on space rockets in a document captured from the Peenemunde Germans, he "wrote a four- or five-page memo proposing a Navy satellite development but it was scorned."[6]

In February 1946, the U.S. Army Air Corps asked the major air frame companies to submit secret proposals for the design of an "earth orbiting satellite." Douglas Aircraft was notified on July 1 that its design was judged the winner. Today's aerospace companies will find it hard to believe that the Air Corps evaluation was completed in less than four months! What's more, Douglas was funded for the study at the "unheard of amount in those days" of $1 million, but their contract was switched to the newly formed Project RAND (for Research and

Development) in Santa Monica, California.[7] RAND's eventual report on "Preliminary Design of an Experimental World-Circling Spaceship" predicted, with keen perception, that "The achievement of a satellite craft by the United States would inflame the imagination of mankind, and would probably produce repercussions in the world comparable to the explosion of the atomic bomb."[8] Alas, nothing followed the study, and so it would not be a U.S. satellite that would generate those repercussions.

In Russia, as mentioned earlier, Konstantin Tsiolkovsky had shown mathematically in 1903 how a device launched at a certain velocity would achieve Earth orbit. Then in 1948, forty-five years later, the visionary Mikhail Tikhonravov had made the case to Korolev for developing just such a device. At first he was unable to get support for the concept. His presentation to a meeting of the Academy of Artillery Sciences was treated skeptically. Golovanov quotes the remarks of the Academy president, Anatoli Blagonravov, at the meeting: "The topic is interesting. But we cannot include your report. Nobody would understand why. . . . They would accuse us of getting involved in things we do not need to get involved in. . . . " However, what Blagonravov said officially was not what he thought instinctively. This courteous, mild-mannered, chain-smoking, white-haired former general, who would become in later years one of the chief spokesmen for the Soviet Union in the United Nations Committee on the Peaceful Uses of Outer Space, was bothered by the wary reception of Tikhonravov's ideas. "There was no way he could escape the thought that this ridiculous report was in fact not very ridiculous at all." Blagonravov, risking the derision of his colleagues, put the report back on the agenda, thereby giving Tikhonravov—and Korolev—license to study possible satellite designs.[9]

Some five years later, towards the end of 1953, having redesigned the R-7 rocket to carry a heavier payload, Korolev had drafted a proposed decree for the Central Committee of the Communist Party that included the possibility of using the vehicle to launch a satellite. However, while the draft "was making its way to the top," mention of the satellite was struck out.[10] Not until May 26, 1954, did Korolev formally propose the satellite launch to Dimitri Ustinov, Minister of Armaments.[11] By then the R-7 was capable of propelling an H-bomb warhead of 5 tons—the actual size had not yet been determined—over an intercontinental ballistic trajectory. It could easily orbit a satellite of some 1.5 tons. According to Korolev's deputy, Vassily Mishin, Korolev had to propose a sputnik launch as part of the test program of the ICBM program.[12] In any case, Korolev's proposal to Ustinov is so delicately phrased that R-7 is not mentioned at all, but merely referred to as the

" . . . new article which permits speaking about the possibility of designing an artificial Earth satellite within the next few years. By a certain reduction of the weight of the payload it will be possible for the satellite to achieve the necessary velocity of 8,000 m/sec."[13]

As Roald Sagdeev, longtime head of the USSR Space Research Institute, put it, "Korolev and his colleagues could have only had a vague idea of how heavy the final reentry vehicle [for the ICBM warhead] should be. Sakharov [Andrei] was still far from knowing how to make this deadly weapon relatively compact and easily portable. Since rapid progress was required, rocket designers adopted a worst-case strategy and started to develop an ICBM that, as it was discovered later, had a substantial excess of launch capability, or throw weight."[14]

Meanwhile more substantive thinking on possible satellite designs had been resumed in the United States. The most expedient design approach came from von Braun's team at the Army's Redstone Arsenal in Huntsville, Alabama. Starting with a meeting in early 1954 with George Hoover of the Office of Naval Research, von Braun and his colleagues eventually came up with Project Orbiter, which would have been an Army-Navy-Air Force design using already developed Army Ordnance weapons technology to put a small satellite into orbit.[15] But the Eisenhower administration had reasons to choose a different approach. Eisenhower wanted to keep the military out of the IGY program, which was dedicated to scientific purposes. That was the surface reason. The other one, more telling, was—ironically—based on military strategy. He wanted to be consistent with his well-publicized "Open Skies" stance at a time when U-2s and spy satellites were being developed to begin reconnoitering the USSR. Eisenhower reasoned that a satellite put up as part of the IGY program would strengthen the freedom of the skies policy and would be less likely to disturb Nikita Khrushchev's sensibilities about overflight than one sponsored by three military services. Also, Soviets were likely to launch an IGY satellite themselves. And so a project named Vanguard—which proved to be an embarrassing choice of names—was chosen to carry the U.S. banner into the space age. Based on a sounding rocket developed by the Naval Research Laboratory, but under the auspices of the National Science Foundation, it was going to be a riskier venture than Orbiter because it called for substantially modified first and second stages, a new third stage, and new rocket engines and guidance technology.

On July 29, 1955, the Eisenhower administration announced that the United States would launch Vanguard for scientific purposes during the 1957–1958 IGY. A few days later, at the Sixth Congress of the International Astronautical Federation in Copenhagen, a delegation of Soviet scientists, appearing at IAF for the first time, revealed at a press conference that the USSR, too, might be in the game. Leonid Sedov, head of the Soviet delegation, and the newly appointed chairman of an Academy of Sciences Commission on Interplanetary Communications,[16] choosing his words carefully, said:

> From a technical point of view, it is possible to create a satellite of larger dimensions than that reported in the newspapers which we had the opportunity of scanning today. The realization of the Soviet project can be expected in the comparatively near future. I won't take it upon myself to name the date more precisely.[17]

But Sedov, it seems, was speculating, since no official decision had yet been made that there would be a Soviet satellite in IGY. In fact, not until January 30, 1956, would the Council of Ministers issue a decree authorizing its development.

It was at this time that Korolev was able to arrange for Mikhail Tikhonravov and his team, which had been working on a satellite concept at Special Design Bureau Number 385 in the Ural Mountains, to join him. One of the members of that team, Konstantin Feoktistov, recalled:

> We wanted to build a satellite but Korolev had that responsibility. Tikhonravov was transferred to Korolev's bureau in early 1956 after the Party and the Government had authorized Korolev to proceed with the development of a satellite. But the rest of us in the group had to apply to Korolev individually. However, Korolev relied on the advice of Tikhonravov, his old friend and, by the end of 1957 I was chosen, although it was difficult to leave Number 385.[18]

Even the January decision, however, was not followed by sufficiently aggressive support for the satellite development in the opinion of Korolev and his close ally, Mstislav Keldysh. Time to beat the Americans had flown by and the Soviet establishment was not yet revved up. Nine months later, on September 14, in what must have been exasperation, and probably with Korolev's prodding, Keldysh appeared before the presidium of the Soviet Academy of Sciences to state his case.[19] He first reviewed patiently, like a good teacher, the physics of placing a satellite in orbit. Then he covered the pioneering scientific measurements that the Soviet satellite would make—the Earth's magnetic fields, the "ionic composition of the upper layers of the atmosphere," the "corpuscular radiation of the Sun," cosmic radiation, possible micrometeorites.

He must have caused eyes to widen a bit when he said " . . . we are considering placing a live organism in the satellite—a dog. It turns out that the perception of a dog is the most similar to the perception of a man—biologists so consider it. The dog will live there in the absence of a gravitational field, in conditions of irradiation, of cosmic radiation. The dog will be undergoing all sorts of dangers, because if the satellite is hit by a large meteoric particle, it will broach the satellite. . . . "

By then he surely had their full attention, but he went on. "We, of course, can't stop at the task of creating an Earth satellite. We, naturally, are thinking of further tasks—of space flight. The first project along these lines, I believe, will be to fly around the Moon and photograph it from the side which is always hidden to us." Then came hard criticism: "We have come up short in a whole series of tasks in the Academy of Sciences, and are lagging now. Back in August we were to turn in the dimensions of the equipment and their mode of attachment to the rocket. . . . We delayed this work, which resulted in a notable delay in the planning and development of the satellite itself."

He called out specific industry laggards: "In general, the radiotechnological industry isn't helping us enough. . . . They are sluggishly regarding the creation of this satellite. . . . We have already committed one breach in delaying the delivery of size specifications and other information to the Korolev Design Bureau." Finally, the ultimate motivation came out baldly: " . . . it would be good if the Presidium were to turn the serious attention of all its institutions to the necessity of doing this work on time . . . we all want our satellite to fly earlier than the Americans'."[20]

Keldysh and Korolev had not only the Academy of Sciences and industry to motivate, they had to deal with the objections of the military generals, who feared that the satellite project would slow down the development of the R-7 ICBM. The fear was understandable since R-7's first five launch attempts had failed.[21] But foot-dragging by the support institutions was only one of Korolev's problems. In spite of Keldysh's speech, the satellite that was supposed to have the honor of being first continued to fall behind schedule. In fact, it would lag so badly that it would become *Sputnik 3*—a huge, sophisticated satellite, eventually launched on May 15, 1958. *Sputnik 3*'s 1,327-kilogram payload included virtually all of the instruments called out in Keldysh's speech.

"But," said Gyorgi Grechko, one of the engineers who worked at the Korolev design bureau during those days, and who later became a cosmonaut, "these devices were not reliable enough, so the scientists who created them asked us to delay the launch month by month. We thought that if we postponed and postponed we would be second to the U.S. in the space race, so we made the simplest satellite, called just that—*Prostreishiy Sputnik*, or 'PS.' We made it in one month, with only one reason, to be first in space."[22]

It was at this time, on August 21, 1957, that the R-7, in its sixth attempt, propelled a dummy H-bomb warhead all the way to Kamchatka, some 6,000 kilometers. With that success, the confident Korolev made his move to beat the Americans with his "PS." His next obstacle was a skeptical State Commission for the R-7 ICBM. Discussion of Korolev's satellite proposal before that body, according to a 1992 report by a journalist in a Moscow magazine, was "sharp, the opponents arguing primarily about the tight timing," and "complete agreement was not achieved." Korolev had to go back to the commission a second time. This time he tried a different ploy. Why not, he challenged the members, put the question of authorization to the presidium of the Central Committee of the Communist Party—in the context of whether or not the USSR should try to be the first country in the world to launch a satellite? The commission blanched. "Nobody wanted to be scapegoat." And so the project proceeded.[23]

With the commission's nervous acquiescence, Korolev bulldozed the development, in a little more than a month, of a plain, polished 83.6-kilogram sphere containing only a radio transmitter, batteries, and temperature measuring instruments, with the intent to place it in orbit on a rocket that had failed in five of its first six launch attempts. It was a very hectic month, and while the satellite was simple, the attention given to its manufacture was unsparing, especially by Korolev himself.

"I coordinated the production, testing, launch preparations, and the launch itself," recalled Oleg Ivanovsky, looking back thirty-six years during a 1993 interview. Ivanovsky had been deputy to Mikhail Khomyakov, *Sputnik 1*'s principal designer. He recalled that there were problems:

> For example, there were two peculiarities which satellites had that missiles did not. One was that the satellite required precise thermal control and the other was that vacuum sealing was used to assure reliable performance. We had to find new techniques of manufacturing the surfaces in order to achieve the necessary optical and thermal qualities. We had no experience in this work. We needed vacuum chambers. I recall one episode when we had to persuade the production shop that the satellite was a new item, not a missile.
>
> Korolev, with his iron character, was able to influence the attitude of people. The Party directed that new paint be put on the factory walls. Korolev put the satellite on a special stand, draped in velvet, in order that the workers would show reverence towards it. He supervised the carrying out of the production schedule every day personally.[24]

One of the metalworkers who was assigned to the *Sputnik 1* manufacture was Gennadi Strekalov, later to become a cosmonaut. "My teacher in metalworking did the finishing," he told me. "Two half spheres were stamped, then machined, then the masters did the finishing."[25] Strekalov, who, in 1995, flew to the Mir space station with American astronaut Norman Thagard and then participated in the first Mir-Shuttle rendezvous, is as proud of his work on *Sputnik 1* as he is of his four orbital flights.[26]

"Korolev came over to the shop and insisted that both halves of the sputnik's metallic sphere be polished until they shone, that they be spotlessly clean," recalled Konstantin Feoktistov, who would be the first engineer-cosmonaut to go into orbit in the three-man *Voskhod 1* seven years later. "The people who developed the radio equipment were actually the ones demanding this. They were afraid of the system overheating, and they wanted the orbiting sphere to reflect as many rays of the Sun as possible."[27]

An idea of how intense were the preparations for the launch of the "simplest satellite," and how sensitive Korolev was to the event's historical importance, comes from the recollections, on the thirtieth anniversary of the launch, of one of the design team, Mikhail Floriansky:

> The jettisoning of the nose cone and the process of separation of the sputnik from its carrier was being tested at the assembly shop late in August, 1957. It is not a complex procedure, but fraught with possible surprises. Everything was going on normally—or so it seemed—when Korolev all of a sudden subjected the plant's chief engineer to a terrible dressing down. Korolev was berating him—what for!—for the poor quality of the surface of the mockup of the sputnik! The quality of the surface is really important in flight because the heat conditions of the sputnik depend on it, but why the dressing down now, when quite a different process was being tested?
>
> Korolev said angrily:
>
> "This ball will be exhibited in museums!"[28]

It was Korolev's aesthetic as well as engineering sense that had led him to insist on the ball shape for *Sputnik 1,* although one of the early designs proposed was for a cone-shaped structure. "Today, after decades have passed," recalled colleague Mark Gallai in a 1980s interview, "we simply cannot imagine the first sputnik to be anything other than what it was: an elegant ball . . . with an antenna thrown back like a galloping horse."[29] There were two flight-ready spheres built, one for the launch and another one for ground testing, developing the welding, and other fabrication techniques. The second one would later be launched too, with the carrier for the dog Laika on *Sputnik 2.*

A recollection of the preparations for *Sputnik 1,* and the launch itself from Baikonur, comes from Colonel Mikhail Rebrov:

> People in the "space room" worked in white smocks, performing each operation with the greatest thoroughness. The rocket was assembled in the big hangar. Silence fell when the Chief Designer appeared. At the time Sergei Korolev was exacting and more strict than ever. . . .
>
> Only two days were left. The carrier rocket was rolled out to the launch pad in the early morning of October 2, 1957. Korolev walked in front, together with all the other chief designers. They walked in silence the entire 1.5-km-long way from the assembly-testing building to the pad. No one will ever know what was going through Sergei Korolev's mind at the time. Later on, when the sputnik was installed in orbit, and its call sign was heard over the globe, he said:
>
> "I've been waiting all my life for this day!"
>
> The moment of the blast-off has been described many times. Then the rocket got out of the radio zone. The communication with the sputnik ended. The small room where the radio receivers were was overcrowded. Time dragged on slowly. Waiting built up the stress. Everyone stopped talking. There was absolute silence. All that could be heard was the breathing of the people and the quiet static in the loudspeaker. . . . And then from very far-off there appeared, at first very quietly and then louder and louder, those "bleep-bleeps" which confirmed that it was in orbit and in operation.
>
> Once again everyone rejoiced. There were kisses, hugs and cries of "Hurrah!" The austere men, who were greeted out of space by the messenger they had made, had tears in their eyes.[30]

Hardly tearful, more like rueful, was the reaction of the American space community. Even though the possibility that the Soviets had been making plans for launching a satellite was known in the United States, and not only to government insiders, the fact of the launch was a deep jolt to the space professionals.[31]

At that time, I was executive secretary of the professional society for space engineers—the American Rocket Society (ARS). *Astronautics* magazine, the main ARS publication, could not resist reminding its readers that, "A little over two years ago, when the government's guided missile policy committee decided against the von Braun-Medaris[32] satellite proposal in favor of Project Vanguard, there were four dissenters who voted to send the idea on to the National Security Council for further consideration. One of them was the 'Lone Eagle,' Charles A. Lindbergh, the last one to make aviation headlines of the same magnitude as 'Sputnik.' "[33]

Grim determination characterized the American rocketeers. "They [the Russians] must not be allowed to win this game—a game with far-reaching political, social and economic consequences," *Astronautics* editorialized.[34]

In the same issue, a very insightful interpretation of the technological significance of the Russian feat by Martin Summerfield of Princeton University had an upbeat aspect, but was somber in its reflection on the advanced state of Soviet space technology.

> The success of the Russian "Sputnik" was convincing and dramatic proof to people around the world of the real prospects of space travel in the not distant future. The fact that a 23-in. sphere weighing 184 lb has been placed in an almost precise circular orbit indicates that a number of important technological problems such as high thrust rocket engines, lightweight missile structures, accurate guidance, stable autopilot control, and large scale launching methods have been solved, at least to the degree required for a satellite project.[35]

I got a phone call at my home in Princeton about 7:00 P.M. on Friday evening, October 4, from the *New York Times* aeronautics reporter, Richard Witkin. Had I heard? What is the reaction of the U.S. rocket community? My response is not even in my memory. But the impact of the launch on the United States, as well as on my own career, would be powerful indeed. ARS at the time had a membership of about 5,000 engineers and scientists, most of them working on missile programs, although a few dozen were on Project Vanguard. By 1962, just seven years later, the membership quadrupled to 20,000, a growth so rapid that industry and government pressures caused a merger of ARS with the Institute of Aeronautical Sciences, the society for aeronautical engineers, into what is today the 35,000-member American Institute of Aeronautics and Astronautics (AIAA).

The news of the launch in the world's leading newspapers got Second Coming treatment. The *New York Times,* receiving the story in the late afternoon of Friday, October 4, printed the next morning a rarely used three-line head in half-inch capital letters, running full length across the front page:

> SOVIET FIRES EARTH SATELLITE INTO SPACE;
> IT IS CIRCLING THE GLOBE AT 18,000 M.P.H.;
> SPHERE TRACKED IN 4 CROSSINGS OVER U.S.

Other world newspapers gave the event similar play. Then the interpretation began. The *Manchester Guardian* needed only a couple of days to begin to speculate apocalyptically on what the Russians might now do. An October 7 editorial titled "Next Stop Mars?" read, "The achievement is immense. It demands a psychological adjustment on our part towards Soviet society, Soviet military capabilities and—perhaps most of all—to the relationship of the world with what is beyond." Some of the *Guardian*'s speculation was downright clairvoyant. "We must be prepared to be told what the other side of the Moon looks like [*Lunik 3* produced the photos only two years later], or how thick the cloud on Venus may be [revealed by the Soviet Venera and U.S. Mariner series starting in the 1960s]." Accurately, the *Guardian* pointed out that, "The Russians can now build ballistic missiles capable of hitting any chosen target anywhere in the world." That, certainly, was true, but it would be some years before improvements in guidance technology made the capability an actuality. It certainly did not follow, as the *Guardian* stated a few sentences later, that "Clearly they have established a great lead in missile technology." This was one of the earliest inaccurate predictions of a missile gap.

French reaction was equally ebullient. "Myth has become reality: Earth's gravity conquered," bannered *Le Figaro*, which went on to report the "disillusion and bitter reflections" of "The Americans [who] have had little experience with humiliation in the technical domain."[36] For three weeks the world could hear the beeping of *Sputnik 1* before the radio died out, and it orbited more than 1,400 times before burning up in the atmosphere after three months in space.

Much has been written about the effect of the *Sputnik 1* launch on the world scene. Many American space enthusiasts, stricken with gloom at the time, now reflect that it might have been the best thing that could have happened to awaken the need for an aggressive space program. Only four years later President John F. Kennedy would call for what became the Apollo program. An enormous infrastructure of space research and development centers, test and launch facilities, and supporting industry and university programs, would come into being.

There were those who reacted negatively to the surge in space technology and its consequent spurring of the growth of high-tech weapons. As former Ambassador to the Soviet Union George Kennan put it in his memoirs, "It [*Sputnik*] caused Western alarmists, such as my friend Joe Alsop, to demand the immediate subordination of all other national interests to the launching of immensely expensive crash programs to outdo the Russians in this competition. It gave effective arguments to the various enthusiasts for nuclear armament in the American military-industrial complex. That the dangerousness and expensiveness of this competition should be raised to a new and higher order just at the time when the prospects for negotiation in this field were being worsened by the introduction of nuclear weapons into the armed forces of the Continental NATO powers was a development that brought alarm and dismay to many people besides myself."[37]

While all this introspection was taking place in the West, the creators of sputnik were unable to be interviewed, take bows, be photographed, get medals. Sergei Korolev was literally back at the office, because Khrushchev, realizing what a hot property he had, gave him new orders. Do something bold, Sergei Pavlovich, to celebrate the upcoming fortieth anniversary of the Revolution! It's only a month away. Years later Mishin recalled that he, Korolev, and others who had launched *Sputnik 1,* after a five-day rest at Bulganin's big dacha in Sochi, were ordered to fly back to Moscow immediately. Cosmonaut Grechko tells this story.

> I heard this from Korolev himself with my own ears. After *Sputnik 1* Korolev went to the Kremlin and Khrushchev said to him,
>
> "We never thought that you would launch a sputnik before the Americans. But you did it. Now please launch something new in space for the next anniversary of our revolution."
>
> The anniversary would be in one month! I'll bet that even with today's computers nobody would launch something into space in one month. It was, I think, the happiest month of his [Korolev's] life. He told his staff, and his workers, that there would be no special drawings, no quality check, everyone would have to be guided by his own conscience. The engineers would make drawings and give them directly to the workers. And we launched on November 3, 1957, in time for the celebration of the Revolution.[38]

The payload of *Sputnik 2* weighed 508 kilograms, more than six times the weight of *Sputnik 1.* A shroud housed a carrier containing the world's first space passenger, the mongrel dog Laika, plus a duplicate of the *Sputnik 1* sphere. Laika is reported to have barked and eaten food during his lonely sojourn but, alas, he died when the capsule overheated after failing to separate from its booster, thereby rendering the thermal control system inoperative. Animal groups protested, but the Soviets made Laika into a martyr for a noble cause. Veterans of the *Sputnik 2* project regard it as an even more significant achievement than *Sputnik 1.* Not a single engineering task had been performed on it until after *Sputnik 1* went up.

With the new triumph, Khrushchev could not resist escalating his needling of the Americans. In a speech at the fortieth anniversary of the Revolution, on November 6, he said, "It appears that the name Vanguard reflected the confidence of the Americans that their satellite would be the first in the world. But . . . it was the Soviet satellites which proved to be ahead, to be in the vanguard. . . . In orbiting our earth, the Soviet sputniks proclaim the heights of the development of science and technology and of the entire economy of the Soviet Union, whose people are building a new life under the banner of Marxism-Leninism."

Those derisive comments proved even more galling to the Americans a month later when, on December 6, the first attempt to launch the Vanguard satellite was an ignominious failure before the world's television cameras. The three-stage rocket was designated *TV-3,* for *Test Vehicle 3,* and it had been originally scheduled to be just that, a test. But under sputnik pressure, the "test" was moved up several months and made into a full-fledged attempt at a satellite launch. But the vehicle got only a few feet off the ground before sagging back, buckling, bursting into a huge conflagration, and tossing its tiny 1.47-kilogram payload, still transmitting, some yards away.

Pravda delightedly reproduced the front page of the London *Daily Herald,* which showed a photo of the Vanguard being readied on the launch pad next to one of the explosion. Superimposed above the immense *Herald* headline, which read "OH, WHAT A FLOPNIK!," was *Pravda*'s comment, "Reklama i Deistvitelnost," or "Publicity and Reality."

Following the *Sputnik 2* launch, a team composed of the von Braun group from the Army Ballistic Missile Agency—for the launch vehicle—and one from the Jet Propulsion Laboratory in California—for the space capsule—the latter headed by William H. Pickering, was given the nod to put up the first U.S. satellite. On its initial try the team launched the 14-kilogram Explorer 1, on January 31, 1958. Fortunately, Pickering and Ernst Stuhlinger, one of von Braun's staff chiefs, conscious of the odds against Vanguard's success, had persuaded James Van Allen of the University of Iowa to make the package of scientific instruments being readied for Vanguard compatible with *Explorer 1*'s Jupiter C launch vehicle.[39] This turned out to be particularly fortuitous because *Explorer 1,* with only one sixth the payload weight of *Sputnik 1,* scored a major scientific coup when its instruments sensed a pattern of radiation around the Earth leading to the discovery of the now famous Van Allen belts. It is an interesting footnote on scientific history that *Sputnik 1* could have made the discovery if it had installed simple instruments. *Sputnik 2,* in fact, did have the instruments, wrote Van Allen: "a pair of Geiger-Mueller tubes . . . which operated properly and yielded data for seven days"[40] but the Soviet scientists failed to develop the data necessary to interpret the discovery.

Oleg Ivanovsky, who worked on *Sputniks 1, 2,* and *3,* speaks with deserved satisfaction of the Soviet achievements of those months in late

1957 and early 1958. But there was trouble, too. "We had our first space failure," he told me, with the difficult *Sputnik 3*. "It was April 27, 1958, and it was caused by a rocket engine failure. The rocket went up about 12 or 15 kilometers and the satellite fell separately. There was a search for the satellite. I remember that the pilots conducting the search were not allowed to know what they were looking for. 'Just search the area for anything unusual,' they were told, 'and don't attract the camels.' It was crazy secrecy. Finally one pilot came back and said he had seen something that sounded to us like the satellite. We sent out a rescue team in an armored vehicle. When we got it back some of the instruments could still operate."[41] Ivanovsky then showed me proudly a copper wire that he had recovered from one of the instruments that kept the satellite beeper from operating prematurely.

When the 1.3-ton *Sputnik 3* was launched successfully, on May 15, 1958, it caused even more anxiety in the West. Any doubt that the Russians would soon have the capability to send an ICBM to the United States was demolished. Lyndon Johnson, then Senate Majority Leader, had demanded a congressional investigation of the impact of *Sputnik 1* only a few days after its launch. On the Senate floor in January he had made recommendations originating in his Preparedness Subcommittee to "Start work at once on the development of a rocket motor with a million pound thrust," "Put more effort in the development of manned missiles [satellites]," and "Accelerate and expand research and development programs, provide funding on a long-term basis, and improve control and administration within the Department of Defense or through the establishment of an independent agency."

During the same months in early 1958, President Eisenhower, with his science advisor, James Killian, also concluded that a civilian space agency was needed, and directed Hugh Dryden, head of the National Advisory Committee for Aeronautics (NACA), to prepare legislation that would create such an agency on that relatively small organization's structure. The result was Public Law 85-568, signed on July 29, 1958, by the president, calling for the creation of the National Aeronautics and Space Administration (NASA). It was quite an about-face for a president whose staff members had belittled *Sputnik 1* as "a silly bauble" and "a neat scientific trick" and who, himself, had said that it had not bothered him "one iota."[42]

Sputnik 3's large load of scientific instruments was designed to measure micrometeorites, density of the upper atmosphere, cosmic rays, solar radiation, the presence and effect of high energy particles, and the Earth's own radiation environment.[43] It could have performed a tour de force of scientific research in virgin territory. The *Manchester Guardian* reported that, "This impressive list [of instruments] is a telling demonstration of the fact that the latest Russian sputnik has been launched for strictly scientific purposes." Unfortunately, to the great embarrassment of the Soviets, the huge vehicle missed one more chance at what would have been its most significant achievement—mapping the radiation belts. *Explorer 1*'s instruments, which had revealed the existence of the radiation from the belts, had at first been overwhelmed by its intensity, and it took some time for the Van Allen analysis team to understand what it had measured. *Sputnik 3* could have mapped the belts systematically, but failed to do so because of a defective tape recorder. The recorder, designed to store and transmit to Earth the information collected by the instruments when it was out of direct radio contact, had failed utterly to transmit the necessary data. Roald Sagdeev recounts what happened:

> A scientific team landed at the Baikonur Cosmodrome for the final integration and testing of hardware on Sputnik 3. Korolev invited everyone for the last briefing before the final okay was to be given and the countdown started. . . . It was the first impressive collection of scientific instruments, each of which was reported to be functioning normally.
>
> However, trouble was soon discovered in some of the supporting hardware. The problem was with the more or less routine tape recorder, whose function was to accumulate data from the different experiments and to prepare messages for the ground station. The spacecraft, revolving around the globe, would only be in contact with the ground station during periods of "direct radio visibility." Simply speaking, the ground station would be unable to sense the signals from the spacecraft when it was behind the horizon. . . . With such a crucial role, members of the scientific team were extremely worried about the troubled tape recorder and they recommended postponing the actual launch to give the technicians a chance to fix it. However, the tape recorder's ambitious engineer, Alexei Bogomolov ["I too often had to depend on his hardware," Sagdeev footnotes acidly] did not want to be considered a loser in the company of winners. He suggested that the testing failure was simply caused by electromagnetic interference from the multiplicity of different electrical circuits in the test room. He boldly proposed to launch Sputnik 3 on time.
>
> To the great disappointment of the scientific team, Korolev accepted Bogomolov's suggestion. . . . During the flight, however, it was confirmed that Bogomolov had been dead wrong. His tape recorder did not work. Consequently, the scientific information gathered was limited by the area of direct radio visibility. . . . Each scientific group had results, but because of the recorder failure they had to guess whether the phenomena discovered were of local or planetary significance.[44]

The most disappointed scientist, says Sagdeev, was Sergei Vernov, a renowned physicist. The detectors on *Sputnik 3* sensed extremely high levels of radiation, but was it local or did it exist around the Earth? Some six weeks earlier, on March 26, *Explorer 3* had been launched (*Explorer 2* had failed to orbit), carrying the first tape recorder ever launched on a satellite. As Van Allen wrote, it "functioned beautifully in response to ground command and fulfilled our plan of providing complete orbital coverage of radiation intensity data."[45]

The Soviets, without tape-recorded data, were hog-tied. As Van Allen recalled, his team was at first puzzled that they "were encountering a mysterious physical effect of a real nature."[46] They "worked feverishly in analyzing the data from Explorers 1 and 3 [by primitive hand reduction of pen-and-ink recordings] and organizing them on an altitude, latitude and longitude basis."[47] In an interesting sidelight on the whole episode, at first there was suspicion that the intense radiation was coming from a Soviet nuclear test. Subsequent analysis, however, proved that it was "geomagnetically trapped corpuscular radiation" distributed in a "belt" around the Earth. At a conference in the summer of 1958, the name "Van Allen radiation belt" was applied for the first time.[48] More confirmation of the origin of the radiation, as well as the discovery that there was also an outer belt, came two months later—again from the Americans—when *Explorer 4* mapped the belts from July 26 to September 19, 1958.

"On a purely observational basis," wrote Van Allen, "the Sputnik 3 data actually represented discovery of the earth's *outer* radiation belt inasmuch as they were acquired before those of Explorer 4 and Pioneer 3."[49] However, Vernov and colleagues had not yet interpreted the observational finding, although in what seems to have been hindsight, Vernov published a photo of what is now known as the Van Allen belts in *Pravda* on March 6, 1959, alleging that the data were based on findings that he had reported at an IGY conference in August 1958. However, colleagues of Van Allen, who heard the paper, maintain that the fragmentary data available at that time from *Sputniks 1, 2,* and *3* could not have formed the basis for such a finding, and, what's more, no such finding was reported in the paper.[50]

Afterwards, a Russian joke circulated that the belts were to be called the Van Allen-Vernov radiation belts. "What did Vernov do? He discovered the Van Allen belts."

Looking back, it is remarkable how intense, frustrating, and vigorously competitive was the space activity during 1958, the first full calendar year of the Space Age. There were twenty-two launch attempts, seventeen by the catch-up minded Americans, and five by the Russians.

Testimony to the difficulty of going up the learning curve: Thirteen of the seventeen U.S. attempts and four of the five Russian attempts were failures, or partial failures. Five of the U.S. failures were of Vanguard, although it also had its first success as well. *Sputnik 3*, even with its tape recorder disappointment, merits classification as the only Russian success of that year:[51]

1958: The First Full Year of the Space Age

January 31	USA	*Explorer 1* (14 kg), America's first satellite, discovers the Van Allen radiation belts
February 3	USSR	First try to launch *Sputnik 3* fails
February 5	USA	A second Vanguard try fails
March 5	USA	*Explorer 2* fails to orbit
March 17	USA	*Vanguard 1* (1.47 kg) successfully orbits, establishes the pear-shapedness of the Earth
March 26	USA	*Explorer 3* orbits, collects radiation and micrometeoroid data
April 28	USA	Another Vanguard fails to orbit (third failure)
May 15	USSR	*Sputnik 3* (1,327 kg) orbits, carrying large array of scientific instruments, but tape recorder fails, so it can't map Van Allen belts
May 27	USA	Vanguard fails for the fourth time
June 26	USA	Vanguard fails for fifth time
July 26	USA	*Explorer 4* orbits and maps Van Allen radiation belts for 2 1/2 months
August 17	USA	An Air Force Pioneer attempt at a lunar probe fails
August 24	USA	*Explorer 5* fails to orbit
September 23	USSR	First lunar probe attempt fails
September 26	USA	Vanguard fails for the sixth time
October 11	USA	*Pioneer 1* travels over 113,854 km into space, setting a distance record, but fails to reach the Moon
October 12	USSR	Second lunar probe attempt fails
October 23	USA	*Beacon 1* fails to orbit
November 8	USA	*Pioneer 2* lunar probe fails to orbit
December 4	USSR	Third lunar probe fails to orbit

December 6	USA	*Pioneer 3* goes 102,333 km, finds two separate bands of Van Allen radiation belts, but fails to reach the Moon
December 18	USA	Project Score becomes world's first communications satellite, broadcasts taped messages for 13 days

As this list makes clear, in the very first year of the Space Age concerted attention, by both the USA and USSR, went to the Moon as a destination. But the Earth's only natural satellite proved to be most elusive as both countries failed with a total of seven attempted lunar probes during the year.

With 1959 approaching, the efforts by the two rivals to reach the Moon would become more intensive, and the sphere of competition would also extend to Venus and Mars.

8

Unmanned Firsts: Hitting the Moon and Venus

Going to the Moon and the planets, first robotically, and then with cosmonauts, had long been a prime objective of the Chief Designer. As early as 1955, two years before the *Sputnik* launch, Korolev had raised the possibility of developing a three-stage version of R-7 that could carry a payload to the lunar surface. Historian Gyorgi Vetrov:

> I was lucky to find documents covering an August 30, 1955 meeting that Korolev had with Vassily Mikhailovich Ryabikov, who had a consulting status with the State Gosplan as well as a Military Commission job. Also present were Keldysh and Alexander Grigorevich Mrykin, representing the Ministry of Defense. The issue of launching of *Sputnik* was discussed. At the same meeting the issue of the development of a three-stage vehicle for the Moon launch, based on the two-stage R-7, was discussed, as well as the feasibility of a Moon launch in general. *Sputnik* had not yet been decided![1]

No go-ahead was authorized at the time, of course, but three years later, in 1958, with the successes of the *Sputniks* behind him, Korolev wrote a formal paper pointing out confidently that "The level of development of technology, achieved at the present time, makes it possible to carry out rocket flight to the moon."[2]

The paper proposed the first robotic Moon missions—impacting the surface, photographing the back side, and exploding a flash that would be visible from Earth (an idea also conceived by Robert Goddard). Although the document deals primarily with scientific investigations by robotic spacecraft it states flatly that a main objective is to prepare the way for manned exploration. "[These] first studies of the Moon and interplanetary space," Korolev wrote, "at distances from Earth that reach 400–500 thousand km will also create the necessary prerequisites/premises for the penetration of man into interplanetary space, the Moon and the planets."[3]

Where did Korolev get his ideas for such bold initiatives? Once again a big influence was Tikhonravov, who had inspired a team of bright young engineers at the Artillery Institute. One of them, Gleb Maximov, recalled the period.

> I had been working with Tikhonravov's group in 1952 and 1953, doing research on ballistics of various satellites in various orbits. The Korolev Bureau was busy working on ballistic missiles, but Korolev

> himself was interested in the possibility of space exploration using ballistic missile technology. Korolev came to Tikhonravov's institute [the Artillery Institute] and talked to his deputy, Tyulin. Tyulin told him about my data on artificial satellites, so he came to see me. I sat next to him like I'm sitting next to you and I showed him my graphics. I was young and emotional and I had no idea who Korolev was. I said several times, "Do you understand?" Suddenly I felt a kind of invisible wall.
>
> Korolev looked at me like a snorting bull and said, "I do understand—everything." I was 25 or 26, Korolev about 46."[4]

Eventually the Tikhonravov group was transferred in toto to OKB-1. "There were two different types of designers in the Korolev Bureau," said Maximov, "the creators of preliminary design, concerned with the main characteristics of the spacecraft [I was one of these] and those who created the drawings for production."[5]

Maximov's description of what it was like to work under Korolev reveals the man's sometimes quixotic style, but also his ability to get top effort out of the design team:

> Korolev had a very complex nature. He understood that designers can make mistakes, and as long as the designer acknowledged his fallibility he was understanding. But he didn't like people who tried to cover up their mistakes, even if they were small details. . . .
>
> He was a man of strong will, but he was very careful in choosing the technical means to solve problems. We had a lot of meetings with different organizations. Korolev "sucked out the problem," looked at it from various sides. But after choosing the technical means he became very decisive. Everyone in the design bureau was afraid of him although he never fired anybody.
>
> He was actually very warmhearted. For example, during tests at Tyuratam we drank tea in his small house and discussed problems. He had a penchant for punishing by shouting, swearing, and this was not always justified. The next day he would see you and be embarrassed. He wouldn't apologize directly but would wordlessly convey his feelings with maybe a pat on the back.
>
> We worked nine- to eleven-hour days. We even looked over the factory work on the hardware. I've been an engineer for forty-one years but those ten years with Korolev were the most difficult and the happiest years of my life.[6]

With all the careful preparation, the first robotic efforts to reach the Moon were not easy for Korolev, nor were they for his U.S. competitors who were gearing up for similar missions.

"My father's passion for reaching the Moon was so vivid that he called the first moon ship *Mechta* [dream]" said Korolev's daughter,

Natasha.[7] The dream would be very difficult to turn into reality. The first three Mechtas, or Lunas as the Soviets called them officially, failed to reach the Earth orbit from which they would have transferred to a lunar trajectory. Following what became standard Soviet practice, none of the failures were given names. Then *Luna 1,* launched on January 2, 1959, got a name because—although it missed its target by 5,998 kilometers—it became the first manufactured object to go into orbit around the Sun. Also, it achieved other substantive results—its instruments determined that the Moon had no magnetic field, and it released a cloud of sodium vapor at 100,000 kilometers that was visible over the Indian Ocean to astronomers. Maximov described Korolev's approach to the first Luna missions.

> Most people use the simplest path in difficult situations—it's usually the optimal way—but Korolev used all the paths, even the most difficult, in order to get redundancy. For example, he ordered a number of people to solve the same problem of getting the right trajectory for the Lunas to hit the Moon. We used radio data to adjust the optimal trajectory. But in those days it was difficult to understand radio data, so he wanted another method for confirmation. We discussed the possibility of using a small nuclear bomb to explode so we could see the flash and measure the time it took for the flash to appear.
>
> Thank God we didn't pursue that one.
>
> Zeldovich [Yaakov], who was in charge of the nuclear program, proved that because of the lack of an atmosphere on the Moon the bomb would create only a small cloud of dust. We had another variant—a "sodium comet"—a container with 1 kilogram in a "thermit"—a kind of napalm ignited the sodium producing a big vaporous cloud of bright yellow. Korolev liked this idea and used it, and the foreign and USSR astronomers could see the clouds. The radio data correlated well, however, and Bernard Lovell heard the radio transmission at Jodrell Bank, so we didn't need the sodium vapor.[8]

One of those disappointed by the failure of *Luna 1* to hit the Moon was Arvid Pallo, who had worked for Korolev in the years up to his arrest, then came back to the bureau in time for development of the first lunar spacecraft. "I did design and testing [for the lunar impact missions]," Pallo said, "and integrated the various systems. I supervised all of the testing at the factory and the launch site, and carried out the final inspection."

Korolev's consciousness of the tremendous significance of the *Luna 1* mission is made clear by one of Pallo's assignments. "I designed the sphere containing memorabilia that would have been left on the Moon. There were two things in it, based on the ideas of Tsiolkovsky. We immersed several layers of materials in a fluid of the same density. In the middle, also immersed so as not to be damaged, was a piece of material inscribed 'USSR.' There was also a spherical explosive device containing seventy-two stainless steel segments. Stamped on one side of each was the launch date, and on the other side was the heraldic symbol of the USSR [hammer and sickle in a wreath under a star]. We designed the sphere so that on impact all seventy-two pieces would fly radially at 2 to 6 kilometers per second velocity."

But this time the impact was not to be. "I'll have to look at my notes to determine what the reason was for the failure. It was either a targeting error or an incorrect trajectory," said Pallo wistfully in an interview at age eighty-one.[9]

It would be some thirty years before the Soviets would admit that *Luna 1* was meant to hit the Moon. Ignoring the failure, Nikita Khrushchev described the mission as if it were part of a long-range Soviet strategy. In a speech to the 21st Party Congress, on January 27, 1959, he crowed grandiloquently,

> In the first days of the new year, 1959, the first year of the Seven Year Plan, Soviet scientists, designers, engineers, and workers achieved a new exploit of worldwide importance, having successfully launched a multistage cosmic rocket in the direction of the moon. . . . Even the enemies of socialism have been forced, in the face of incontrovertible facts, to admit that this is one of the greatest achievements of the cosmic era. . . .

Luna 2, launched on September 12, 1959, did impact the Moon, enabling the Soviets to claim one more first—the first spacecraft to reach another celestial body—and the Soviets quickly released enough detail about the launch to allay Western skepticism about the credibility of the achievement. On September 14, Hugh Dryden, NASA deputy administrator, prepared a statement for President Eisenhower's morning report, which was also released to the press, confirming the fact that tracking data from Jodrell Bank (UK) and Fort Monmouth (NJ) Observatory, and analysis of the signals from the spacecraft

> indicated that the probe vehicle accelerated as it came under the influence of the moon's gravity field. The acceleration combined with trajectory data confirm the Soviet claim of impact. NASA joins with the world's scientific community in acclaiming this as a truly great engineering achievement. It is further evidence of the excellence of the Soviet capability in the propulsion field. The mass that was sent into space orbit required a multistage vehicle, the first stage of which would have a thrust which was probably greater than 500,000 pounds.[10]

But what kind of guidance did the vehicle have?

"Hopefully," said Dryden's statement, "future reports from the Russians will tell us whether guidance was a function of early staging or some form of mid-course or terminal guidance."

The fact is that the Russians had not yet developed mid-course guidance technology but this was a sensitive subject for the Americans, since precise guidance was needed to send ballistic missiles to their targets, and the United States wanted its citizens to know that its ICBMs—which did have mid-course guidance capability through inertial systems developed primarily by Stark Draper and his colleagues at MIT—were quite up to their job, even if its space vehicles weren't.

"Our military vehicles," said Dryden, using blunt Cold War language, "for example, have demonstrated guidance capacity to pinpoint target areas at distances of many thousands of miles. In many respects this is favorable comparison when one compares the several square miles of a city at the end of a 6,000-mile ballistic flight with the 2,160-mile diameter of the moon, after a 240,000-mile flight." But the United States, Dryden continued, "still lacks the propulsion capability to place this guidance equipment into payload vehicles designed for space exploration. This large propulsion capability is the characteristic feature of the Russians' 5-year lead."[11]

Luna 3, sent into space only three weeks later, flaunted that lead. Launched on October 4, 1959 (October 5 in Baikonur), on the second anniversary of *Sputnik 1,* it was an especially elegant mission. The 278.5-kilogram spacecraft swung gracefully around the Moon and, with the Sun behind it, photographed the far side, never before visible to Earthlings. The technology that made it possible could not help but impress the Americans—solar cells for power, photoelectric cells to orient the spacecraft properly with respect to the Sun, gas jet stabilization. For forty minutes the spacecraft took photos, covering some 70 percent of the far side. The film was developed automatically onboard, scanned by a television unit, and transmitted by radio to the ground. The resultant images were not of high quality, but they were precious—the very first from another celestial body.[12] Boris Rauschenbakh claims that: "*Luna 3* was the first spacecraft with an attitude control system. We were ahead of the Americans and had nothing to learn from them."[13] That statement, however, is contested by Daniel DeBra, a guidance and control expert from Stanford University, who points out that the U.S. Discoverer series of reconnaissance satellites, launched earlier in 1959 than *Luna 3,* used active, jet-powered attitude controls during the coast phase of their orbital entry. The Discoverer system used horizon sensors and gyroscopes in a "gyro-compassing mode" to orient the satellites' cameras as they photographed areas of Earth.[14]

An ironic note in connection with *Luna 3*'s pictures. A. C. B. Lovell of Jodrell Bank, whose tracking of Soviet satellites had helped give corroboration to their successes, found himself persona non grata when he deciphered the *Luna 3* signals, converted them to photos, and published them in the British newspapers one day before the Soviets themselves did. Unfortunately, he had incorrectly estimated the scale for the dimensions and so the photos were distorted. It is said that Lovell's nomination for foreign membership in the USSR Academy of Sciences, some years later, was voted down because of the incident.

Like virtually every project in the Korolev design bureau in those days, the preparations for *Luna 3* were very intense right up to the moment of launch. " . . . only at the cosmodrome were we able to finally develop and adjust the system of electrical power supply," wrote Arkady Ostashev, a designer.[15] "We worked around the clock, and many specialists did not leave their test stations for several days in a row." The vehicle lifted off on schedule but before long it was obvious that radio communications were not coming through properly.

What Korolev then decided is testimony to both the primeval state of his electronic communications and the supreme conviction that he, himself, must take charge of whatever crisis loomed. He decided that he, personally, along with Mstislav Keldysh and other high officials, would best head for the Crimea and be there by 4:00 P.M. when *Luna 3* came into radio range with the control center at Ai-Petri.[16] At about 10:00 A.M. on October 6, having just returned to Moscow from Baikonur, recalled Ostashev, the group set out for the Crimea. First there was a careening car ride to Vnukovo airport, followed by a flight to Simferopol in the Crimea. At Simferopol airport the group boarded a helicopter for the trip to Ai-Petri. But soon after the helicopter passed over the Crimean mountains the pilot came into the passenger cabin. "It is sleeting in the region of Ai-Petri," Ostashev remembered him saying. " . . . the visibility is practically zero and it is not advisable to land." After first considering a landing anyway, Korolev decided it would be more prudent to head for Yalta airport where cars—arranged through a radiogram to local Communist Party officials—would speed them the 10 miles to the control center.

"All seven of us got into one Zim. . . . They gave us a driver who . . . was simply a virtuoso. 'Well, dear boy,' Korolev told the driver, 'show us what you and the Zim can do . . . we are really in a great hurry.' " The driver never reduced his speed along the precarious, sharply curving road, leading the passengers to worry about a crash with an oncoming car at virtually every turn.[17]

Arriving at the control center, Korolev called the staff together and discussed the probable reasons for the poor communications. Was it a problem of incorrect directionality of the antennas? Perhaps, but the design was made at the Korolev design bureau, not here, said the technicians. Not a good answer for the resolute Korolev, who remarked, according to Ostashev, that, " . . . the reason for the malfunction must be sought in our common and as yet poor skill in controlling the flight of such stations." He then talked individually to each of the control center operators, "creating an atmosphere of confidentiality and simplicity necessary for people to speak out frankly." What he learned was that "there was no proper verification of the correctness of performing [each] completed operation before performing the subsequent one." In other words, poor quality control.[18]

The next person to catch hell was Ostashev, himself. An analysis of the telemetry showed that the thermal regulation system was not working properly and the spacecraft was heating up. Ostashev's recollection of the dialogue: " 'Why are we overheating?' barked Korolev."

"I don't know yet," Ostashev's answer, was not what the Chief Designer wanted to hear. He then fulminated: "If you as a test specialist believe that questions of planning and ideology of the system are not your problem, you are deeply mistaken. Where were you when the AIS [Automatic Interplanetary Station] was being prepared for flight? Why didn't you find out the capacities of the thermoregulation system? This is your gross error which cannot be allowed in the future. And now I am waiting for proposals from you. Involve all of us here, anyone in Moscow or in the entire Soviet Union. . . . But remember—the fate of the results of the first AIS flight is now in your hands."[19]

By the evening of October 6, Ostashev had worked out a method of cooling the spacecraft by taking two clever steps. First he rotated the spacecraft into different positions with respect to the Sun along the longitudinal axis, being careful not to hinder the charging of the chemical batteries by the solar cells. Second he programmed various heat-generating devices to turn off whenever they were not functioning. The temperature dropped from 40°C to 27 to 30° over the next few days.[20]

But it wasn't only the engineers and technicians on the scene in the Crimea who incurred Korolev's wrath. He had plenty in reserve for those back in Moscow who were maintaining constant calculations of the spacecraft trajectory.

One of the veterans of that duty, Efraim Akim, recalled the experience in 1993. Still youthful-looking and healthy, although probably in his sixties at the time of the interview, Akim has a high position in the Keldysh Institute of Applied Mathematics. In 1959, though, he was a young ballistician, very much in awe of Korolev and of his own direct boss, Keldysh, both of whom were in the Crimea trying to keep track of the capricious *Luna 3*.

> Two different ballistic centers, the one at NII-88 and ours [at the Keldysh Institute] were performing calculations at night. We never slept at launch time in those days but on this occasion we had done much work, had solved problems, run many calculations. It was the kind of heavy pace that never happens today. We calculated one more maneuver and I decided to give our people a rest. We brought a few folding cots in. But there wasn't enough space for all of our people in the computer building. I went to the next building for a rest. Korolev telephoned from the Crimea to talk about the next maneuver. Some young female computer operator picked up the phone—no one else was there. She didn't know who Korolev was. Remember he was anonymous to all but the insiders. She told him,
>
> "Akim is asleep and there is nobody available for you to talk to."
>
> You can imagine Korolev. After we came back from our rest in about an hour I phoned the Crimea. I was told that Korolev had been furious, had made unbelievable noise, and it was only thanks to Keldysh that he was calmed down. Keldysh told him it was impossible that everyone went home, that they were surely where they were supposed to be. At the moment of my call Korolev himself was resting and couldn't come to the phone. When Keldysh and Korolev came back to Moscow there was much talk about this incident. Keldysh had smoothed things over but Korolev never forgot the matter.[21]

Although it might not have been Korolev's style to forget such letdowns on the part of his staff, he could be depended on to reward good work as well, giving out bonuses and recommending prizes generously. Oleg Ivanovsky, who was one of Korolev's designers of the Luna spacecraft, is proud of the fact that he was given a Lenin Prize after the *Luna 3* mission, "although nobody knew who we were," and he has very warm memories of another generous act by SP stemming from the same mission.

> At the factory one day, back in 1960, somebody found me and said I should call Korolev urgently. I called and he said,
>
> "What are you doing? If you can, please come to my office." Usually he just said, "Come."

> When I got to his office he stood up and gave me a thin blue book. In it were the first photos of the back side of the Moon. The inscription, dated December 31, 1959, was,
>
> "To Oleg, with good memories of our work together."
>
> Then he gave me two bottles of wine for the New Year. It seems that a group of people in Paris had talked about whether anybody would ever see the far side of the Moon. One guy said it's impossible and bet 1,000 bottles of wine that it would never happen. Then when *Luna 3* did it, his friends made him come across. They sent the wine to the Soviet Academy of Sciences accompanied by a letter explaining the bet. I was too young and stupid to save the two bottles. We drank the wine and threw the bottles away.[22]

A few weeks after the *Luna 3* mission, a Soviet delegation presented copies of the photos of the Moon's dark side, as well as a model of the spacecraft, to John Stapp, president of the American Rocket Society, at the annual ARS meeting in Washington. That event is especially vivid in my memory since it was I who met, at New York's Idlewild Airport, the Russians who would act as "front men" for Soviet space at international gatherings for years to come—Leonid Sedov and Anatoli Blagonravov. I then flew with them to the nation's capital for the ARS meeting. That was my first of some thirty-eight years of meetings with Russian space scientists and engineers. Alas, I was never to meet Korolev.[23]

The Americans, meanwhile, had been experiencing a succession of Moon-shot failures. The first stumble was on August 17, 1958, when the initial stage of an Air Force Thor-Able launcher malfunctioned and the Pioneer spacecraft never made it off the pad at Cape Canaveral. *Pioneers 1, 2,* and *3* likewise failed to make it to the Moon, although *Pioneer 1* set a distance record by traveling some 113,854 kilometers into space, and *Pioneer 3* provided important data on what were later designated the Van Allen radiation belts. All told, there were seven straight Pioneer failures up through 1960.

In the middle of these, NASA's first administrator, T. Keith Glennan, issued prickly statements to the effect that the United States was methodically developing its own space exploration program and was not in any kind of a race. His draft of the portion of Dwight Eisenhower's January 1960 State of the Union speech dealing with space made the point that "we are not going to attempt to compete with the Russians on a shot-for-shot basis in attempts to achieve space spectaculars."[24] Looking at the record, it is no wonder that he offered that kind of rationalization. Here are the Russian and American Moon missions between 1958 and 1960.

Moon Attempts 1958–1960

	US	USSR
8/17/58	Pioneer (Thor-Able) lunar probe fails	
9/23/58		Lunar probe fails
10/11/58	*Pioneer 1* fails to reach Moon	
10/12/58		Lunar probe fails
11/8/58	*Pioneer 2* lunar probe failure	
12/4/58		Lunar probe fails
12/6/58	*Pioneer 3* fails to reach Moon, but discovers outer radiation belt	
1/2/59		*Luna 1* misses Moon by 5,998 km
3/3/59	*Pioneer 4* misses Moon by 60,000 km, goes into solar orbit	
6/18/59		Lunar probe fails
9/15/59		*Luna 2* impacts Moon
10/4/59		*Luna 3* photographs far side of Moon
4/12/60		Lunar probe fails
9/25/60	Pioneer lunar probe fails	
12/15/60	Pioneer lunar probe fails	

Glennan's words notwithstanding, the subsequent efforts by the United States not only to reach the Moon, but to match the USSR in exploring new celestial arenas—Mars and Venus—were intensive. Pushing hard on the Russian side, according to some reports, was the newly converted space aficionado, Nikita Khrushchev, who is said to have called for the planetary shots, starting with an attempt to reach Mars in October 1960, just in time to herald his visit to the United Nations General Assembly. Not likely, says Nikita's son Sergei, now a professor at Brown University. "In those days if a space shot was successful, the credit was claimed by the space people. If it failed, it was said to be Khrushchev's fault for ordering it."[25] In any case, while space triumphs offered nice prestige bonuses for the USSR, Khrushchev's main preoccupation continued to be military strength.

The level of activity at the Korolev design bureau in this period was frenzied. On the agenda were spacecraft for the Moon, Mars, and

Venus; communications, reconnaissance, and weather satellites; and the Yuri Gagarin flight, which was only months away. All of these missions would depend on versions of the R-7 launch vehicle with new upper stages. The missions designed for lunar flyby, lunar impact, and fly-around (for photographing the far side) used a third stage powered by the RO-5 engine developed by the Korolev and Semyon Kosberg design bureaus.[26] The engine, the first to be ignited in space, had a thrust of 5.6 tons, using liquid oxygen and carbon-hydrogen propellants. It was developed in just nine months.[27]

The lunar soft landings required a different third stage, powered by a Kosberg four-chamber rocket of 30 tons thrust, plus a new fourth stage. The latter, propelled by a Korolev rocket engine of 7 tons thrust, also used liquid oxygen and carbon-hydrogen propellants. It was the first "closed loop" rocket and also the first "launched in weightlessness."[28]

This same fourth-stage engine was necessary for the planetary missions as well. Low-thrust solid rockets provided the initial acceleration needed to reliably ignite the fourth-stage engine in zero gravity. The fourth stage then propelled the spacecraft on a heliocentric orbit to Mars or Venus. This launch vehicle was later used to place the first Molniya[29] communications satellites into highly elliptic orbits over the northern latitudes of Russia.[30]

New technology and new facilities, which were surely very expensive, had to be developed to cope with the demands of this new generation of spacecraft. Korolev had numerous people to convince of the feasibility of his ambitious planetary exploration strategies. One of them was his long-time ally, Mstislav Keldysh. "Keldysh was against the development of such craft [for exploring Mars and Venus], and even called them a 'rocket trick,' " remembered Gleb Maximov. But eventually Korolev's conviction that he could be successful convinced Keldysh and from then on he got the physicist's fervent support.[31]

Cost was no problem in those days. In fact, no one really knew how many rubles were being spent. Among the technological challenges were how to make the instrumentation airtight; how to develop thermal environment chambers so as to simulate the heat regime that the spacecraft would experience; how to develop onboard power supplies for operating the instruments and for controlling the spacecraft orientation over long periods of time; how to make mid-course trajectory corrections by radio communications over great distances.[32]

Korolev ordered a new Center for Deep Space radio communications to be built in Yevpatoriya in the Crimea. Because this was the USSR's only such center it was unable to communicate when the Earth rotated and occluded the spacecraft. So there were limited opportunities to communicate trajectory corrections. Also, it was necessary to make corrections enabling the spacecraft to arrive at the planet when it had radio visibility with Yevpatoriya.[33] The U.S. Deep Space Network gave the Americans much more flexibility and communications coverage. Begun in 1958, the DSN by 1965 had round-the-world coverage, including communications stations in Goldstone, California; Auroral Valley, near Canberra, Australia; and Johannesburg, South Africa.

The Mars and Venus destinations would prove much more difficult to reach than the Moon, and would be just as spiritedly contested with the United States. Korolev was ready with two spacecraft for most of the astronomical windows for both planets right up to his death, and the Americans made many of the windows as well. The first to fail were the two Soviet Mars probes attempted in October 1960. Neither spacecraft made it to Earth orbit due to third-stage engine failures. Then seven straight Venus probes—five by the Russians and two by the Americans—failed in one nineteen-month stretch between February 1961 and September 1962.

It must have been especially frustrating for Korolev to have had so many failures though he had personally monitored launch preparations meticulously. One such instance was recalled by Maximov. It was in 1961, during a pressure chamber checkout of a planetary spacecraft. He had gotten an evening phone call from an engineer who reported that the craft showed an air leakage rate three to four times what the blueprints called for. Was it OK to fly? Maximov recalls:

"Having not quite awakened, and therefore being overly calm, I multiply the measured loss by the time of flight, divide it by the free volume of the compartment, and compare the obtained value in pressure drop with the range of pressure in the compartment . . . the result is positive, I give the 'OK' to the head designer, and go back to sleep. . . . "

The next morning Maximov encountered Korolev, who asked, " 'Who gave you the right to approve the craft for launch?' . . . I explained my computation.

'What is the leakage allowable in the blueprints? Was this condition fulfilled before?'

Yes, it was. . . .

'So, why didn't you wonder why it isn't being fulfilled now?' "

It turned out that several bolts on the spacecraft's airlock had not been tightened sufficiently. This time Maximov got off with a dark look.[34]

Although sometimes, as with Maximov, Korolev was hard-bitten with his staff during this period, he made an effort to thaw what had been a cold relationship with his daughter, Natasha.

"When I was taking my practical courses at Khotkovo, near Abramtsevo," she told me, "he visited me there. He was very pleased when I became a doctor. I had graduated with a gold medal cum laude, from Elementary #243. Then I went to the First Moscow Medical Institute and became a lung surgeon. . . . I visited my father's house in 1962 when Andrei [Korolev's first grandchild] was three. My father dreamed of playing with Andrei when he got older. "We will be good friends," he said.[35]

From the time of the divorce from Xenia in 1948, Natasha and her father had not been close. "Xenia and Korolev's mother took the marriage to Nina hard," I was told by Golovanov. "The young Natasha had a negative attitude towards her father. She didn't have contact with him, didn't even invite him to her wedding. Their relationship was reestablished only when Andrei was born, in the last years of his life."[36]

One of the most interesting—and, in retrospect, chilling— incidents of this period occurred during Korolev's attempt to send the world's first payload to Mars in October 1962, at the very time when President Kennedy and Chairman Khrushchev were having a standoff over the Cuban missile emplacements. Here is a firsthand account from Korolev's deputy, Boris Chertok:

> I was at Baikonur readying a spacecraft for a launch to Mars. Suddenly, Colonel Anatoli Kirillov [deputy chief of the Baikonur test range] told me to get the launch vehicle off the pad so that they could replace it with an ICBM, that there was some kind of national emergency crisis. All of the phone links were being used by the military, so I couldn't get through to Korolev, who was at his home in Moscow, ill with a cold. I told Kirillov how difficult it would be remove the Mars vehicle, that there was a limited window for the launch. He said he'd get court-martialed if I didn't move the Mars vehicle away. . . . Kirillov's team began to do all of the checks of the ballistic missile at its assembly site. But only Korolev would be able to get through to Khrushchev to countermand these terrible instructions. . . .
>
> There were soldiers with machine guns all around "in case of attack by U.S. paratroopers" we were told. So we all went to Korolev's house. I remember that we ate a watermelon and thought we might be waiting for a U.S. thermonuclear strike. Well, it would finish off Keldysh, Smirnov and Voskresensky, who were all there. But after we ate the watermelon, a smiling Anatoli Semyonovich Kirillov appeared at the door and said that Moscow had told him to abort the ballistic missile preparations.[37]

In another account of the incident, V. M. Bryushinin recalls that the preparations were sufficiently serious that all industrial personnel had been cleared from the checkout facility and the launch pad. It was not until Kirillov reported that the "the issue has been settled via diplomatic channels" that anxieties subsided, and the Mars launch was able to proceed.[38]

A little-known sidelight is that when the Mars spacecraft was finally launched on October 24, 1962—in the middle of the crisis, which was not settled until October 27 when Khrushchev announced that he would dismantle the Cuban missiles and send them back to the Soviet Union—it exploded into so many pieces that it nearly caused radar observers at the U.S. Ballistic Missile Early Warning System to think that a Soviet nuclear attack was underway. Fortunately, BMEWS computers, which assess trajectory and impact points, proved the alarm to be false within a few seconds.[39] Although it seems incredible that a launch pad for a Mars vehicle would also be needed for an ICBM launch, that was, indeed, the case. There were only two pads at Baikonur and one at Plesetsk and they had to double as ICBM and spacecraft launchers. In September 1961, based on pictures from the first five reconnaissance satellites, the CIA estimated that there were only ten to twenty-five ICBMs in the whole USSR at that time.[40]

One week after the missile scare, on November 1, 1962, a second spacecraft, designated *Mars 1,* made the world's first Mars flyby, although it was mute by the time it went past the Red Planet, because its attitude control system failed and the spacecraft lost its communications lock with Earth after traveling 106 million kilometers. It had held some sixty-one communications sessions back to Earth over two- and five-day intervals until March 21, 1963—no mean feat.

The competition with the United States for planetary firsts grew more and more intense over the next several years, with each nation scrutinizing, and trying to anticipate, the other's efforts. A CIA report transmitted to the Lyndon Johnson White House on June 1, 1964, noted that *Zond 1,* which had been launched two months earlier, on April 2, would reach the vicinity of Venus on July 20.

"We cannot yet tell whether it will impact on Venus or fly by, perhaps ejecting an instrumented probe to explore the planet's atmosphere as it goes by," said the report. In fact, *Zond 1* came within 100,000 kilometers of Venus, but its radio failed and no data were returned.

However, by the end of the 1960–1965 period, the Soviets were able to claim not only the first Mars flyby but also the first Venus impact, with *Venera 3.* They also had a space technology first with *Zond 2,* launched on November 30, 1964. This was the first spacecraft to use electric thrusters for attitude control.

As for the Americans, after a *Mariner 1* attempt to reach Venus failed, *Mariner 2* made it to within 35,000 kilometers of that planet.

Mariner 3 missed Mars, but *Mariner 4* became the first substantially successful Mars mission when it took spectacular TV pictures of the cratered surface from a distance of 9,844 kilometers in July, 1965.

It would not be until 1971, more than five years after Korolev's death, that the Soviets would be able to claim their next planetary first—the first spacecraft to impact Mars. There were, in fact, two Soviet spacecraft—designed by the Lavochkin design bureau, to which Korolev had transferred planetary exploration responsibility—that hit Mars within one week in that year. Capsules released by *Mars 2* and *Mars 3* crashed into the Martian surface on November 27 and December 2, respectively, and neither vehicle was able to transmit pictures back to Earth.

This was an especially exciting phase of the USA-USSR planetary competition because the American *Mariner 9*, developed by the NASA-Cal Tech Jet Propulsion Laboratory, had reached Mars two weeks earlier, on November 14. Its mission was not to land on Mars but to orbit the planet, taking photos, and it orbited patiently throughout a dust storm, then took spectacular pictures of the pockmarked surface until January 1972.

Mars and Venus Attempts in Korolev's Time

	USA	USSR
10/10/60		Mars probe, third stage fails
10/14/60		Mars probe, third stage again fails
2/4/61		*Sputnik 7*, actually a Venus probe, suffers fourth-stage failure
2/12/61		*Venera 1* is launched from the parking orbit of *Sputnik 8*, but the Sun orientation system fails and radio contact is lost after 4.7 million miles; spacecraft goes into solar orbit
7/22/62	*Mariner 1* Venus probe fails	
8/25/62		Venus probe fails
8/27/62	*Mariner 2* flys within 35,000 km of Venus, enters solar orbit	

9/1/62		Venus probe fails
9/12/62		Venus probe fails
10/24/62		Mars probe fails
11/1/62		*Mars 1* loses communications link with Earth after 106 million km enroute to Mars when attitude control engines fail
11/4/62		Mars probe fails
3/27/64		*Kosmos 27,* Venus probe, fails
4/2/64		*Zond 1* misses Venus by 99,779 km, fails to return data
11/5/64	*Mariner 3* misses Mars by wide margin	
11/28/64	*Mariner 4* passes by Mars (7/14/65) at a distance of 9,844 km and takes 21 TV pictures, the first close-range pictures of another planet	
11/30/64		*Zond 2* passes 1,609 km from Mars, but fails to return data
11/12/65		*Venera 2* passes 23,950 km from Venus (2/27/66) but fails to return data
11/16/65		*Venera 3* impacts Venus (on 3/1/66), first spacecraft to reach the surface of another planet, but returns no data
11/23/65		*Kosmos 96,* probable Venus probe, is left in parking orbit

In 1961, the second year of these planetary tries, both countries resumed their Moon shots. On the American side, the first Pioneer launches had been conducted by the Air Force. Then, when NASA was created in October 1958, its Jet Propulsion Laboratory, operated in

Pasadena by California Institute of Technology, took over the program. But for some time there would continue to be disappointment. The Ranger program succeeded Pioneer, and *Rangers 1* and *2*, which were to be orbital test vehicles, never made it to orbit. *Ranger 3* began 1962 by missing the Moon by almost 37,000 kilometers.

Ranger 4, on April 23, 1962, became the first U.S. spacecraft to impact the Moon, but its instruments failed. *Ranger 5*, October 18, 1962, missed the Moon by 700 kilometers. *Ranger 6*, January 30, 1964, hit the Moon but its television didn't work.

"We called the Rangers 'American kamikazes,' " said Oleg Ivanovsky, who was one of Korolev's designers of the Luna spacecraft.

But the Korolev team had its own miseries with the Moon, experiencing miss after miss, although they were always reluctant to acknowledge their failures, or even their tries. "There were about eighty lunar missions over the years, forty by the United States up through Apollo, and forty by the USSR," said Ivanovsky, "but only twenty-four of our forty were announced. There were a lot of failures on both sides."[41]

As the table below indicates, the Americans finally succeeded in reaching new heights on the learning curve with *Rangers 7*, *8*, and *9* in 1964 and 1965. Those missions provided some 17,000 astounding pictures of the lunar surface. Their quality was far superior to the pictures made by the Russians, as the latter had not yet developed the technique of digital signal processing. By using that process to enhance *Ranger's* analog data, the images of lunar features transmitted over some 400,000 kilometers were breathtakingly sharp. The Soviet record of Moon missions during the period was dismal, with eleven out of twelve failures. Only the *Zond 3* mission, which took some twenty-five pictures of the far side of the Moon, can be considered successful.[42]

Moon Programs 1961–1965

	USA	USSR
8/23/61	*Ranger 1* fails to orbit	
11/18/61	*Ranger 2* fails to orbit	
1/26/62	*Ranger 3* misses Moon by 36,800 km	
4/23/62	*Ranger 4* impacts Moon, instruments fail	
10/18/62	*Ranger 5* misses Moon by 725 km	
2/2/63		Lunar probe fails to orbit
3/3/63		Lunar probe fails to orbit

4/2/63		*Luna 4* misses Moon by 8,499 km
1/30/64	*Ranger 6* hits Moon, TV fails	
3/21/64		Lunar probe fails
4/20/64		Lunar probe fails
6/4 or 6/18/64		Lunar probe fails
7/28/64	*Ranger 7* returns 4,308 photos before impact	
2/17/65	*Ranger 8* impacts Moon (2/20), returns 7,137 photos	
3/12/65		*Kosmos 60,* probable lunar probe failure
3/21/65	*Ranger 9* impacts Moon (3/24), returns 5,814 photos	
5/9/65		*Luna 5* hits Moon but fails to make soft landing
6/8/65		*Luna 6* misses Moon by 161,000 km
7/18/65		*Zond 3* flies by Moon, takes photos
10/4/65		*Luna 7* crashes on Moon, fails to make soft landing
12/3/65		*Luna 8* crashes on Moon, fails to make soft landing

The last of the Luna soft landers to fail in Korolev's lifetime, *Luna 8,* would be especially frustrating. "It was just a few weeks before his death," remembers Mikhail Rozhdestvensky. "Korolev was in charge of launch operations himself at TsUP (Flight Control Center in Kaliningrad) at that launch."[43] That very fact reveals the Chief Designer's obsession with complete control over as many aspects of the mission as he could muster. In the United States it would have been like NASA Administrator James Webb directing launch operations.

Korolev once again blew his top because "There had been breaks in the communications to the spacecraft," said Rozhdestvensky, "and Korolev shouted at some Army general that, 'If you don't fix this in ten minutes I will make you a soldier.' I had never heard anything like this."[44]

It seems especially poignant that Korolev's death occurred only seventeen days before the very successful *Luna 9* mission, launched on January 31, 1966. Although the spacecraft was built by the Lavochkin bureau, the design was from Korolev's engineers. "Gyorgi Nikolaevich Babakin, the chief designer of Lavochkin at the time, was given complete documentation for the spacecraft, and all of the results of the experimental testing," recalls Gyorgi Vetrov.[45] The transfer, however, was not casual. Rozhdestvensky remembers that "I was present at the meeting [when Korolev handed off the responsibility for *Luna 9* to Babakin]. Korolev paraphrased a famous literary quotation by Taras Bulba to his son, from the Gogol novel, *Dead Souls,* 'It was I who gave you birth. If you fail it will be I who will end your life.' "[46] Those were the words of a man who quite obviously regarded the whole Soviet space program as his province.

Lavochkin took over not only the robotic Moon missions, but the planetary missions as well. For Korolev it was, at long last, an admission that his projects had simply become too many and varied to be managed effectively. The failures in the Luna series, for example, were probably the consequence of inadequate testing. Lavochkin, with long experience in series production of military aircraft, " . . . did more thorough ground testing," said Rozhdestvensky. "We cleaned up a lot of bugs. We used vacuum chamber tests, drop tests, worked on the landing controls."[47]

The result was an impeccable *Luna 9* mission. Brian Harvey, a British expert on the Soviet space program, analyzed the mission vividly from Russian sources.

> . . . 46 seconds before impact, the retrorocket fired, exactly on schedule. It halted Luna 9 dead in its tracks. The little 99.8 kg capsule separated and bounced along the rocky, dusty lunar surface like a heavy-bottomed toy. It settled. On the top part, four petal-like wings unfolded. Four minutes later, four 75 cm aerials poked their way out of the dome, slowly extending.[48]

In seven separate communications, lasting more than 8 hours, *Luna 9*'s tiny TV camera, rotating 360 degrees around its landing area in the Ocean of Storms, sent back to Earth excellent photos of the boulders, rocks, and Moonscape. There had been much debate in those days in the United States about the possibility that the Moon would swallow a spacecraft in dust. The Russians had the same concern. "The astronomers," recalls Boris Rauschenbakh, "couldn't reach agreement, so there was a meeting to discuss the problem. After some considerable discussion, Korolev, having listened carefully, said, 'The Moon's surface is solid.' It was intuitive. One of the eminent scientists stood up and said, 'No serious scientist can support such a statement.' Korolev wrote down in a notebook, 'The surface is solid,' signed it and gave the notebook to the scientist. 'Here, you have a signature.' He had guts and instinct, and he was right."[49]

An American mission incontrovertibly confirmed the finding. *Surveyor 1,* launched four months later than *Luna 9,* on May 30, 1966, made the first controlled soft landing on the Moon, reached out with a digging device and determined that the lunar surface was, indeed, solid.

Radio Moscow declared the *Luna 9* mission, which proved to be one of Korolev's last legacies, "a major step towards a manned landing on the moon and other planets."[50]

Now we must back up from 1966 to the preparation of the cosmonauts for that landing, starting in 1961 with still another world-shocker, the launch into orbit of Yuri Gagarin.

9

Gagarin First, Shepard an Anti-climactic Second

"A BLACK Zil 110 drives up and this guy in a navy blue overcoat and hat, with a big, powerful neck, steps out. Under his brimmed hat you could see penetrating black eyes. To turn around he didn't just turn his head, he turned his whole body. His eyes got livelier when he saw us." So recalled Alexei Leonov, who would be the first human to walk in space, of his first impression of Sergei Korolev. It was April 8, 1961, only four days before Yuri Gagarin would become the first man ever to orbit the earth.[1]

"He [Korolev] was then fifty-four, and he was looking at his future—about twenty of us, tough, well-conditioned military fighter pilots, experienced in Migs 15, 17, 19," Leonov told me.

> He said, "Sit down, my little falcons [*sokoliki*]." He took up a list of who was who: Bykovsky, Gorbatko, Gagarin. . . . He looked at each man. When he came to Gagarin he looked him over carefully. You could see he liked Gagarin—blue eyes, nice, bright smile, a relaxed way of talking. He looked at him as if they were the only two in the room. Then he went on to the others. He said, "Patriotism, courage, modesty, iron will, knowledge, and love of people, cosmonauts must have those qualities." Mind you he didn't say "love of Party, or love of the Government." He was thinking of all mankind. When he left we all grouped around Gagarin and said, "He chose you." In fact, Korolev later said that first meeting had determined his selection. We had already been a half year in training so he probably already knew a lot about Yuri Alexeevich from reports."[2]

On April 12, Gagarin would orbit Earth a single time, becoming the first human in space and frustrating the hopes of America's Alan Shepard to have the honor. Shepard would have been able to make the claim by going up on a suborbital flight three weeks earlier. He was ready, but Wernher von Braun wasn't, having decided that the *Mercury-Redstone* vehicle needed one more test after (in January) having traumatized a chimpanzee named Ham by subjecting him to far more acceleration, and more than 100 miles greater distance, than had been planned. So a March 24 launch became another test, rather than a Shepard launch.[3] A deeply disappointed Shepard wrote years later, about his Russian rivals, "We had 'em by the short hairs, and we gave it away."[4]

But even if he had gone up in March instead of when he did—on May 5—it's unlikely that the world would have accorded him the plaudits that Gagarin got. Shepard's flight was for 15 minutes. The spacecraft reached 186 kilometers altitude and covered a distance of 489 kilometers. Gagarin completed one orbit of the Earth in 1 hour, 48 minutes. If Shepard had been launched first, Korolev was ready to state that a suborbital flight was not really a space flight and had even assigned one of his top engineers, Boris Rauschenbakh, and a staff assistant, Vladimir Shevalyov, to prepare a document arguing the point.[5]

In any case, Gagarin's spectacular achievement made the argument moot. Once again the world's headlines trumpeted a Soviet space first.

The New York Times again ran three lines of huge type across the front page:

SOVIET ORBITS MAN AND RECOVERS HIM;
SPACE PIONEER REPORTS: 'I FEEL WELL';
SENT MESSAGES WHILE CIRCLING EARTH[6]

The next day the *Times* editorialized that the flight "will be hailed as one of the great advances in the story of man's age-old quest to tame the forces of nature."[7] But this time *Pravda* was even more excited, declaring it a GREAT EVENT IN THE HISTORY OF MANKIND[8] and then devoting its entire six pages to the feat for three straight days. Besides the facts about the launch—devoid, of course, of any mention of the Chief Designer, Sergei Pavlovich Korolev, or his team members—there were exultant proclamations by the Communist Party about the clear superiority of their system over capitalism, detailed accounts of Gagarin's life, with emphasis on his humble beginnings as child of a farmer and a dairy maid, reactions from the world as well as from Soviet citizens, and so on and on.

The *New York Times* reported a few days later that the reaction of "official Washington" was "little more than a yawn that the inevitable had occurred." The Soviets, the article said, were capitalizing mainly on space spectaculars while the United States was methodically developing space science and technology—and had, in fact, launched some thirty-seven satellites to the USSR's twelve, while discovering the Van Allen belts, the pear-shape of the Earth, and the extent of its magnetic fields.[9] *Times* reporter John Finney called attention to what would be a continued Russian advantage in rockets—the Soviet launch vehicle could put up a 5-ton capsule while the thrust of the U.S. *Atlas* limited the size of the *Mercury* spacecraft to only 1 ton. Readers were reminded, however, that America was already developing a much larger engine.[10] This was the 1.5 million-pound-thrust F-1 rocket engine, already some three years into development, which would prove to be the crucial factor in the coming race to put a man on the Moon. Some say that the inability to fund and develop a comparable engine doomed the Russians' Moon race effort.

In the meantime, however, Korolev's now reliable R-7 rocket, which had powered the first ICBM and the first satellite, and which also launched Yuri Alexeevich, enabled the Soviets to continue to ride high.

Korolev had a massive supporting cast of organizations developing the infrastructure required to ensure safe space travel for humans. He was careful, for example, to consult Soviet space medicine specialists on the possible effects of weightlessness on man.

"Korolev was very conscientious in protecting the cosmonauts medically," said Oleg Gazenko, one of the USSR's top space doctors. "He was very annoyed by medical restrictions, but he observed them. . . . We saw no insurmountable barriers as a result of our [rocket] tests with living animals—rats, dogs, mice, rabbits, and guinea pigs. . . . Living creatures were weightless for as long as 12 minutes, so we realized there would be no deaths. . . . But Korolev was not interested in the data on these experiments. He always wanted 'Yes' or 'No' on humans."[11]

Dr. Gazenko, interestingly, was one of the few Soviet space professionals who met Wernher von Braun, whom he found to be—in contrast to Korolev—deeply interested in space medicine details.

"In 1965," he said, "I spent several days in Greece with von Braun after the International Astronautical Congress in Athens. A wealthy Greek shipping magnate, Typaldos, took me, von Braun, Hermann Oberth, Leonid Sedov, and several others for a cruise on a big yacht. Von Braun questioned me at great length about the space medicine data that we had accumulated."[12]

Space medicine, not incidentally, was one of the few disciplines in which there was a relatively frequent and productive exchange of data between the Soviets and the Americans, even during some of the most frigid periods of the Cold War.

Preparations for sending cosmonauts into space by experimentation with animals had been going on in the USSR since as early as 1951. Vassily Mishin wrote that the dogs Dezik and Tsygan were sent to 100 kilometers altitude in that year, and that the same pod used for carrying them was later used to house Laika in *Sputnik 2*.[13] "I was surprised that no Americans knew of this medical research," said Gazenko. "Under Yazdovsky[14] there were six or seven of us in the beginning. Then by 1956–57 it went up to thirty or forty doctors and technicians, maybe half doctors. By 1960–61 it was one or two hundred." In one test the dogs Albina and Tsyganka, wearing space suits, were ejected from a capsule at 85 kilometers and recovered safely.[15]

The Americans actually had a four-year lead in this research, having started sending mice up in V-2 rockets in 1947. The U.S. specialists continued their research with monkeys in the 1950s, concluding—just as the Russians did—that neither weightlessness nor the high acceleration expected during these flights would adversely affect humans.

Real differences in the approach of the two rivals to human space flight occurred in the design of the spacecraft and its life-support system. NASA experts decided on the *Mercury* capsule—a truncated cone with a blunt nose—having determined from intensive research that the shock wave formed on the nose would absorb 90 percent of the heat of reentry from space.

Korolev had been thinking of the design of a manned spacecraft since as early as 1948, recalled one of his early design engineers, Viktor Sadovii. It was in that year that he hired an engineer from the Yakovlev aircraft design bureau, A. V. Afanasyev, to work on the "pilot's cabin in one of the rockets."[16] However, serious design studies on what was to become the *Vostok* spacecraft, which would carry Yuri Gagarin, did not begin until 1958. Once again, the design choice was a spherical shape as with *Sputnik*. Korolev's reasoning was that the sphere, being symmetrical, would be more stable dynamically. A conical shape, he knew, would pitch and yaw, thereby requiring complex attitude control devices, and Soviet attitude control technology was not as advanced as that in the United States.

One of Korolev's early decisions was whether or not to put the *Vostok* cosmonaut in a space suit.

"A space suit caused a great weight penalty, but we designers, and the doctors, wanted one," said Gai Severin, Russia's preeminent maker of ejection seats and space suits for more than thirty years.[17]

On the other hand, the *Vostok* engineers felt that depressurization of the spaceship was less probable than other problems.

"So," recalled Severin, "a debate took place in mid-1960 at a large meeting, and after hearing the arguments the decision was made—to use a space suit. But there was little time before the scheduled flight."

"Korolev said, 'I'll give you as much weight as you need, just make the space suit in time.' So we made it in nine or ten months. We accepted a simpler design than that used for *Mercury,* which used a single-joint space suit that was an integral part of the life-support system. We had to make an independent system because the onboard life-support system had already been developed."

As for the life-support system, the Americans, in hindsight, chose unwisely for the breathing medium. The Russians decided to approximate the Earth's air for *Vostok*—creating a one-atmosphere environment of 80 percent nitrogen and 20 percent oxygen. The Americans decided on pure oxygen at a pressure of one third of an atmosphere. The *Vostok* system did have a disadvantage. It exposed the cosmonaut to decompression—the "bends"—if he had to switch to a space suit in mid-flight. In the U.S. system, the pressures in the cabin and in the space suit were the same, so there was no decompression problem as long as the astronauts pre-breathed oxygen prior to flight to remove the nitrogen from their bloodstreams. However, the absence of nitrogen, and the presence of 100 percent pure oxygen in the cabin, presented a fire danger.[18] The danger never led to a problem through seven *Mercury* missions between 1961 and 1963, and then 10 Gemini missions in 1965 and 1966. However, during the *Apollo 1* test, on January 27, 1967, astronauts Gus Grissom, Roger Chaffee, and Ed White died in the cabin from fire inhalation. An investigation board hypothesized that an electrical arc had jumped between two portions of a wire whose insulation had been worn off by door openings, igniting the oxygen-rich atmosphere.[19]

There was another significant divergence in the U.S. and USSR approaches to human space flight. The Russians preferred to depend on automated systems, with the cosmonaut in a passive role. The Americans, however, gave the astronaut a more controlling role, allowing him to override automated systems.

"We believed that we could do everything automatically, even with the cosmonaut present. Gagarin had nothing to do," Boris Rauschenbakh told me in a 1991 interview. "The U.S. put man in control first."[20] In fact, some of the procedural measures taken to limit the cosmonauts' role would probably have caused mutiny if imposed on the free-wheeling American astronauts. With such restrictions on his role, it is not strange that Yuri Gagarin did not know he was to be the first cosmonaut until four days before his flight. John Glenn, by contrast, got his assignment on November 29, 1961, almost three months before his February 20, 1962, flight. What's more, he had known a year earlier that it would be either he, Alan Shepard, or Gus Grissom.

Despite the light piloting role, however, there was much concern about how Gagarin might handle himself mentally. Mark Gallai, a former Russian test pilot who was put in charge of cosmonaut flight training by Korolev, described these measures, which were supposed to deal with possible psychological instability:

> All of the test pilots believed that these concerns were stupid. Many pilots had flown in the stratosphere at night, or in heavy cloud conditions, but the spacecraft designers decided to install a logic lock that would prevent the cosmonauts from using manual controls. There was a special panel with six buttons on it. The buttons had to be pushed in a coded sequence for the manual controls to work. Three of

> the six digits would be sent to the cosmonaut by radio. None of us pilots liked this system and we made a great noise about it. It was our feeling that the possibility of a pilot going crazy was much smaller than the possibility of failure in the radio communication.
>
> Korolev didn't like the system either, but he decided to accept it to quiet the physicians, then he would simply give the cosmonaut a sealed envelope with the three digits. This became a formal procedure. Suppose a cosmonaut opened the envelope and made a mistake punching the buttons and returned successfully, or went to another orbit. Who would punish him? Anyway, both I and Ivanovsky, separately, gave Gagarin the digits. Later Korolev admitted that he also gave them to Gagarin, who never let on.[21]

The differences in design philosophy between the USA and USSR led naturally to a difference in the qualifications criteria of the cosmonauts/astronauts. Gallai told me candidly that, "The U.S. selected as astronauts professional test pilots with high technical education and strong flight experience. Consequently, they were about ten years older than our cosmonaut candidates, who were young fighter pilots. Our main demand was perfect health. Experience has shown that the U.S. approach was more logical, and now the Russians are going the same way. Intellectual skills are more important than perfect health."[22]

My interview with Gallai took place in his Moscow apartment. Although his full head of hair had turned gray, he was still an active and healthy-looking man at age seventy-nine. He looked quite Western, dressed casually in a plaid shirt, jeans, slippers, argyle socks, wearing a pilot's watch. Although he didn't seem to have the swagger of a test pilot, there were several suggestions of that stereotype around the apartment, including a large calendar featuring a voluptuous nude, and several cabinets containing models of airplanes he had tested, photos of his colleagues, and a small bust of Korolev.

Gallai was an ideal Korolev choice to train the cosmonauts, because of an unusual engineering and piloting background. A specialist in flight dynamics, with a Candidat degree—"like your PhD"—he had test-flown 124 different types of aircraft, including fighters, bombers, and transports. "Before the war I did special testing on flutter vibration," he said. "I was one of the very few who actually experienced flutter. I was young and stupid." He had met Korolev in the 1930s when both were glider pilots. "Of course, no one had space flight experience," he told me, "so Korolev chose test pilots to train his cosmonauts."

It was not until six months before Gagarin's flight that the cosmonaut training began. Gallai continued:

> . . . we started with the first six cosmonauts—Gagarin, Titov, Nikolaev, Popovich, Bykovsky, and Nelyubov—he was later fired by Kamanin for getting drunk, a decision that I feel was too severe. The choices were made by the space medicine expert Yazdovsky, with perfect health being the criterion.
>
> . . . The physicians had selected twenty cosmonauts in the first group, but only the six named above were picked for training, those not taller than 167 cm (about 5′6″) and not weighing more than 65 kg (about 143 lbs). If the criteria had been different, certainly Komarov, who was very intelligent, would have been in the group. He had Air Force Academy flight experience. He had a great influence on the design of *Vostok* and *Voskhod*.[23]

Besides being older, the first seven American astronauts averaged more than three inches taller and more than 20 lbs heavier than their Soviet counterparts. Their mean height was 177 cm (5′ 9 1/2″) and their mean weight 55.3 kg (166 lbs).[24]

Some insightful background on the design of *Vostok,* as well as on Korolev himself, was given me by one of the Chief Designer's key engineers—Konstantin Feoktistov. Feoktistov, who was eventually picked by Korolev to fly as a cosmonaut on the *Voskhod* vehicle he had helped design, is a calm, confident man who looked younger than his 65 years at the time of my interview.[25] He is not given to easy smiles or laughter. He had a harrowing experience during the Great Patriotic War when, as an Army scout, he was captured by the Germans, put in front of a firing squad with other soldiers, and shot. But he was only wounded, and somehow escaped. After the war he graduated from the Bauman Institute as an engineer. He still works as a consultant to RSC Energia and lives in the special complex of attractive apartments built for civilian cosmonauts near Korolev's museum-home. Feoktistov said that he had first worked for Korolev, for a short time, in 1951, and "didn't like him in the beginning. He gave one of those inspirational speeches which seemed to me to be full of clichés."

Feoktistov went back to Special Design Bureau No. 385 in the Urals to resume working for the visionary Mikhail Tikhonravov, who had wanted to build a satellite himself. But the government gave the satellite task to Korolev, who promptly persuaded his old friend Tikhonravov to come with him to Podlipki. That was in 1956. The next year Feoktistov, who had just earned his Candidat degree, at age twenty-five, came to the Korolev bureau, too. He was given a choice—automated interplanetary vehicles or manned spacecraft. He chose manned spacecraft and became

Korolev's principal designer of the family of vehicles that would lead the Soviet exploits in the first years of human exploration starting with the *Vostok* in April 1958.

Ever sensitive to the competition from the Americans, Korolev asked Feoktistov and the other designers what they could do to shorten the time needed to prepare for a manned flight. "We worked out fifteen points," he told me, "that would shorten the time."

The main point was to use a standard aircraft ejection seat, slightly redesigned, for the escape system should there be an explosion at launch. A catapult device might have landed the cosmonaut in the flame deflector, "so we proposed to catch him in a big net. The risk was that the crew compartment would follow an unguided trajectory. So, we designed the system so that the spacecraft would separate from the booster and descend by means of an aerodynamic drag device and a parachute. I proposed testing the ejection system myself, but Korolev rejected the idea."

Feoktistov's opinion of his legendary boss, which had started out negative, remained somewhat ambivalent but "eventually he won my respect and he was the right man for the enterprise he ran. We had a very good relationship."

Another of Korolev's engineers who had a key role in developing *Vostok* was Oleg Ivanovsky. Sitting comfortably in his Moscow high-rise apartment in 1991, clad in a colorful plaid shirt and jeans with the "Lee" label showing on his rolled-up cuffs, he looked like a Westerner on vacation. The fact is that he spent more than forty years in the Soviet space program doing pioneering work with Korolev on the *Sputniks, Vostok,* and the missions to the Moon, Mars, and Venus. He was still working at the time of the interview, as head of the museum at the NPO Lavochkin design bureau. Full of vitality, although perhaps in his early seventies, he recalled:

> I went to work on *Vostok,* got to know all of the astronauts beginning in September 1960. It was then that the major attention shifted from the spacecraft to the pilots. We called the design area the "Tarzan site" because of all the wires hanging around during the electrical tests, like vines in a jungle.
>
> Design and production of *Vostok* involved many people. The age of personal invention was gone. We had to do man-rating, using classical schemes for this. We worked "15 days a week." Sometimes we slept in a corner of the factory. Nobody thought about pay. Bonuses were gratuitous, not expected. Maybe we would get two or three months salary.[26]

Throughout the *Vostok* design cycle, Korolev took a strong personal role. "He was very strict, sometimes crude," said Gallai. "Many people feared him. He wanted people to fear him. But he never did irreparable damage to anyone. I didn't fear him, not because I was so brave but because I was not his subordinate. I reported to Nikolai Sergeevich Stroyev, Chief of the Flight Test Institute." But those who were in Korolev's own organization often felt the sting of his temper and sometimes his irrational behavior. Gallai remembered one incident in which Ivanovsky was the butt of that behavior, only about two weeks before the Gagarin flight.

> Korolev came to the assembly building at Baikonur and shouted to Ivanovsky, his leading designer,"You no longer work for me. You're fired."
>
> Ivanovsky said, "OK" and continued his business. Two or three hours later Korolev appeared, forgot about the earlier incident, and for some other reason, said to Ivanovsky,"I'm making a notation." This meant that he was giving Ivanovsky a kind of demerit.
>
> Ivanovsky said, "You have no right to do that."
>
> Now it's impossible to imagine that anyone could say this to Korolev. Korolev lost his breath. "I have no right! What do you mean?"
>
> "I no longer work for you. You just fired me," said Ivanovsky.
>
> Then, in an untypical voice, Korolev said, "You are an SOB!"
>
> Life went on. Ivanovsky was one of the few people who didn't fear Korolev.[27]

Unmanned tests of the *Vostok* began somewhat inauspiciously, with the launch of the first unmanned prototype, on May 15, 1960. On the sixty-fourth orbit, a "command was given from earth to ignite the retroengine, but a fault, not previously apparent, in one of the attitude-control systems deflected the braking pulse from the calculated direction and the spacecraft could not be returned to earth."[28] Another *Vostok* prototype flight, not known to Westerners until mentioned in a 1994 publication, carried the dogs Chaika (also known as Bars) and Lisichka on July 28, 1960. The dogs were killed when the launch vehicle exploded.[29] As Feoktistov noted, the August 19 flight of the dogs Belka and Strelka was successful. The animals, which were recovered by parachute after eighteen orbits, were the first living creatures successfully returned to Earth.[30]

On September 19, a few weeks after the Belka-Strelka recovery, the Central Committee of the Communist Party was sent a request for authorization of the launch of a human into space. It was signed—in characteristic share-the-responsibility-the-glory-and-the-possible-blame fashion—by no less than ten people: Ministers Dimitri Ustinov, and Ridion Malinovsky, Marshal Mitrofan Nedelin, Chief Designers Korolev,

Valentin Glushko, Mikhail Ryazansky, Nikolai Pilyugin, Vladimir Barmin, Viktor Kuznetsov, and Vice President of the Academy of Sciences Mstislav Keldysh. The note, in the usual stilted language for such communiques, stated that:

> The successful launch, flight into outer space and landing of the spaceship [the *Vostok* prototype carrying Belka and Strelka] raises anew the question about the time of performing a flight of a human into outer space. . . . Development of planned technical decisions present an opportunity to create a spaceship [the *Vostok* 3A object] and to resolve the issue of flying a human into outer space on this object in 1960. . . . On this basis, the following proposals are being suggested: . . . to perform a flight of a human in December of 1960.

Not even the ballistic missile explosion at Baikonur a month later, on October 24, which killed some 165 people, including one of the signatories, Marshal Nedelin, failed to halt the plans to go ahead, although it caused a delay and a shift to an alternative launch pad. Another event that might have given pause to Korolev and his team—not to speak of the cosmonauts—was the ultimate sacrifice of the dogs Pchelka and Mushka in another *Vostok* prototype on December 1, when a control error caused the cabin to burn up on entering the atmosphere at a steeper angle than planned.

Still another try, on December 22, carrying the dogs Damka and Krasavka, did not achieve orbit because of a premature shutdown of the third-stage engine, but the dogs were recovered in the reentry vehicle.[31] Two subsequent flights, however, were successful. The dog Chernushka, accompanied by a dummy, was recovered on March 9, 1961, and Zvezdochka, also traveling with a dummy, was recovered on March 25, so the stage was set for a human, although, based on the spotty record of success, the cosmonauts would seem to have had every right to be apprehensive.[32]

If there was apprehension, it was perhaps mitigated by two *Vostok* features designed by Severin—an improved escape mechanism for launch mishaps, and an ejection and soft-landing system for the end of the journey. Severin, who was only twenty-two when he graduated from Moscow Aviation Institute in 1948, garnered considerable experience designing and building catapult ejection systems for such Soviet aircraft companies as Sukhoi, Tupolev, and Mikoyan before working on spacecraft.

A dark-skinned, rather lean and handsome man whose black hair has turned mostly gray, Severin is still the dominant figure in his field, and is working on the design of a space suit that he hopes will be used by astronauts and cosmonauts called on to assemble the International Space Station starting in 1997. He recalled his space projects in the late 1950s:

> I suggested the idea for the cosmonaut rescue systems for the *Vostok*. . . . One of Korolev's characteristics was a patience with young engineers. He would listen to their ideas, whereas some others would not. . . . When I suggested the soft-landing system there was overwhelmingly negative speculation by some of the senior design people. Korolev was one of the few people to support the idea, which we eventually developed with mockups. [33]

Aside from design challenges, Korolev had some bureaucratic wickets to go through in order to get official approval of the cosmonaut launch. On March 30, a note went to the Party Central Committee indicating full readiness for the flight, again signed by a multitude of ministers and designers, including Korolev. It said, "[We are] reporting that . . . a large volume of scientific, research and testing work, both in ground and flight conditions, has been performed. . . . The results of this work performed for development of the construction of the ship-satellite [*korabl-sputnik*], of the means of recovery on Earth, and of cosmonaut training make possible the performance of the first flight of a human into outer space. Six cosmonauts are prepared for flight. . . . "[34]

The note went on to state that the flight would be for one orbit and would land in the USSR "on a line running through Rostov, Kuibyshev, Perm . . . " and that precautions were carefully taken should the landing system fail, in which case "the craft can descend by natural braking in the atmosphere over two to seven days, with touchdown between the latitudes of 65° north and south." The cosmonaut would have "appropriate instructions" in case of a "forced landing in foreign territory" and besides the capsule's ten-day supply of food and water there would be another three-day "portable emergency supply," plus a radio and transmitter. The spacecraft "is not provided with an emergency self-destruct system for the reentry vehicle."[35]

Knowing the Party's reluctance to publicize Soviet space missions unless and until they were successful, the next admonition might have seemed audacious, but its logic was compelling.

> We consider it advisable to publish the first TASS report immediately after the satellite-spacecraft enters orbit, for the following reasons:
>
> (a) if a rescue becomes necessary, it will facilitate rapid organization of a rescue;
>
> (b) it precludes any foreign government declaring that the cosmonaut is a military scout.

. . . In the TASS reports, it is suggested that the satellite-spacecraft be called "Vostok." . . . [36]

The *Vostok* name had been secret, since it applied to both a reconnaissance satellite—later named Zenit—and a manned spacecraft, so some did not want the name for the Gagarin ship revealed. It would just be another sputnik. Rauschenbakh, however, recalled that it was Tikhonravov who insisted passionately that this momentous flight not be simply the next in the sputnik series. His view prevailed.[37]

The note was endorsed, on April 3, by the Central Committee with a decree entitled "About the launch of a space ship-satellite." The next day, April 4, Korolev reported readiness for the flight to a governmental commission in Baikonur itself.

Five cosmonauts—Gagarin, Titov, Nelyubov, Nikolaev, and Popovich—arrived at Baikonur on April 5. Three days later, on April 8, came the meeting described above when Gagarin was selected. Apparently the selection was a fait accompli, since the recommendation of Gagarin, with Titov as backup, had already been made by Lieutenant General Nikolai Kamanin, then chief of the Soviet space program commission. There was a "ceremonial" session that evening, filmed by Soviet reporters, at which Gagarin and Titov were introduced, and it was announced that on April 10 at 7 A.M. the launcher would be rolled out and given a "dry" checkout, and on April 12 the launch would take place.

At 1 P.M. on April 11, the twenty-seven-year-old Yuri Alexeevich Gagarin was driven to the pad for an orientation. According to one report, after he and Korolev spent an hour together at the top of the pad going through procedures, Korolev was exhausted, and had to return to his tiny wooden cottage to take medication for his weak heart. He recovered, though, and by morning was fit for the big day. Gagarin, bunking with Titov in an almost identical adjoining cottage, had slept soundly.[38]

Here are excerpts[39] from Gagarin's personal account of the flight, as presented to the State Commission in the city of Kuibyshev, the day after he was recovered near the Volga town of Engels, about 700 miles north and east of the Baikonur launch site. It is a fascinating recap, especially that portion dealing with the end of the flight when an anomaly occurred, which, if it had been revealed at the time, would have gotten almost as much attention as the successful landing. At the end of the first and only orbit the retropack failed to separate on signal—a failure of the automated system that Gagarin could not override, as an American astronaut would have been able to. Gagarin was in deep trouble, his spacecraft spinning wildly, and only after ten long minutes did he get out of it. He could easily have been killed. This experience was hidden until these excerpts were published on the thirtieth anniversary of the flight, in 1991.

The account begins with preflight preparations after Gagarin had been awakened at about 5:30 A.M. on April 12:

> I myself felt good, because I had relaxed and slept well beforehand . . . the members of the military team helped me put on my spacesuit. They did a good job, adjusted it, pressurized it. Then they placed me in the test chair and checked . . . ventilation and communications systems. . . .
>
> Then there was the trip to the launch site in the bus. I, my cosmonaut friends (my replacement was Gherman Stepanovich Titov), and the directors traveled to the launch pad . . . they took me up to the capsule . . . in an elevator. I was placed in the seat by a team directed by Oleg Genrikhovich Ivanovsky. All . . . the attachments were connected well. . . . The communications were two-way and stable. . . .
>
> Then hatch No. 1 was closed. I heard it close, I heard the wrenches clink. Now they begin to open the hatch again. I look, and the hatch is removed. I knew something was not in order. . . . [40]

The problem was that the hatch contacts had not indicated closure. Korolev himself, again dominating the crucial actions, this time by serving as capsule communicator, told Gagarin what had happened. The hatch was reopened, the contacts checked and confirmed as operative, and the hatch closed again.[41]

Gagarin was asked by Popovich at 8:14 if he was getting bored. He replied, "If there were some music, I could stand it a little better." At 8:19 he was pleased to acknowledge that "They gave me love songs." His report at Kuibyshev continues:

> Then they announced T minus 15 minutes. I put on my sealed gloves. I closed the helmet. T minus 5 minutes. T minus one minute and launch. Before that, you could hear the tower being taken away. There were some light blows to the structure of the rocket. The rocket seemed to rock a little.

At 8:41 he heard the valves working and eventually, as he described it:

> A faint noise started. . . . When the engines entered their main, primary stage, the noise intensified, but it wasn't so sharp that it deafened me or interfered with my work. . . . I was prepared for much more noise. Then the rocket smoothly, lightly rose from its place. I didn't even notice when it started. Then it felt like there was a slight shiver in the structure of the rocket. The vibrations were high frequency, low amplitude. . . .

> I prepared myself for ejection. I'm sitting and observing the liftoff process. I hear Sergei Pavlovich reporting. . . .

It was then that Gagarin shouted "*Poyekali!*" (Let's go!) a cry that would be quoted for years to come in the Soviet media.

The launch was at 9:06. Three minutes later, Gagarin remembered, he was feeling "fine." His account continued:

> . . . The g-load is increasing steadily, but it's completely manageable, as in normal airplanes. About 5 g's. At that g-load, I reported and communicated with the ground the whole time. It was somewhat difficult to talk, since all the muscles of my face were drawn. There was some strain. The g-load continued to rise, then reached its peak and began to steadily decrease. Then I felt a sharp drop in the g-load. I felt as if something had separated from the rocket. I felt something like a knock. Then the noise dropped sharply. The state of weightlessness began to emerge, although the g-load was about 1 at that time. Then the g-load came back and began to increase. I began to be pressed to the seat, and the noise level was substantially lower. At 150 seconds, the nose fairing separated. The process was very crisp.

He then gave the first-ever of what would be many subsequent descriptions of the Earth from space by subsequent cosmonauts and astronauts—waxing more enthusiastic than poetic.

"I see the clouds. The landing site. . . . It's beautiful, what beauty!"

At 9:12 the second stage of the rocket shut down, and then the third stage

> shut down abruptly. The g-load increased a little, and I felt a sharp knock. After about 10 seconds, separation occurred. I felt a jolt. The craft began to rotate slowly. . . .
>
> The Earth began to pass to the left and up, then to the right and down. . . . I could see the horizon, the stars. . . . The sky was completely black, black. The magnitude of the stars and their brightness were a little clearer against that black background. . . . At the very surface of the Earth, a delicate light blue gradually darkens and changes into a violet hue that steadily changes to black.
>
> . . . In my flight over the sea, its surface appeared gray, and not light blue . . . uneven, like sand dunes in photographs.
>
> . . . I could eat and drink. I noticed no physiological difficulties. The feeling of weightlessness was somewhat unfamiliar. . . . Here, you feel as if you were hanging in a horizontal position in straps.
>
> . . . I released the tablet with a pencil, and it floated in front of me. Then I had to write the next report . . . but the pencil wasn't where it had been. It had flown off somewhere. . . . I closed the journal and put it in my pocket . . . because I had nothing to write with.

> . . . The entry into the Earth's shadow was very abrupt. Up until now, I had at times observed intense illumination through the windows. I had to turn away from it or cover my face. . . .

At 9:57 he told the communicator, "I'm in a good mood. I am continuing the flight. I am over America."

At 10:13 he reported that the flight was going "smooth as silk" as he was given the first of several commands to prepare for reentry.

What happened next, however, not only disturbed the "smooth as silk" flight but almost caused a disaster, even though Gagarin says it "was not an emergency situation":

> . . . at precisely the appointed time, the third [reentry] command was issued. . . . I felt the braking rocket kick in . . . the g-load began to rise a little, and then weightlessness came abruptly back. . . . The braking rocket operated for exactly 40 seconds . . . as soon as [it] shut off there was a sharp jolt, and the craft began to rotate around its axis at a very high velocity . . . about 30 degrees per second, at least. I was an entire "corps de ballet": head, then feet, head, then feet, rotating rapidly. . . . First I see Africa . . . then the horizon, then the sky. I only barely managed to hide my eyes from the Sun. . . . I waited for the separation. There wasn't any. I knew that, according to plan, that was to occur 10–12 seconds after the braking rocket switched on. When the braking rocket shut down, all the lights on the stage separation monitoring console went out. . . . There was no separation. Then the indicators on the . . . console began to light up again. . . . There was no separation whatsoever. The "corps de ballet" continued. I decided that something was wrong. . . . I estimated that all the same I would land normally . . . the Soviet Union was 8,000 km long, which meant I would land somewhere in the Far East.
>
> . . . I reasoned that it was not an emergency situation. I transmitted the all-normal signal with a key. . . . The separation occurred at 10 hours 35 minutes [10:35 A.M.]—and not at 10 hours 25 minutes, as I had expected.
>
> . . . The craft's rotation was beginning to slow, but it was about all three axes. The craft began to oscillate about 90 degrees to the right and left. . . . Suddenly, a bright crimson light appeared along the edges of the shade. . . . I felt the oscillations of the craft and the burning of the coating. I don't know where the crackling was coming from: either the structure was crackling, or the thermal coating was expanding as it was heated—but it was audibly crackling. One crackling lasted about a minute. . . . I felt that the temperature was high. . . . The g-load was small, about 1–1.5. Then the g-load began to steadily increase.
>
> . . . It felt as if the g-load was 10g. There was a moment for about two–three seconds when the indicators on the instruments began to become fuzzy. Everything seemed to go gray.

> . . . When the g-loads had fallen completely, which apparently coincided with passage through the sound barrier, I began to hear the whistling of air.
>
> . . . I'm awaiting the ejection . . . at an altitude of approximately 7,000 meters hatch No. 1 was shot off. . . . I'm sitting there thinking, that wasn't me that was ejected, was it? Then I calmly turned my head upward, and at that moment, the firing occurred, and I was ejected. It happened quickly, and went . . . without a hitch. I didn't hit anything. . . . I flew out in the seat. Then a cannon fired, and the stabilizing chute deployed.
>
> I was very comfortable in the seat, as if I were in a chair. It felt as if I were rotating to the right. I immediately saw a large river. I thought, that's the Volga. . . . When I was parachute training, we had jumped many times over this very site. . . . I recognized a railroad, a railroad bridge over the river, and a long spit of land extending far into the Volga. . . . I was landing in Saratov [about 200 miles south of Kuibyshev].
>
> Then the reserve parachute deployed . . . and hung in mid air. It didn't open. Only the pack opened. . . .
>
> As I descended, I noticed that to my right there was a field camp on a terrace. There were many people and machines there. It was right next to the road. The road led to Engels. Then I could see a rivulet flowing in a gully. . . . I saw some women there tending a calf. Well, I think, now I'm probably going to fall into that very gully, but there's nothing I can do. It feels as if everyone is looking at my pretty orange canopies. Then I see that I am going to land in a plowed field. . . . I hit with my feet. The landing was very soft. . . . I didn't realize I was already standing. . . . The rear parachute fell on me, the front parachute fell in front of me. . . . I was alive and well.
>
> . . . I went up on a knoll and saw a woman and a little girl coming toward me. . . . I saw the woman slow her pace, and the little girl broke away. . . . I then began to wave my arms and yell, "I'm one of yours, a Soviet, don't be afraid, don't be scared, come here." It's hard to walk in a space suit, but I was doing it anyway. . . . I went up to her and said that I was a Soviet and that I had come from space.[42]

Clearly, Yuri had the right stuff, having coped courageously with what must have been a frightening crisis with the retropack. A British expert on Soviet space history, Phillip Clark, years later found an article from a Russian publication that revealed that separation of the 2-metric-ton retropack from the 2.5-metric-ton descent module did not occur until the heat of reentry melted the umbilical cord. "To say that Gagarin was lucky to survive might be putting things mildly."[43] Russian space historian Maxim Tarasenko is less apocalyptic and more explicit about what happened.

> The principal connection—four metal bands, strapping "the ball" to the conical equipment section, jettisoned normally. However, the plate of cable connectors linking the two sections of the spaceship failed to disconnect from the recovery vehicle and the two compartments remained loosely connected by these cables. Because the recovery vehicle reentered with a lighter equipment section trailing in the back, the recovery vehicle itself began to rotate, pretty much like a yo-yo, and that continued until the cables burned through and broke. That event was noticed in previous missions and repeated afterwards. It was not deadly dangerous, but it did increase the g-loads affecting the cosmonaut, particularly in the lateral direction.[44]

Eventually, in any case, a search helicopter from the nearby military garrison found Gagarin and flew him to Engels, where he received phone calls of congratulations from Brezhnev and Khrushchev. Then he was taken to Kuibyshev.

Gagarin's mode of parachute landing was to cause him some embarrassment, according to Gallai. "At a press conference in Moscow two days after the flight he was asked, 'How did you land?' Some bosses had told him to answer that he had landed within the capsule. If he was to qualify for establishing a record for the flight, according to the rules of the Federation Aeronautique International, he should have landed in the same craft he took off in. This was the first of many times that he was asked the same question and it put him in a very difficult position. It's easy to understand his dilemma. . . . He always avoided an answer, although later it was admitted by officials that he had ejected from the spacecraft and landed by parachute."[45]

For Korolev, the Gagarin triumph was a big milestone on a quest that still had a long way to go. He had been thinking of human space flight for some two decades, since the '30s, and actively developing the technology for almost ten years. It was particularly galling for him, therefore, to be unable to participate prominently in the revelry that enveloped Moscow.

"When Gagarin came back after his flight, Korolev went to meet him with Khrushchev at Vnukovo," I was told by Shevalyov. "He had arranged [for the design bureau] to buy a Chaika automobile from some foreign embassy for 40,000 rubles. [Chief Designers rated only less impressive Volga cars.] But when Khrushchev embraced Gagarin, Korolev was kept off to one side. Then when the cavalcade returned to Moscow, the fan belt broke on Korolev's Chaika and he had to be moved to a more modest car so he was again in the background. He was not allowed even to wear his medals."[46] A photo shows the anonymous Chief Designer well off to the side in a group that includes Nikita and

Nina Petrovna Khrushchev and Anastas Mikoyan feting Gagarin and his wife at a reception.

Back at OKB-1, however, Korolev observed what would become a tradition for every cosmonaut returning to Earth. In the open square in front of the design bureau, the entire working force met Gagarin for "friendly socialization. . . . It is difficult to overestimate the importance of these meetings for instilling labor enthusiasm and pride in one's profession in the collective," recalled Pavel Kostyukevich, a deputy in the design bureau's Party committee for scientific research.[47]

In the United States, as George Low, then NASA's chief of manned space flight, recalled, " . . . there was sudden interest again, in this country, in manned space flight."[48]

Low and the newly appointed NASA administrator, James Webb, as well as Webb's deputy, Robert Seamans, had been testifying on the state of their efforts before the House Committee on Science and Astronautics the day before the Gagarin flight. They scratched their plans to show a film, on the next day, April 12, of the recovery of the chimpanzee Ham, which had taken place two and a half months earlier. The poor chimp was in a traumatized state after having experienced an excess of g's. "We thought," remembered Low, "it would not be in our best interest to show how we had flown a monkey on a suborbital flight when the Soviets had orbited Gagarin."[49]

President John F. Kennedy, less than three months in office, extolled the Soviet feat graciously, "We, all of us, as members of the race, have the greatest admiration for the Russian who participated in this extraordinary feat." He then turned to the tough job faced by the United States:

> I indicated that the task force which we set up on space, way back last January—January 12th—indicated that because of the Soviet progress in the field of boosters, where they have been ahead of us, that we expected that they would be the first in space, in orbiting a man in space. . . . We are carrying out our program, and we expect to hope to make progress [note the weak wording from the normally crisp young chief executive] in this area this year ourselves.[50]

To a reporter's question on when the USA might "perhaps surpass Russia in this field," Kennedy remarked that "the news will be worse before it is better."[51]

Indeed, it got worse. Gus Grissom's suborbital flight on *Mercury-Redstone 4,* on July 21, was resoundingly topped by Gherman Titov's seventeen-orbit flight in *Vostok 2* on August 6, although Titov experienced a "difficult period of adaptation because of weightlessness sickness," he told me in an interview.[52]

NASA did not even put up an unmanned *Mercury,* the vehicle that would orbit John Glenn the following year, until September 13. On November 29 another *Mercury* carried the chimp Enos, who was recovered after two orbits. As mentioned earlier, it was after this flight that Glenn was designated to be the first American to circle the Earth.

On February 20, 1962, Colonel Glenn, launched by *Mercury Atlas 6,* made three orbits in *Friendship 7,* which now sits splendidly in the National Air and Space Museum. The American catch-up plan was underway.

The Vostok-Mercury Manned Flight Competition[53]

1961		
April 12	USSR	*Vostok 1.* Yuri Gagarin, first man in space, orbits the Earth once, lands in a parachute.
May 5	USA	*Mercury-Redstone 3.* Alan Shepard flies 486 km, reaches 187.5-km altitude in ballistic trajectory.
July 21	USA	*Mercury-Redstone 4.* Gus Grissom flies ballistic trajectory to 190-km altitude. Space capsule sinks.
Aug 6	USSR	*Vostok 2.* Gherman Titov completes 17 orbits in 25 hrs 18 min.
1962		
February 20	USA	*Mercury-Atlas 6.* John Glenn does 3 orbits in 4 hrs 55 min.
May 24	USA	*Mercury-Atlas 7.* Scott Carpenter does 3 orbits, splashes down 420 km beyond target area after reentry errors.
August 11–15	USSR	*Vostoks 3* and *4.* Andrian Nikolaev does 64 orbits, and Pavel Popovich, launched a day later, does 48. The two spacecraft come within 5 km of each other. Cosmonauts land in parachutes 193 km and 6 min apart.
October 3	USA	*Mercury-Atlas 8.* Walter Schirra does 6 orbits in 9 hrs 13 min in "textbook flight," splashes down only 7.24 km from recovery ship.

1963		
May 15–16	US	*Mercury-Atlas* 9. Gordon Cooper does 22 orbits in 34 hrs 20 min, splashes down 6.4 km from ship after manually controlling reentry.
June 14–19	USSR	*Vostoks 5* and 6. Valeri Bykovsky does 81 orbits, passes within 5 km of Valentina Tereshkova, launched two days later, who does 48 orbits.

There was speculation in the United States on the technical strategy, as well as the propaganda implications, of the Soviets in carrying out the Nikolaev-Popovich group flight in *Vostoks 3* and *4*. "What was so special about it that could not have been done by, say, the Americans?" Vassily Mishin was asked in a 1990 interview. His answer:

> In order for craft to be able to rendezvous in orbit, they had to be outfitted with approach equipment, the likes of which did not exist on either the Vostok or Mercury craft. But we were working on the Soyuz, while the Americans did so on the Gemini. Future craft were capable of performing those types of maneuvers in space. But the side that could develop its own craft first would become the leader. . . .
>
> The group flight . . . well, a day after launch, the first craft was over Baikonur. If the second craft were launched now with great precision, then they would turn out to be next to each other in space. And that's what was done. Nikolaev's craft was launched on 11 August 1962, and Popovich's craft on 12 August. The craft turned out to be 5 kilometers from each other! Well, since, with all the secrecy, we didn't tell the whole truth, the Western experts, who hadn't figured it out, thought that our Vostok was already equipped with orbital approach equipment. As they say, a sleight of hand isn't any kind of fraud. It was more like our competitors deceived themselves all by their lonesome. Of course, we didn't shatter their illusions.[54]

Mishin has been very candid in recent years about such gamesmanship. Another baldly politically propagandistic example was the launch of the USSR's second cosmonaut, Gherman Titov.

Mishin: "Khrushchev himself set the date for Gherman Titov's launch. Several days after Titov's flight, they built the wall in Berlin. Western experts believe that the effects produced by the flight gave the GDR [German Democratic Republic] government a lot of moral and political support during the preparations for and the execution of that action."

The Tereshkova flight was also blatantly propagandistic. Mishin said that Khrushchev, in one of his speeches, claimed that Valentina's flight, the first for a woman in space, "demonstrated to the entire world the equality of men and women in our country."[55] This was especially ironic because, in Mishin's words, "Tereshkova turned out to be at the edge of psychological stability. It would seem that her flight, on the contrary, should have discredited N. S. Khrushchev."[56]

But it was quite clear that what was happening in space during these Cold War years was of equal interest to both Kennedy and Khrushchev. The stakes for performance went far beyond the mere attainment of the space exploration goals of Korolev and NASA.

Consequently, there was plenty of support, circa 1963, for racheting up the effort on both sides to compete in the next phase of the competition.

Voskhod: A "Circus Act" 10

"ONE PERSON alone, in a single-seated spacecraft, will never undertake interplanetary travel such as flight to the moon, not to mention Mars. And so the next step [after *Vostok*] was to design a new spacecraft, the *Voskhod,* with room for a crew."[1]

So wrote four of the Soviet Union's leading spacecraft designers, looking back, in 1967, on ten years of the Space Age. In fact, however, *Voskhod* (means "rise," as in "sunrise" which, of course, occurs on the eastern horizon) was hardly a "new spacecraft." Although it had more capability than *Vostok*—that is, it had a redundant solid-propellant braking system and weighed 600 kilograms more—*Voskhod* was *Vostok,* rigged up to be able to sardine three cosmonauts into a spacecraft designed for one. Undertaking the jerry-built project bespeaks of the high-stakes gamble that Korolev was willing to take to outpace the Americans. The Soviets were scrambling. Korolev's bureau had been working hard on the *Soyuz* manned spacecraft, which would be a truly significant step forward technologically, and which, with some adaptations, would be able to carry cosmonauts to the Moon. Indeed, uprated versions of *Soyuz* are still carrying cosmonauts to orbit today. But *Soyuz* was not ready to compete with the two-man American *Gemini,* whose design had been well-known since the end of 1961 from the open U.S. literature.

Once again Nikita Khrushchev butted into the system, according to Vassily Mishin. "Khrushchev phoned Korolev and ordered the launch of three cosmonauts right away," recalled Mishin. This would not be easy. Mishin:

> But fitting a crew of three people, and in spacesuits, in the *Voskhod* was impossible. So—down with the spacesuits! And the cosmonauts went up without them. It was also impossible to make three hatches for ejection. So—down with the ejection devices. Was it risky? Of course, it was. For approximately 20 seconds of flight prior to insertion into orbit, the crew did not have any means of escape in the event of an emergency.[2]

Korolev, says Mishin, took the risk, but first extracted a trade-off. In a personal meeting with his leader in February 1964, Korolev accepted the *Voskhod* task, but on condition that the N-1 program, aimed at a manned lunar landing, would get stronger backing than it had been getting. Khrushchev agreed to the deal, although "This agreement was never spelled out openly, nor was the staff of OKB-1 ever told that 'Khrushchev personally ordered us to do this or that.' That was not the practice at the time," according to Gyorgi Vetrov.[3]

Sergei Khrushchev regards these recollections as dubious. They are examples, he told me, of placing hindsight blame on his father. "Soviet leaders," he said, "could not give such orders, since they did not know what was technologically possible, although my father might have urged Korolev to do something when he heard about *Gemini.*"

At any rate, Korolev barreled ahead with *Voskhod.* For this project, however, he wanted to include an engineer in the crew. His choice was Konstantin Feoktistov, who was given a major role in the design of the spacecraft. "There were ten candidates who passed the medical tests," Feoktistov recalled. "The final decision was made by Korolev, and he picked me."[4]

One must admire the courage of the young Feoktistov, then in his early thirties. He was being asked not only to adapt a one-man spacecraft to accommodate three cosmonauts, and do it at minimum increase to the weight of the capsule, but to fly the hazardous mission himself.

> I began working closely with Korolev himself on the first concept—spherical shape, two compartments, one for the crew and one for the other systems. By May we gave Korolev our proposals. He was highly inspired by them and he ordered us to deliver full design reports. We delivered our proposals about two months later, in August, and then we got the green light to go ahead and build the spacecraft. The U.S. was developing *Gemini* at the time, so we knew we were in a space race. There were about fifty design engineers working on the project; and then later a lot of technicians were needed to develop the instruments, the electrical systems, etc.; then hundreds of engineers to build the whole system. At the time I had very close connections to Korolev, although they were purely administrative. I would see him every week, sometimes more often.[5]

According to Mishin, it was Feoktistov, himself, who suggested the idea of eliminating the space suits and ejection systems.[6] Without space suits the crew would depend on an absolutely leak-proof capsule to maintain a breathing atmosphere. The life-support system provided only a means of changing the composition of the gases—removal of the carbon dioxide exhaled by the cosmonauts—but did not allow for replenishment of any lost gases.[7]

How to deal with the absence of a system for ejecting the cosmonauts at landing time? The solution was to work out a soft-landing system for the capsule itself. For this Feoktistov got help from Gai Severin.

"We worked closely with Feoktistov on the design of *Voskhod,*" said Severin.[8] What he came up with was a second parachute as well as small retrorockets that would slow the capsule on reentry to one meter per second at touchdown.[9] It was a dicey solution.

"The skeptics," said Severin, "were worried that, since rocket engines are ignited at landing, the parachutes might catch fire. Korolev recognized that there were risks, that there might be failures during the test period, but he was courageous enough to make the decision to go ahead with the tests. We knew that we could show that the danger of fire was small, but that even if there were fire, the cabin serves as a heat shield."[10]

The fact is, risky or not, the Russians pulled it off. After one unmanned test flight—*Kosmos 47*—only six days earlier, *Voskhod,* on October 12, 1964, launched Feoktistov, Vladimir Komarov, and the first doctor to go into space—Boris Yegorov—into a flight of sixteen orbits. This was an incredible seven months after the project was okayed.

"The world applauded again," said Mishin. "It was as if there was, sort of, a three-seater craft and, at the same time, there wasn't. In fact, it was a circus act, for three people couldn't do any useful work in space. They were cramped just sitting! Not to mention that it was dangerous to fly. But, in the West, they drew the conclusion that the Soviet Union possessed a multi-seat craft. It would never have entered anyone's mind there that we would send a crew into orbit without the appropriate means of rescue. It was good that everything turned out all right. But what if it hadn't?"[11]

The Korolev team later rationalized the achievements of the "circus act" flight. "A broad program was carried out," they wrote, "testing the three-man capsule, observing the efficiency and teamwork of the crew, collecting technological and scientific data, and gathering a wide range of biomedical information applicable to long-duration space flight, with the direct participation of the scientist and the doctor. For the first time, investigators observed and worked in space."[12]

It was ironic that Khrushchev, who had ordered the flight, was deposed on October 14, the day after *Voskhod* returned to Earth.

There were only two *Voskhod* flights, but both had spectacular impact. Only five months after *Voskhod 1* came another show-stopper, the seventeen-orbit flight of Alexei Leonov and Pavel Belyaev, on March 18, 1965, in *Voskhod 2,* in which Leonov performed the world's first walk in space—a 20-minute EVA (extra-vehicular activity). This was just five days before Gus Grissom and John Young became the first U.S. pair to go into orbit, on *Gemini 3.*[13]

The pace for developing the *Voskhod 2* mission was just as frenetic as for *Voskhod 1.* "Originally," said Vetrov, "no one had planned for *Voskhod 2* to be launched so soon. The first plans for EVA were much more cautious and gradual, with dogs leaving the cabin before an actual human. But American plans with EVAs on *Gemini* forced speedier action."[14] The tension was not lessened by the explosion of a *Voskhod 2* unmanned prototype on February 22, 1965, caused by botched signals from the ground. A worried Nikolai Kamanin recorded in his diary that Korolev called him to his sickbed on February 26 (doctors had diagnosed the Chief Designer as having a "focal pulmonary inflammation" causing a fairly high fever) to express his anxiety over the upcoming manned flight after the prototype's demise. He was assured that the signal botch-up would have been overridden by the cosmonaut. But Korolev was also concerned that the base ring of the airlock, which would be left protruding after the airlock itself was released following Leonov's ingress from space, might cause the spacecraft to spin during reentry. He directed that an unmanned *Zenit* spacecraft be sent up, carrying the ring, to see how it behaved. *Zenit* went up on March 7 and touched down on March 15, the airlock ring causing no excessive spinning.[15]

The design and construction of the airlock itself was an example of the jerry-built approach, albeit imaginative, which characterized the *Voskhod* program. Rather than install an egress hatch that would open, exposing both crew members to space, as was done with *Gemini,* Korolev had Severin design a simple 1-meter cylinder that stovepiped at an angle into *Voskhod 2*'s side. Made of a double-thickness of a rubberlike material, covered with protective fabric, it accordioned under the shroud in the space capsule on takeoff. It was equipped with a regulation system for controlling air pressure from normal Earth atmospheric pressure to that of the vacuum of space.[16]

"Gai Ilyich," Korolev directed Severin, "demonstrate to us the durability of your airlock chamber." Severin, "a tall, stately, athletic looking man who was still young, easily jumped up and hung on the end of the cylinder." Mock-ups of *Voskhod 2,* with the airlock, stand proudly today in the museums of both the Korolev design bureau and Severin's Zvyezda design bureau.

Also in a place of honor in Severin's museum is the space suit designed for Leonov. He recalled how the suit was developed under great time pressure:

> We had only about nine months, starting in May or June 1964, to design and make the suit for Leonov's space walk.
>
> We suggested [a suit] based on the old *Vostok* design. Korolev approved our suggestions. We knew that Ed White[17] was getting ready. Unlike White, who just opened the hatch and looked around, Leonov went through a depressurization system. . . . We had very modest

production facilities at Zvyezda, so some of the pieces of equipment were made by Korolev. We had a series of failures. Our plan envisioned a full repetition of the EVA system we had undergone for the unmanned test for the Moon flight. We deployed the space suit in an inflatable chamber, depressurized the chamber, opened the hatch, recorded how the space suit performed, then detached the pressurization chamber and allowed it to burn up because the landing capsule must be spherical on reentry.

We made an additional chamber and space suit and launched it about a month and a half before the Leonov flight. We launched the spacecraft, it entered orbit, the pressurization chamber was deployed, the space suit pressurized, then the communications dead zone started, so we had to wait for the next orbit to see if it was OK. The spacecraft never appeared again. Nobody could understand what happened. It turned out that the spacecraft had been destroyed. The main command signal and the backup signals from the ground stations combined to trigger the self-destruct signal, which was to be activated in case the spacecraft was headed for a landing in an unfriendly territory. So we still had not tested our space suit. The question in front of Korolev was what to do? There were two alternatives: we could repeat the test program, but that would take six months; or we could take all of the responsibility and go ahead with the real experiment. Korolev allowed me to make the decision, and then to defend it before the State Commission. There were two groups organized for analysis. We analyzed the data from the aborted experiment, compared them with the results of our own experiments in the pressurization chamber, and I concluded that there was no reason for delay. The partial data proved that we could proceed.

Late at night I called Korolev at his house in Baikonur and reviewed the results. He agreed with my suggestion to ask the State Commission to allow us to go ahead, but he said the Commission would accept my recommendation only if I could prove convincingly that it would work. Korolev gave me the chance to defend my position. I must emphasize that every Chief Designer must be himself responsible for his projects, so obviously he was placing a great deal of faith in me. The State Commission accepted my recommendation, thank God.[18]

The crew for *Voskhod 2*—Belyaev as crew commander and Leonov the pilot-cosmonaut for the EVA—was not finally confirmed until three days before the flight itself. There had been some apprehension about Belyaev's fitness for the mission when he did poorly on altitude chamber tests two months earlier. As Kamanin explains, however, "Leonov told me what had happened. During the training session, Belyaev began to gasp for breath because oxygen was not coming into the altitude chamber. But he displayed admirable composure, found the problem, and corrected it. Those responsible for the malfunction were the factory specialists supporting the equipment, but Belyaev did not want to 'tell' on them and took all the responsibility for what had happened."[19]

An exultant Kamanin recorded in his March 18 diary that, "Today has been a hectic, unforgettable day. For the first time in history, a man walked in outer space. . . . " After donning a suit and prebreathing oxygen to purge his lungs of nitrogen, Leonov entered space through the cylindrical airlock. TV pictures of the walk were widely broadcast. What was not revealed until years later, and was not included in Kamanin's diary, was that Leonov almost could not reenter the airlock. His space suit had expanded and it took him twelve minutes of struggle before he could depressurize the suit and squeeze himself back through the opening.

A second anomaly on that flight occurred when the automatic reentry system failed. Alexei Yeliseev, one of the engineer-cosmonauts, was in the control center at the time. "No one understood what the problem was," he remembered.

> There were many guesses, frantic proposals—everyone had clearly begun to get nervous. . . . And here Korolev took the supervision into his own hands. He established quiet and asked everyone to sit down. Then he calmly listened to the work supervisor in charge of the control system. He asked . . . about the possible causes and . . . a suggestion for further action.[20]

Korolev's decision was to perform a manual descent. He told the crew of the decision himself, said Yeliseev,"in such a calm and assured tone that on board the craft and on Earth a normal businesslike atmosphere was again restored." The spacecraft flew an extra—seventeenth—orbit, to give Belyaev time to prepare for manual retrorocket firing. He asked Leonov to check the spacecraft's attitude, and the extra time needed for the check caused the spacecraft to overshoot its intended landing area by about 2,000 kilometers. *Voskhod 2* set down in deep snow near Perm, wedged between two large fir trees about 3 or 4 meters above the snow. The cosmonauts opened the hatch but stayed inside. A recovery helicopter reached them after two and a half hours but could not land safely in the deep snow and thick growth of trees. They spent the night in the capsule in bitter cold, while wolves howled around them. In the morning they got out of the capsule somehow and a helicopter flew over them and reported seeing them chopping wood for a campfire. Korolev sent his own team of rescuers who helicoptered to the site, went down a rope ladder, and skied about 200 meters to the capsule. The rescuers and the cosmonauts then skied to the helicopter, where they had to spend a second night in the forest, although this time with food, tents, and warm clothes.[21] "Sergei Pavlovich was constantly in communication with the head of the search service. . . , " recalled Yeliseyev, and when the rescue was complete, had enough of a sense of humor to say, "And now bring a half a kilo of Validol [a tranquilizer] for the State Commission."[22]

In an interview in 1991, fifty-seven-year-old Leonov—affable and communicative, and probably 40 pounds heavier than he was as a cosmonaut—recalled his famous flight, and remembered also a 1962 meeting with Korolev when he was first shown the *Voskhod* that he would fly:

"Korolev told us," he said, "that 'Any seaman on a ship has to know how to swim and so each cosmonaut has to know how to swim and do construction work outside of his vehicle.' He looked at me and said, 'Now *orlyonok* [eaglet], put on a space suit, and go through all the procedures for the engineers.'

"I went through the procedures and then Korolev left. Gagarin said, 'Congratulations, now your selection has been made.'

"But actually Korolev had chosen me in advance, just as he had Gagarin."[23]

Intensive work, under great pressure from Korolev, was done in 1964–1965 on a *Voskhod 3* mission that never materialized. It would have involved a test of an artificial gravity system consisting of rotating "orbital blocks" tethered by flexible cable. One of the engineers involved in the project, which was directed by Pavel Tsybin, was L. B. Vilnitskii. "Our section," he wrote, "had to develop complex and large mechanisms: winches, drums with an entire kilometer of cable, fittings, drives, etc. All this had to be done in 2–3 months instead of the 1 to 1 1/2 years it should have taken in normal operations. . . . Sergei Pavlovich . . . began to literally beg and plead for us to maximally accelerate our development."

But the development of the system presented problems that proved insurmountable. This is not surprising, since artificial gravity systems have still not been developed in either the United States or Russia.[24]

It was not until *Gemini 4,* on June 3, that an American astronaut, Ed White, flying with Jim McDivitt, would top Leonov's feat with a twenty-one-minute EVA. White maneuvered himself with a handheld nitrogen-powered thruster. The Americans, too, had trouble with automation. As a result of a computer failure, McDivitt, making a manual reentry after four days in orbit, was one second late in activating the retrorockets. The spacecraft, after sixty-six orbits over a ninety-nine-hour period, landed in the ocean about 64 kilometers off course.

It was this flight, almost four times as long as any previous one, that caused particular exultation among the NASA medical specialists, and probably the Soviet doctors as well. Dr. Charles Berry, the astronauts' flight surgeon, said after the flight that, "We knocked down an awful lot of straw men. We had been told we would have an unconscious astronaut after four days in weightlessness. Well, they're [sic] not. We were told that the astronaut would experience vertigo, disorientation, when he stepped out of the spacecraft. We hit that one on the head."[25]

Unlike the built-on-the-cheap *Voskhod,* however, *Gemini* was a major step ahead in spacecraft technology. Bell-shaped like *Mercury,* it was almost three times as heavy at nearly 4 metric tons. It was, however, more than a ton lighter than *Voskhod.* The space for the two astronauts was comparable to that in "the front seat of a small sports car." It could maneuver with its own onboard propulsion and had a guidance and navigation system and a rendezvous radar.[26] Its design, of course, was geared toward developing technology and techniques that would contribute to the *Apollo* lunar landing program. Although not officially approved until December 1961, *Gemini* had been intensively worked on throughout 1961, its progress sped up by President Kennedy's announcement, on May 25, 1961, that a landing on the Moon by an American astronaut "before this decade is out" was a national objective.

That meant taking steps that would ensure a victory over the Russians in less than nine years. Already well underway was the *Saturn* launch vehicle whose first stage would be powered by the giant F-1 rocket engine. But a savvy group of engineers at NASA Langley Research Center, including the clever and tenacious John Houbolt, came up with a shortcut for the mission itself: rendezvous two spacecraft in lunar orbit, thereby greatly reducing the initial launch mass, as well as the time to develop the needed technology.

Gemini became the vehicle for learning how to achieve rendezvous. This was not easy. McDivitt and White, in *Gemini 4,* had trouble with a practice rendezvous because, according to Andre Meyer Jr., of the *Gemini* project office, they didn't understand station-keeping operations.

"Adding speed also raises altitude, moving the spacecraft into a higher orbit than its target . . . "wrote Meyer. "As the *Gemini 4* crew observed, the target seemed to gradually pull in front of and away from the spacecraft. The proper technique is for the spacecraft to reduce its speed, dropping to a lower and thus shorter orbit, which will allow it to gain on the target. At the correct moment, a burst of speed lifts the spacecraft to the target's orbit close enough to the target to eliminate virtually all relative motion between them. Now on station, the paradoxical effects vanish, and the spacecraft can approach the target directly."[27]

There were, all told, ten two-man *Gemini* flights over a period of twenty months, from March 1965 to November 1966, and they were crucial to the eventual *Apollo* program.

Gordon Cooper and Pete Conrad, in *Gemini 5,* had no ill effects from eight days of weightlessness, the time required for going to the Moon and back.

Gemini 7, with Frank Borman and Jim Lovell in crew, set a trip longevity record that would stand for five years—fourteen days in orbit—and actually flew before *Gemini* 6. The latter's flight was delayed when an Agena vehicle, which was supposed to serve as a rendezvous target, exploded six minutes after launch. Instead, *Gemini 6,* with Wally Schirra and Tom Stafford, went up on *Gemini 7*'s eleventh day in orbit. Schirra performed the first real rendezvous when he brought his ship, with no less than 35,000 thruster firings, to within 2 meters of the Borman-Lovell craft.

Pavel Popovich tried to one-up Schirra in an interview with an *Izvestia* reporter on December 21, 1965, a few days after the *Gemini* 6-7 rendezvous. Speaking of the *Vostok 3* and *4* flights three years earlier he said, " . . . our ships came to within 5 kilometers distance in space. Thus, in principle, the American experiment of an orbit rendezvous repeats in some degree what we did." He then conceded that "techniques had advanced a great deal" and graciously complimented Schirra for his skill.[28]

Schirra, though, at a press conference a few days later, denigrated the comparison. "If anybody thinks they've pulled off a rendezvous at three miles [5 kilometers], have fun! This is when we started doing our work. I don't think a rendezvous is over until you are stopped—completely stopped—with no relative motion between the two vehicles, at a range of approximately 120 feet [37 meters]. That's rendezvous!"[29]

At the end of 1965 it must have been debilitating for a tired, ailing Sergei Korolev to realize how substantial a series of achievements the Americans had accomplished with the *Geminis*. The outlook for the Russians in the Moon race was getting gloomier and gloomier.

Here is the scoreboard for the 1964–1965 laps in the race:

Voskhod vs Gemini

1964		
October 12–13	USSR	*Voskhod 1*. Feoktistov, Komarov, Yegorov. First three-man flight, 16 orbits
1965		
March 18–19	USSR	*Voskhod 2*. Belyaev and Leonov. 17 orbits, 26 hrs. Leonov does first EVA, 20 min
March 23	USA	*Gemini 3*. Grissom and Young perform first manned orbital maneuvers, orbit 3 times in 4 hrs, 54 min
June 3–7	USA	*Gemini 4*. McDivitt and White do 66 orbits. White does 21-minute EVA
August 21–29	USA	*Gemini 5*. Cooper and Conrad orbit 128 times in 190 hrs, 54 min
December 4–18	USA	*Gemini 7*. Borman and Lovell do 220 orbits in 330 hrs, 36 min; serve as target for *Gemini 6* rendezvous
December 15–16	USA	*Gemini 6*. Schirra and Stafford approach within 1 foot of *Gemini 7*, land after 17 orbits, 25 hrs 54 min

The Americans were on a roll. Korolev, pumping hard to keep up, was not even free to focus full-time on the race. He was still a major contributor to ballistic missile development, and to unmanned lunar and planetary exploration. Simultaneously, he had been working on a new genre of applications satellites—for spying and for communicating. He would cope with that responsibility remarkably well, by using some of his manned spacecraft technology.

Spy Sats and Com Sats

11

The David-and-Goliath nature of Korolev's contest with the United States comes into sharp perspective when one realizes that his design bureau, alone in the mid-sixties, was pitted against a huge array of American aerospace firms, most of them concentrating on one or a few projects in intercontinental ballistic missiles, lunar and planetary explorers, and manned spacecraft. Now add two new fields—reconnaissance satellites and communications satellites—and a new set of American competitors.

Add, in other words, the likes of RCA, Thompson Ramo Wooldridge, AT&T, Hughes Aircraft, Eastman Kodak, Fairchild Camera, Bell Aircraft, and Itek, to McDonnell Aircraft, Douglas Aircraft, Martin Company, North American Aviation, Lockheed Aircraft, General Electric, Boeing Airplane Company, Chrysler Corporation, Curtiss-Wright, Aerojet-General Corporation, Thiokol Chemical Corporation, Reaction Motors, Inc., Bendix Aviation, Convair-Astronautics, United Aircraft, and others.

As might have been expected, the indefatigable Sergei Pavlovich faced up to the new challenge with his customary ingenuity—using to the utmost the technology he had already developed. His initiatives in these fields came as no surprise to the CIA, which predicted, in a report dated December 5, 1962, that, "The first Soviet military space vehicles are likely to be earth satellites used in various support roles—reconnaissance, early warning [EW], geodetic, communications or navigation. We believe that such satellites could be launched at any time; some recent satellites probably have carried out cloud photography and possibly other experimental reconnaissance missions."[1]

Probably, indeed. The fact is that Korolev, although intensely involved in the manned program, had already put up four spy satellites—beginning with *Kosmos 4,* later known as *Zenit* (zenith)[2]—on April 26, 1962—eight months before the CIA report was published.

By 1962 the United States had been flying its own spy satellites, in the highly secret *Corona* program, for almost two years. Not even the name *Corona* was declassified by the CIA until February 1995,[3] although it was leaked in a 1978 book about the family of former CIA head Allen Dulles.[4] In May 1995, *Corona*'s history was divulged in detail at a briefing conducted by the CIA, George Washington University, and the Smithsonian Institution. For many years the name given to the U.S. spy satellites was *Discoverer,* although information on that program was hush-hush as well. Resourceful trade publications, like

Aviation Week, were able to get information on *Discoverer* capsule recoveries. But only a few insiders knew officially of the *Corona* program, never mind the fact that it had a two-year lead on the Russians.

One of those in the dark, John F. Kennedy, would use American secrecy on the subject to his advantage. It was during his 1960 presidential campaign against Richard Nixon that Kennedy cited a "missile gap," alleging that the Soviets had built up a substantial lead in ballistic missiles. President Eisenhower, having access to photos from U-2 spy plane overflights, as well as from the newly operational *Discoverer*, knew that there was no evidence of such a lead, but was unwilling to divulge the source of his knowledge. In fact, in early 1960 the CIA had estimated that by mid-year the Soviets would have only 35 missiles, and 140 to 200 by mid-1961, far fewer than the outlandish estimates made by some of the alarmists. The first *Discoverers* had given fair warning that even these estimates might be high when their photos showed "empty countryside, no missiles," as described by Albert D. Wheelon, a space expert who later had a key role in the CIA.[5] But with such information undisclosed, U.S. critics were free to speculate wildly.

McGeorge Bundy, Kennedy's special assistant for national security affairs, cites a 1959 prediction by Stuart Symington, then a presidential hopeful himself, that "in three years the Russians will prove to us that they have 3,000 ICBM's."[6] Contrast that with Russian historian Maxim Tarasenko's estimate, mentioned earlier, that at the end of 1960 there were only two R-7 pads at Baikonur, and perhaps two—and in no case more than four—at Plesetsk.[7] Tarasenko believes that there were probably only four ICBMs deployed in 1960, although he points out that a Russian Strategic Rocket Forces publication gives a higher estimate—15 ICBMs in 1960 and 200 in 1965. In any case, the real number of Soviet ICBMs was far lower than that offered by the missile gap proclaimers.

Kennedy soon learned the facts about the U.S. lead and became dependent, as president, on the valuable information from the *Corona* program. His successor, Lyndon Johnson, would declare in 1967 that reconnaissance satellite information alone was "worth ten times what the whole [space] program has cost."[8] Wheelon, with justifiable pride, says that, "When the government eventually makes public all the reconnaissance systems developed by our country, the American people will learn of achievements in space every bit as impressive as the Apollo moon landings."[9]

No doubt Nikita Khrushchev developed an equal respect for what he learned from the *Zenit* photography. But for the first several years of *Corona/Discoverer*'s existence, Soviet policy, at the UN and in the press, was to make believe it was not developing its own capability, while protesting U.S. spy satellites as incursions of Soviet sovereignty. When *Zenit* became operational those protests ceased.

The path to operational status, however, was difficult in both the USA and USSR. The first successful mid-air recovery of a camera capsule was on *Corona/Discoverer 14,* built by Lockheed in Sunnyvale, California, for the CIA and the Air Force. That happened on August 18, 1960. After that, between November 1960 and February 1962, eight more *Discoverers* were successfully launched and recovered, either from mid-air catches by aircraft, or from the sea.[10] It should be noted, in gaining an appreciation of the difficulty of developing the technology of launch, on-orbit operations, proper reentry burn, and capsule recovery from either mid-air or the ocean, that there had been numerous failures on *Discoverers 1* to *13* and quite a few more subsequent to the first success.[11] The Soviets, in fact, although two years late getting started, seem to have had less trouble achieving their first capsule recovery than the Americans. Of course, it was easier to recover the Russian satellite, which was parachuted to a landing on terra firma, than to grab the *Discoverer* capsule at altitude by airplane. The CIA had decided that a C-119 aircraft recovery was the more secure way of getting the capsule, worrying that a land recovery could result in impact on a compromising spot, like someone's house.

Embarrassingly, the *Discoverer 2* capsule, launched on April 13, 1959, and ejected on its seventeenth orbit, had eluded capture and was "reliably reported to have fallen on West Spitsbergen Island," Norway, where it was probably recovered by a Soviet mining team working in the area. The seriousness of this mishap is emphasized by the fact that the report of the incident, made after an intensive but fruitless search by the U.S. Air Force, was communicated to the Deputy Secretary of Defense by no less an official than the Chairman of the Joint Chiefs of Staff, General Nathan F. Twining.[12]

No such embarrassment ever happened to the Russians, although if their craft had been recovered by their Cold War foes they would have revealed relatively unsophisticated designs. However, even if the spacecraft were somewhat primitive they did the job, and their development got very high priority.

"Korolev assigned enormous importance to the initial planning stage," wrote Yuri Frumkin, one of the OKB-1 design team, in 1993.[13] The start had been in 1957, meaning that the reconnaissance satellite was one of Korolev's first space initiatives after the R-7 launch vehicle, directly contemporaneous with the *Sputnik* and *Vostok* developments.

The first design was of a conical capsule with a cylindrical equipment section, somewhat like *Discoverer*. But in 1959 Korolev decided instead to base what became the *Zenit* spacecraft on the *Vostok*.[14] Since the start on *Vostok,* which would carry Yuri Gagarin into orbit in 1961, was taking place at the same time, in 1957, why not, Korolev reasoned, use elements of its design for both missions? This might have seemed a strange decision when one realizes that the two missions required totally different interior payloads. But using the same shell and support systems for both made it unnecessary to have two separate design developments going on in the bureau. Also, *Vostok* was designed for automatic flight, without pilot input, making its adaptation for *Zenit* all the easier.

"This decision," according to Yuri Frumkin, "greatly reduced the [development] time of . . . Zenit and substantially increased its in-flight reliability since there already existed the experience for a manned ship on which special, exceptionally high requirements on ensuring safety had been imposed."[15]

Then, wrote Frumkin, the program began to roll rapidly:

> After choosing the variant adopted for realization Sergei Pavlovich became tough and demanding, setting a fast pace which all the participants took up with enthusiasm.
>
> Tasks went on side by side, unsolved problems did not stop the creation of mockups, units, assemblies and apparatuses for the checking and finalization of all components and the space vehicle as a whole. Then the pace slowed down. This was a period of analysis of the results. . . . In general, however, such concepts as a slowdown are difficult to apply when speaking of Korolev.[16]

Once again the Soviet product was bigger and bulkier than its American counterpart. *Discoverer 14* weighed 850 kilograms. *Zenit* was more than five times as heavy, at 4,600 kilograms, about the same weight as *Vostok*.

Zenit's guts, of course, were quite different from *Vostok*'s. The craft carried four cameras, three with 1-meter focal lengths and with 1,500-frame films wound on a cannister. There were also a surveillance radio apparatus, an orientation system, a programming-logic device, a programmed radio link, and a thermostatting device.[17] A 180-kilometer band of Earth could be photographed. Theoretically, then, the whole of the United States could be covered in about twenty-five orbits at an altitude of about 200 to 400 kilometers, although more orbits were needed because of cloud cover, power reserves, and the limitations imposed by film capacity and by some inclinations.[18] Resolution, reported Frumkin, "was so high that it made possible, for example, to determine the number of automobiles in a parking lot."[19]

The orientation system, developed at OKB-1 by a team led by Boris Rauschenbakh, was very sophisticated. Programmed rotations made possible the photographing of regions off to the side of the flight trajectory, which "provided a three-dimensional image of the terrain and made it possible to solve a very complex problem: obtaining a base for mapping many regions and reducing them to a single system."[20]

How delectable it must have been for Soviet officials to examine the photos. "A single flight of five to seven days made it possible to do that [for] which aerial photographic surveys would have required years and incommensurably greater expenditures," wrote Frumkin. The Soviets, for example, were able to reconcile to a unified system of cartographic coordinates maps "of territories located far beyond the boundaries of the USSR." Great for ICBM targeting.

This latter capability proved immensely important, also, in applying the technology of *Zenit* to later generations of observation satellites now used regularly to monitor Earth and ocean resources.

Frumkin, in his 1993 revelation of the tough challenges facing the Korolev team, went into some detail:

> The initial stage, especially everything related to the complex of problems related to photoinformation, proved to be very difficult. Reference is to high quality photoinformation with the identification of objects measuring about 10–15 m. Then the principal difficulties were, first of all, an evaluation of the fundamental possibility of obtaining such information and determining the specifications necessary for such cameras, and second, checking on whether it was possible to create aboard a space vehicle the conditions necessary for their functioning.

New ground had to be explored in optics technology, a field in which the Soviets were far behind the Americans. They would remain far behind in subsequent years as, for example, the U.S. *Ranger, Surveyor, Viking, Mariner, Voyager,* and such observation satellites as *Landsat* and subsequent generations of American reconnaissance satellites produced spectacularly detailed photos of the Earth, Moon, and the planets. But the Soviets went up the learning curve eagerly. As Frumkin recalled:

> In order to obtain a high-resolution photoimage from a flight vehicle which is moving at a velocity of about 8,000 m/s at altitudes 200–400 km it was necessary to formulate the theoretical principles for designing and constructing optical systems with long-focus objectives and large multilayer lenses.

> It was also necessary to overcome other obstacles. For example, in the course of exposure the film had to move at a rate seemingly making it possible to freeze the terrain image. This process is called image displacement compensation ["image motion compensation" in the U.S.].[21] Photograph quality and resolution can be ensured only in a case when the deviations from the stipulated rate of compensatory film movement do not result in a displacement of the "frozen" image by more than 0.01 mm. This can be attained under the following conditions: if, first of all, the altitude of the survey, the velocity of the space vehicle motion and its angular position in space at the time of the photography are known with a high accuracy, and second, if the stipulated rate of film movement in the camera itself is precisely ensured. It was found that both conditions can be satisfied.[22]

Another big challenge was the control of temperature in the camera system, a factor exacerbated by the continual rotation of the spacecraft relative to the Sun, which caused substantial changes in heat flow, especially when the vehicle was in the Earth's shadow. Even a small temperature differential between the glass layers in the viewing port would change their curvature, reducing image quality.

Dealing with these design problems was new to the Korolev team. As Frumkin put it, "These systems did not have analogues in preceding development work. . . . Special gyroscopic transducers were used for the first time in the orientation system . . . and regimes of transducers for constructing the infrared vertical and a number of other system elements were fully developed."

Another complex problem was how to control the camera from the ground. Korolev's engineers were used to one-time start and stop commands but here was the need to transmit a volume of information which, as Frumkin stated, "increased by more than an order of magnitude . . . in particular . . . a large number of complex programs for adjusting each photographic session."[23]

Kosmos 4 was OKB-1's second try to orbit a spy satellite. The first attempt, on December 11, 1961, had failed when the third stage of the booster shut down and the spacecraft never made it to orbit. The *Kosmos 4* flight itself was not without problems. Frumkin reported:

> . . . malfunctions occurred in camera operation and there were errors in the orientation system. Nevertheless, after three days, as was planned, the descent module landed in the stipulated region. Thus began the flight-design tests of the Zenit spy satellite. In 1 1/2 months it was possible to eliminate the shortcomings and already in July of that

same year a new flight [*Kosmos 7*, July 28, 1962] was made during which the onboard systems operated well and the tasks were fully carried out.

Ten test flights took place up through *Kosmos 20*, launched on October 18, 1963. The 65° inclination orbit of the *Vostok*[24] rocket enabled *Zenit*'s cameras to photograph the most interesting parts of Earth. Not covered, though, were the poles, northern Canada and Greenland, most of Alaska and Scandinavia, and the northern latitudes of Russia and Siberia.

The first *Zenits* operated for eight-day periods. Then a ten-day mission was flown in April 1963, and a twelve-day mission in March 1968, the same year in which a maneuverable satellite was launched and two satellites were put in operation simultaneously.

The differences between *Zenit* and *Discoverer* are typical of the differences between USSR and U.S. space technology of the era. *Zenit*, besides being bigger and heavier, was easily mass-produced. A supply of them was ready to be lofted at any time. *Discoverer*, lighter, containing more sophisticated optics and electronics, was tailor-made one at a time at Lockheed.

Discoverer went into a near-polar orbit at 80–83° inclination, riding on a Thor-Agena booster, which gained less thrust advantage from the Earth's rotation than *Vostok*. Its pictures, more detailed, also had the benefit of being taken at virtually the same sun angle on each orbit. *Discoverer* benefited from film and camera technology already in hand from the U-2 and reconnaissance balloon programs. Special high-resolution film had been developed for the U-2 by Eastman Kodak, and newly designed panoramic cameras came from the balloons. Techniques for snagging the parachutes had also been designed for the balloons, although catching the *Discoverer* capsules proved to be a nagging problem.

In 1963 Korolev transferred responsibility for *Zenit* serial production to a branch of OKB-1 in Samara, which eventually became the independent Central Specialized Design Bureau (TsSKB for Tsentralnoe Spetsializirovanoe Konstruktorskoe Byuro). Dimitri Kozlov, another exemplar of the durability of some space leaders, has been in charge of TsSKB—formerly the Kuibyshev Design Bureau—since 1959.

Two years earlier, in 1961, Korolev started a development that was to have import in the USSR at least equal to that of the reconnaissance satellite. It was *Molniya* (literally, lightning), the Soviet Union's first communications satellite. No one realized the value of fast messages to the eleven-time-zone USSR better than Sergei Korolev. Except, perhaps, anyone who has ever had to rely on Russian telephone service. Korolev's mandate was no less than to give the Soviets decent long-distance telephone and television coverage.

Mstislav Keldysh said that, "This achievement alone has huge significance for the development of culture in remote regions and for the facilitation and reduced cost of communications."[25] *Izvestiya,* in 1986, reported that these systems were "three times cheaper than radio relays, and they can be built ten times more quickly."[26]

The fact is, however, that Korolev probably had more trouble getting funding for the *Molniya* concept than he had developing the spacecraft system. One of his young engineers who would later become a cosmonaut, Vitaly Sevastyanov, tells the story of a meeting during which a fifteen-year plan for unmanned and manned spacecraft launches was being discussed before a minister named Psurtsev:

> Korolev introduced me to Psurtsev and said to him, "Sit down and listen." He then tried to persuade Psurtsev to give money for the development of communications satellites. Psurtsev said no, that he would put the money into a cable system. Korolev insisted that new technology should be used. "Give us at least the seed money," he said. This was 1959! He described a heavy communications satellite and Psurtsev finally said OK, he would fund it, and he would open up a department for it in his Ministry. In two days we met with Psurtsev's guys and we went ahead with *Molniya*. Psurtsev got two State prizes for it. It made his career.[27]

In developing an effective communications satellite network for the huge Soviet Union, Korolev was hamstrung by the unavailability, in the early 1960s, of big boosters capable of placing satellites in geosynchronous orbit, from which they could hover over the equator and feed messages and data uninhibitedly.

His clever solution: use a four-stage version of the old reliable R-7. This relatively inexpensive rocket could place the spacecraft into a highly elliptical orbit—400 by 40,000 kilometers—which would enable as few as three satellites, launched on 63–65° inclinations, to provide coverage of most of the Soviet Union. Since a satellite travels more slowly the higher it flies, it takes eight hours to traverse the apogee portion of the orbit, all the while communicating with ground stations.

Molniya must be recorded as one of Korolev's most important legacies. Coverage includes the economically and strategically important far northern parts of Russia as well as the ships in the northern polar region—something that geosynchronous satellites cannot do. Although three *Molniya* satellites can provide the coverage, the Russians usually rely nowadays on four of them, replacing old with new as needed. For the past decade, in fact, the Russians have maintained separate constellations of eight *Molniya 1* and eight *Molniya 3* (advanced design) satellites.[28]

Korolev started his satellite communications program far later than the Americans. He did not even begin to develop the *Molniya* concept until 1961, three years after the United States had already put up its first test communications satellite—called *Score,* which was very simply a "talking" Atlas booster. It contained a tape that transmitted a Christmas message from President Eisenhower. *Echo,* a huge Mylar balloon from which Earth signals could be bounced, was next, in 1960. *Telstar,* the first commercial communications satellite, developed by AT&T, came in 1962, giving the Americans a solid head start in a field that is now a multibillion-dollar-a-year business.

It was in that year that Korolev wrote the first detailed paper on how *Molniya* would work.[29] It may well be that he got the idea for the unusual orbit from papers delivered in 1960 and 1961 by a British engineer, William F. Hilton.[30] Perhaps, however, the concept was derived independently. In any case Korolev decided to test an experimental system carrying a single TV channel or forty to sixty bilateral channels of telegraph and telephone communication between Moscow and the far eastern city of Ussariysk, near Vladivostok. No doubt the prior existence of U.S. satellite communications systems helped him to maintain political support for the concept, but he certainly helped his cause by pointing out that *Molniya* would make possible Vladivostok's rapid communication "with the basic cities of European USSR and Western Europe."[31] He made it clear in the report that this would require a major commitment of resources extending over a considerable period of time. There would be a need for a large number of satellites and launchings, and equipment having lifetimes of 5,000 to 10,000 hours. "At present," he pointed out, "space objects" have "smaller lifetimes."

After two failures, *Molniya 1* made its debut on April 23, 1965. In the previous year the United States had put up its first geosynchronous satellite—*Syncom.* The USSR would not even test-launch its first geosynchonous satellite until *Kosmos 637* was orbited on the SL-12 Proton in 1974. By then, however, a twenty-four-hour system of *Molniya 1*s had been performing well for six years, since 1968.

In 1964, after building eight *Molniya* spacecraft, Korolev turned over responsibility for the system and its future iterations to NPO PM (Scientific Production Association "Applied Mechanics"), directed by a longtime colleague, Mikhail Reshetnev, in Krasnoyarsk, Siberia.[32]

Relief from carrying the load of so many projects was no doubt welcome, although Sergei Pavlovich seemed to bear up under the strain, even though there had already been instances when fatigue had set in, and he had to take time out to rest. His concession to the furious pace was, according to Vladimir Shevalyov, one of his administrative staff, to take a nap after lunch for a couple of hours.[33]

He gave himself few amenities, however. There was a shower in his office and a special canteen where he and his deputies would take meals. But his workday continued to be long and strenuous. One of his secretaries, Antonina Zlotnikova, recalled nostalgically a typical day at the bureau.

> Usually his work day began at 8:30 A.M. Sergei Pavlovich would come into the reception room, shake hands with everyone present, and proceed to his office. Despite the fact that he was a rather corpulent and strong man, his gait was light and soft. . . . [He would be] scheduled literally down to the minute. . . .
>
> [He] was very punctual and demanded the same of his subordinates. He liked his desk very much. He had selected it himself, as well as other furniture in his office. He didn't even allow anyone to move papers around on his desk. He always knew where everything was. He wrote with a common lightweight plastic pen. He dipped the quill in . . . india ink. He signed resolutions in a broad signature with blue pencil. He didn't pay particular attention to his clothing. . . . He liked to have a bouquet of lilacs on his desk . . . he liked to stay late . . . by himself. That is when he laid out the papers on his desk and sang quietly to himself "There's no better flower than when the apple tree blooms."
>
> . . . usually in the evening at about 9:00 o'clock he drank tea with lemon and black bread with a piece of boiled sausage. . . . In general, he loved life and work.[34]

Unlike many of his hard-drinking colleagues, Korolev usually eschewed vodka and other spirits. Shevalyov characterized his boss as having a simple lifestyle, "although hardly an ascetic." He did not pay much attention to material benefits, and dressed unostentatiously. Most Soviet officials wore jackets and ties. Not so Sergei Pavlovich. He disliked ties, and usually wore an open-necked sport shirt to work.

A handsome, well-built man in his youth, with a bull neck and a stern gaze, Sergei Pavlovich was attractive to women, and had an eye for them himself. Antonina Otrieshka confided to me that, "They said I was in love with him. Not so, but everybody loved him and he demanded that people do their work on time and correctly."[35]

To say "everybody loved him" seems the forgiveable but exaggerated claim of a loyal staffer. Korolev may not have "loved" Otrieshka but his fondness for women in general seems an accepted fact. He was not, apparently, a blatant womanizer, but seems to have exemplified the Russian macho male in his penchant for having women on occasion. No lurid tales emerged from the interviews or memoirs, just flat-out statements like this one from Golovanov:

"At least, let's say, he was never an ideal husband . . . He had other girls . . . Nina Ivanovna rejects the notion that he continued to philander but in fact he did."[36]

Vetrov added confirmation of Korolev's liaisons:

> Yes, he liked girls. He knew how to attract women, but in the end nobody was hurt. He had some serious romances, and lots more were rumored. I know very well one of the women who was a serious girl friend. . . . Nina Ivanovna knew about these affairs. She understood them philosophically. . . . You must realize that in the culture of that time the men under Korolev admired him for this.[37, 38]

That admiration might have helped sustain the Chief Designer as the mid-sixties brought him more deeply into competition with the massive industrial might of the United States. Even the unloading of the robotic space explorers and the *Molniya* communications satellite left him with a huge array of work. As he prepared to focus on the central goal of that work, manned exploration of the Moon, he would get more hindrance than help from the bureaucratic Soviet government.

Sergei Pavlovich Korolev at age two in 1909, with grandmother, Maria Matveevna, and mother, Maria Nikolaevna Moskalenko, in Zhitomir, Ukraine. (*Courtesy of Natalya Koroleva*)

At four, everyone's favorite in his grandparents' household. His father had already left the family. (*Courtesy of Natalya Koroleva*)

Growing up in Nezhin—posing with toy rifle, 1912. (*Courtesy of Natalya Koroleva*)

In the cockpit of *Koktebel*, the record-setting glider he designed in 1929. (*Courtesy of Natalya Koroleva*)

Memorial to Konstantin Tsiolkovsky, father of space travel, in Moscow. (*Courtesy of James Harford*)

Korolev, at left, supervises the first successful launch of a Soviet liquid-fueled rocket, August 17, 1933. (*Courtesy of Natalya Koroleva*)

GIRD's sponsor, Marshal Mikhail Tukhachevsky, who was executed in 1937 during the Stalin purges. (*Courtesy of National Air and Space Museum*)

Friedrikh Tsander, the space visionary who worked with Korolev at GIRD. (*Courtesy of National Air and Space Museum*)

Wernher von Braun in 1930, carrying the Mirak rocket to a test at the Raketenflugplatz in Berlin. At left is Rudolf Nebel. *(Courtesy of the National Air and Space Museum)*

The 29-year-old Korolev, with his first wife, Xenia Vincentini, and their one-year-old daughter Natasha in 1936, a year of intensive rocket experimentation at GIRD. (*Courtesy of Natalya Koroleva*)

The author in front of the Moscow apartment at 26 Konyushkovskaya Street from which Korolev was taken in the middle of the night in 1938 and sent to a gulag camp in Kolyma. (*Courtesy of James Harford*)

Korolev in about 1945, just after being released from prison. (*Courtesy of Natalya Koroleva*)

Valentin Glushko, arrested during the purges, later joined Korolev in a sharashka in Kazan. He would become the USSR's leading rocket engine designer. (*Courtesy of National Air and Space Museum*)

Nina Ivanovna Kotenkova became Korolev's second wife in 1948. (*Courtesy of Natalya Koroleva*)

Korolev in Germany in late 1945, with the rank and uniform of colonel, studying the V-2 technology. (*Courtesy of Natalya Koroleva*)

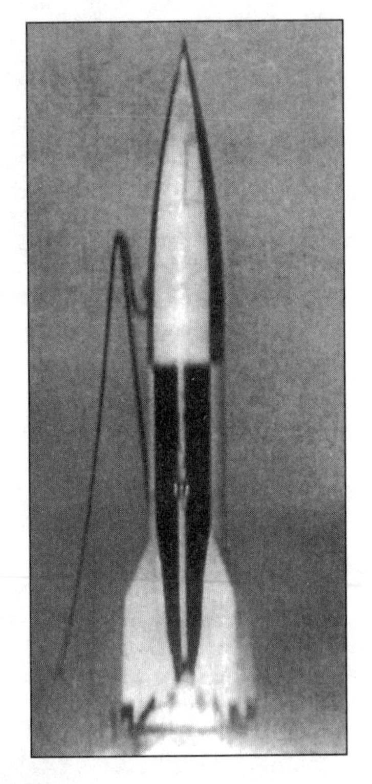

The R-1, the Soviet version of the V-2. (*Courtesy of RSC Energia*)

German V-2s being launched during World War II. (*Courtesy of Ernst Stuhlinger*)

The Council of Chief Designers: Mikhail Ryazansky, Nikolai Pilyugin, Korolev, Valentin Glushko, Vladimir Barmin, Viktor Kuznetsov. (*Courtesy of RSC Energia*)

"This ball will be in museums," Korolev predicted of *Sputnik I.* Representations hang today in many places, including the United Nations Secretariat. (*Courtesy of James Harford*)

Gennadi Strekalov, who helped machine the Sputnik sphere and later became a cosmonaut. (*Courtesy Gennadi Strekalov*)

Following Krushchev's directive, *Sputnik II,* with the dog Laika as sacrificial passenger, was prepared for launch in four weeks to celebrate the 40th anniversary of the Bolshevik Revolution. (*Courtesy of James Harford*)

The RD-107 engine, which was designed by Glushko, has four chambers fed by a single pump. The R-7 launcher is powered by four of these engines plus a central four-chamber RD-108, also designed by Glushko. (*Courtesy of James Harford*)

The Baikonur plaque says "Here the Genius of the Soviet People Began the Bold Assault on the Cosmos." (*Courtesy of James Harford*)

This is the capsule in which Yuri Gagarin made the first Earth orbital flight in 1961. It is in the RSC Energia museum in Kaliningrad. (*Courtesy of James Harford*)

Yuri Gagarin, the first human being in space. (*Courtesy of RSC Energia*)

One of the rare photos of the anonymous Chief Designer at an official gathering, this one a VIP reception honoring Gagarin. From left are cosmonaut Andrian Nikolaev, the Premier, Mrs. Gagarin, Yuri, Mrs. Khrushchev, Anatoli Mikoyan, Korolev, and Nina Ivanovna. (*Courtesy of Natalya Koroleva*)

Korolev, Gherman Titov (second man in space, August 6, 1961), and Mystislav Keldysh, head of the Soviet Academy of Sciences. (*Courtesy of Keldysh Institute*)

A US Air Force C-119 recovers the film package from a Discoverer satellite in mid-air. There were plenty of misses in the early days. (*Courtesy of the National Reconnaissance Office*)

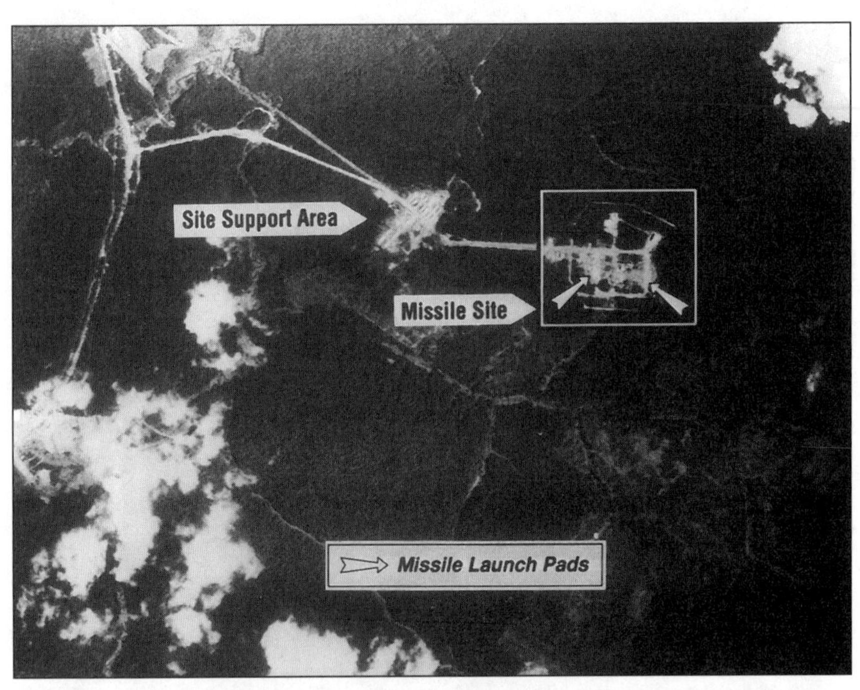

Through photos like this one, the US Discoverer spy satellite confirmed that the number of deployed Soviet ICBMs in 1962 was still small. Yurya is about 500 miles east of Moscow. 1962 was the year when Korolev launched the first Soviet spy sats. (*Courtesy of the National Reconnaissance Office*)

Voskhod 1 looks much like *Vostok* because it was the same spacecraft, stripped to house three cosmonauts without spacesuits or ejection systems. (*Courtesy of James Harford*)

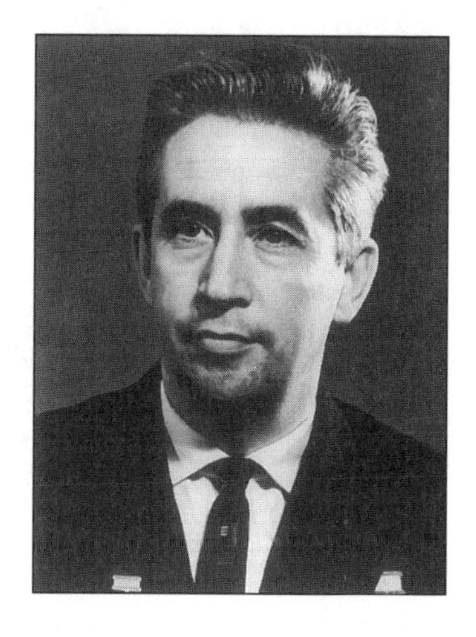

Konstantin Feoktistov, the designer, was also one of the three cosmonauts who flew *Voskhod 1*. (*Courtesy of the National Air and Space Museum*)

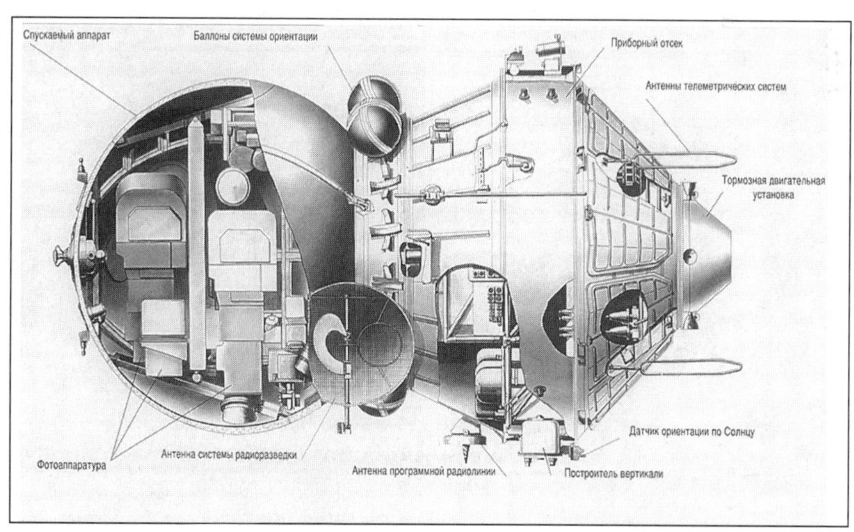

The *Zenit* spy sat used the *Vostok* spacecraft design. This is *Zenit 2*. (*Courtesy of RSC Energia*)

Voskhod 2 deployed a fabric airlock through which Alexei Leonov exited to become the world's first space walker. (*Courtesy of James Harford*)

Leonov and his crewmate, Pavel Belyaev, are greeted by Korolev on *Voskhod 2*'s triumphal return in March, 1965. (*Courtesy of Alexei Leonov*)

Leonov spent 20 minutes outside the spacecraft, 12 of them trying to wedge his way back in. His EVA was not matched until Ed White did 21 minutes in *Gemini 4* more than two months later. (*Courtesy of Alexei Leonov*)

Prototype of *Luna 9* capsule, designed by the Korolev bureau but built by Lavochkin, who took over the robotic spacecraft programs. (*Courtesy of James Harford*)

The four-stage interplanetary version of R-7, now called "Soyuz." Earlier versions launched the first ICBM, and placed the first Sputnik and Yuri Gagarin into orbit. (*Courtesy of RSC Energia*)

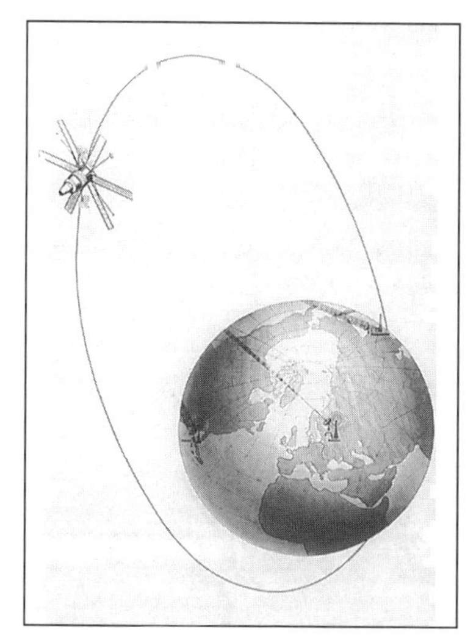

The Molniya communications satellite efficiently covers areas of northern Russia and Siberia with its 1,500–40,000 km elliptical orbit. (*Courtesy of RSC Energia*)

The author examines Block D, one of the upper stages of the N-1/L-3 lunar landing complex, at the Moscow Aviation Institute. (*Courtesy of James Harford*)

The author with Korolev's successor, Vassily Mishin. Mishin is now a professor at the Moscow Aviation Institute. (*Courtesy of James Harford*)

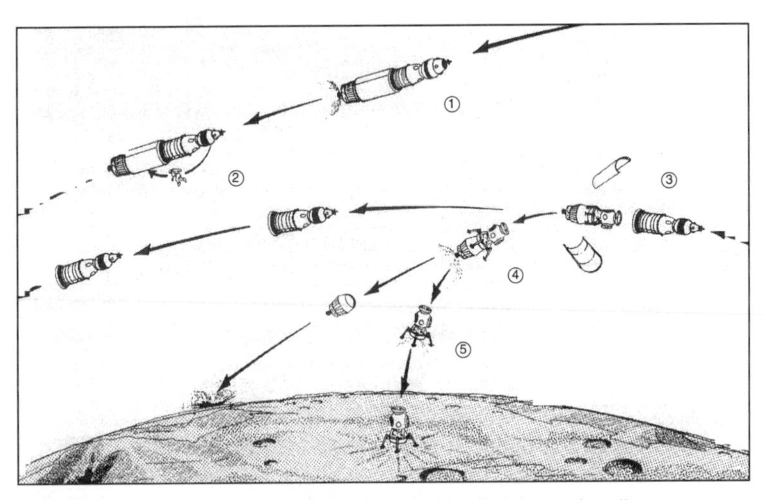

A portion of one of the scenarios for the N-1/L-3 lunar landing mission that never happened. (*Courtesy of Nicholas Johnson*)

Leonid Brezhnev and other officials carry Korolev's coffin to the Kremlin wall, January, 1966. (*Courtesy of Natalya Koroleva*)

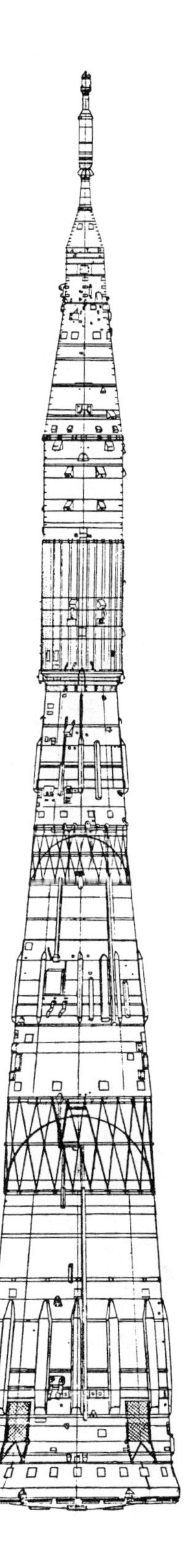

Left: N-1/L-3 weighed 2750 metric tons on the pad, was 105 meters tall, and needed five stages to lunar orbit. The thrust of the 30-engine first stage was 4620 tons. (*Courtesy of Charles P. Vick*)

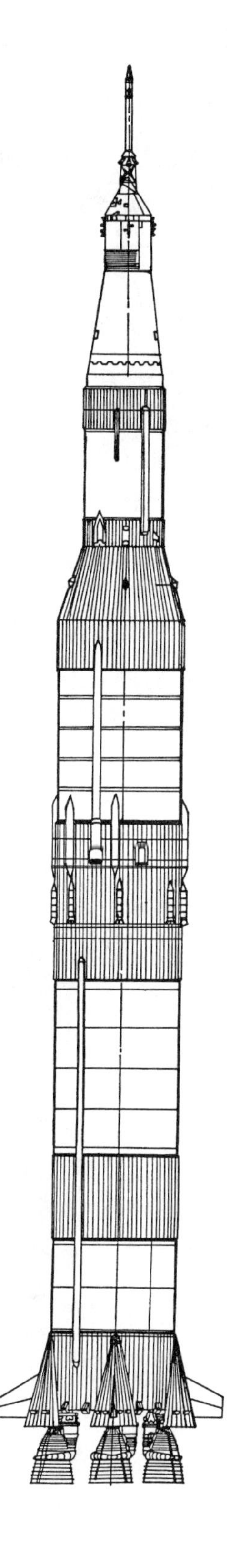

Right: Apollo-Saturn V weighed 2938 metric tons at launch, was 110.6 meters tall, and required four stages to lunar orbit. The thrust of the five-engine first stage was 3447 tons. (*Courtesy of Charles P. Vick*)

There were 30 rocket engines in the first stage of N-1. (*Courtesy of RSC Energia*)

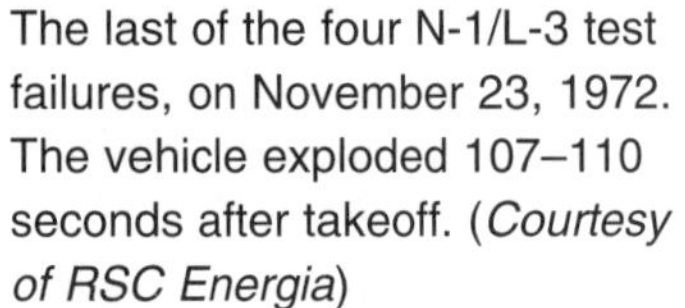

The last of the four N-1/L-3 test failures, on November 23, 1972. The vehicle exploded 107–110 seconds after takeoff. (*Courtesy of RSC Energia*)

Korolev's tomb, not far from Gagarin's. (*Courtesy of Natalya Koroleva*)

Artist's concept of the International Space Station, which is being built by the USA, Russia, Europe, Canada, and Japan. Its first component, a Russian power module, is scheduled to orbit in late 1997. (*Courtesy of NASA*)

12

THE ORGANIZATION: KOROLEV UP AND DOWN

"WE HAD NO idea of what things cost, or how much money came from the government," said Oleg Ivanovsky, who served as one of Sergei Korolev's key designers through the Sputnik, Vostok, Voskhod, Mars, Venus, and Luna programs.[1]

It may be that Korolev knew, but when one looks at the labyrinthine organizational setup he had to cope with, he probably had good reason to keep the numbers obfuscated. Some cost estimates are beginning to emerge from archival documents, or personal recollections of participants, but many of the relevant papers are still in secret Kremlin files, and they remain closed despite agitation from historians, including this one. What does seem clear is that the missile and space program was very expensive when calculated as a percentage of national spending for research and development and—more important—in terms of numbers of engineers and scientists whose talents were monopolized.

In 1992, a year after the downfall of the Gorbachev government, Vladimir Postyshev, one of the new bureaucrats, proposed reorganization of the space establishment, protesting that too much money had been spent on it, and much of that had been wasted in duplication. He wrote that, "Until recently, we were giving space [alone, not counting missiles] 1.5 percent of the gross national product every year, half again as much as in the United States, and six times as much as in France, Japan and Germany."[2] When detailed cost figures are released they should be very interesting, and they ought also to provide a better understanding of how specific decisions were made in the complex government and military-industrial complex that ran the ballistic missile and space program (see Charts A and B, pp. 216–18). Postyshev singles out one instance of apparent egregious waste, although it took place in the post-Korolev era. In the 1970s, rather than choose between missiles designed by two different design bureaus, Leonid Brezhnev ordered both to be funded. "No one could determine which of the two was the winner of the competition. . . . It was easier to dig deeper into the till and stamp out the highest 'A-OK' for both missiles. That, of course, was followed by two gigantic orders, the construction of two production bases, two infrastructures throughout the entire Union with no linkage whatsoever, the training of personnel with different specialties. And money, money, money."[3]

Korolev, although generally frugal with space technology funds, had remarkable budgetary autonomy up to some unknown level. As recalled by Gazenko (Chapter 1), he was able to extract sizeable amounts of cash from his own safe without governmental authorization. Where did the cash come from? What kind of accounting was required? These remain unanswered questions.

The Chief Designer's ultimate bosses were the USSR's supreme leaders—Stalin, Khrushchev, and Brezhnev. The Council of Ministers, chaired by these same men, had formal control of the missile and space program, although Lavrenti Beria, as Minister of State Security, was delegated to bird-dog the ballistic missile projects—until he was executed during Khrushchev's time.[4] It was still another autocrat, the Minister of Armaments, and later chairman of the Military-Industrial Commission of the Council of Ministers, one Dimitri Fedorovich Ustinov, who exercised month-to-month control over the various missile and space programs throughout Korolev's career, and with whom the Chief Designer battled incessantly. But did Ustinov have fiscal authority? To some degree he must have. "Funds were always a problem," Vassily Mishin, Korolev's deputy, recalled, "even though we had a very high priority. Korolev would propose to Ustinov and get as an answer, 'Go ahead and work on it and we'll see.' "

Korolev had many meetings with Ustinov at the design bureau, and sometimes the minister would nitpick to a ludicrous degree. "I remember," said Mishin, "when Ustinov came to examine a new building at the design bureau in 1955. He liked to see the furnishings and even the toilets. He noticed that the toilet cabinet doors went down to the floor and he ordered them to be shortened," presumably to spot goof-offs.

The Stalin directive of 1946, which had put Ustinov in charge of rocket and missile development, had also doled out aspects of missile and space development to a myriad of organizations in different ministries. Involved besides the Ministry of Armaments, which supervised NII-88, and the Ministry of Aviation Industries, which supervised Glushko's OKB-456, there was the Ministry of Electric Industry (the name was later changed to the Ministry of Means of Communication Industry), which had responsibility for the development of guidance and control systems.

There was also a Special Design Bureau in the Compressor Plant of the Ministry of Machine Building and Instrument Engineering, which had been developing unguided solid-propellant rockets since 1941, and which was transformed into the State Union Design Bureau of Special Machine Building. The latter had responsibility for launch complexes, fueling and ground support equipment.[5] It would seem that one consequence of the Stalin decree was to obscure the missions of the designated agencies by giving them long names of indeterminate meaning.

The Ministry of Aviation Industries, besides supervising Glushko's rocket engines in OKB-456, got the substantive job of overseeing the aerodynamic testing of the missiles, including winged missiles.

The Ministry of Shipbuilding was also in the act, having its NII-10, which got the responsibility for developing missile gyroscopes.

That makes seven ministries, plus the powerful Military-Industrial Commission. "The system for administering the space program in the USSR," recalled Postyshev, "was set up . . . in the late 1940's as a copy of the political establishment that existed then. It was a pyramid whose aim was to achieve, at any price, primarily the political objectives that had been established. Under those circumstances—with an intellectual gap in a half-ruined country—that kind of administrative machine, authoritarian to the highest degree, was rather effective and was probably the only kind of system possible."[6]

One must be dubious of the truth of such a statement. In any case Korolev as well as other Chief Designers—including Glushko and Chelomei—not being at the top of the pyramid, had to use their personal persuasiveness, powerful charisma, canniness, and drive to get the multitudinous leaders to authorize their plans. Then, after they had received authorization, they used the system to force decisions.

But perhaps the very complexity made for freedom of action. "Not expecting any directive orders from the leadership, we set our own goals," wrote Konstantin Feoktistov.[7] But Korolev was always careful to validate his goals before what might be called his "shadow cabinet"—the Council of Chief Designers. The original, informal council included the designers with whom Korolev had worked since the days in Germany—besides himself there were Valentin Glushko, Nikolai Pilyugin, Vladimir Barmin, Mikhail Ryazansky, and Viktor Kuznetsov. Later, others were added. But Korolev was very much the most powerful member of the council, and chaired it throughout his career. The Council of Chief Designers does not even appear on the organization charts (pp. 216–18) because it was not an administrative body, but rather an advisory group.

"He would listen to the opinions of his fellow Chief Designers," Yuri Orieshkin, an administrative assistant, told me, "but he would make the ultimate decision, and he didn't care if Pilyugin or even Keldysh agreed with him. If a failure resulted it was his responsibility. For example, on one of the unmanned missions in 1963, Pilyugin's control system failed and the launch vehicle crashed. Korolev took the responsibility. Pilyugin disliked passing the blame but Korolev said, 'I'm the chief designer of the whole system so I am to blame.' "[8]

"None of us [in the Council of Chief Designers] at that time had any scientific degrees," Barmin told an interviewer proudly some forty years later. "We were simply engineers, experienced and knowledgeable, each of whom headed his own design bureau, but, let me emphasize, merely engineers."[9] Those remarks express the pride of aerospace engineers the world over, as well as their resentment that the media habitually credited "space scientists" for achievements that *they* made possible.

The meetings of the Council of Chief Designers, said Barmin, were held in whichever design bureau was responsible for the field under discussion. Engine meetings were in Glushko's OKB-456, ground equipment sessions were at Barmin's. "But S. P. Korolev always presided . . . we always talked things over amongst ourselves and reached a decision that was agreed upon and approved by all. . . . And we never, I should emphasize, never turned the council into a trade-union meeting of sorts, in which the decision was made by a mechanical majority of votes."[10]

Although Barmin, like many of the space veterans interviewed for this book, seemed to give the Council of Chief Designers a kind of be-all and end-all power, he then cited another institution of government bureaucracy as having had a key role as well, a role that seems to have been decisive: the State Commission, another entity that does not appear on the charts. There were numerous such commissions focused on specific projects that had complex and demanding time schedules. As an example, Barmin mentioned the State Commission for Testing Boosters and Launches of the First Satellite, through which "All operations and interagency cooperation were carefully controlled." The chairman of that commission, Vassily Ryabikov, "did much for the formation of domestic rocket-building," said Barmin.

Surprisingly, Barmin seemed to imply that the State Commission used a management approach thought by this writer to have been created uniquely in the U.S. missile and space program: systems analysis.

> And, of course, from the very beginning we used systems analysis in our design work. Moreover . . . the entire organization of operations, ranging from designing concepts to putting something into operation, was united by a single schedule that ran through the project.
>
> . . . All operations were conducted in parallel. The design and manufacture of parts were done while the design sketches were released. Test stands were prepared at the same time that assemblies were being developed. Was there an element of risk? Yes. But the person who does not take a risk is the one who is afraid to take responsibility.
>
> . . . The interrelationship and interdependence of the operations made missing a deadline in any section unacceptable. For that reason, every director and participant—from chief designer to shop master—in

assessing the status of the work in his own area, considered the schedule above all as, by and large, the ultimate end and immediately sounded the alarm if there was a threat of a missed deadline. The feedback worked flawlessly. Any danger of missing a deadline went, like an alarm, immediately to the Council of Chief Designers, to the proper ministry or directing agency.[11]

That last sentence seems to confirm that a lot of bosses were in the act, and perhaps that's why a State Commission ran interference. Yet I kept being told that Korolev manipulated the whole apparatus. In most cases he was apparently able to swing to his point of view whatever commission, council, agency, ministry, party, or, for that matter, premier he had to deal with.

In September 1953, when Korolev first proposed the development of an artificial satellite, he got his approval from Committee No. 2, a special body created in 1946 under the Council of Ministers to supervise all missile development. Korolev made the case that the satellite project would be an integral part of the ICBM development. How he argued that dubious proposition suggests that there was some snake oil in his presentation. When Andrei Sakharov termed Korolev "cunning" and "ruthless," he might well have been talking about such actions as this.

At any rate, he succeeded, and it was to Ustinov, by now head of the Ministry of Defense Industry (a name change from Ministry of Armaments) that Korolev would propose, a year later, on May 26, 1954, the development of a "2–3 ton scientific satellite," as well as a "recoverable satellite," a "satellite with a long orbital stay for 1–2 people" and "an orbital station with regular Earth ferry communication." It seems remarkable that Korolev brought all four of those projects to fulfillment—the first, *Sputnik 3* (although only 1.3 tons), in 1958; the second, *Zenit,* a spy satellite, first flown in 1962; the third, *Vostok* and *Voskhod,* in 1961 and 1965. The fourth, *Salyut,* came about in 1971, five years after Korolev's death. These were truly remarkable achievements in a country operating with severe financial, technological, bureaucratic, and paranoiac problems.

In the United States, while all the funding for missiles and space came from a government that had its own bureaucracy, there was a lot more funding to be had, and there was spirited competition among the aerospace companies for contracts. While there were overruns—sometimes outlandish ones—the request for proposal and bid evaluation phases produced a reasonable degree of cost discipline as well as resourceful design approaches. Numerous aircraft companies expanded their capabilities to go into missiles and space.

The customers for these enterprises were either NASA or the military services, and while budgets for their products ballooned, there was reasonably effective accounting of who spent how much for what, with Congress and the press monitoring the process.

In the Soviet Union, the mobilization for missiles, rockets, and spacecraft development occurred in quite a different way, and with no one watching but a Party preoccupied by many other matters, and an obeisant Soviet press. Exposing irregularities among the secret doings of the ministries, the design bureaus, the Academy of Sciences, and literally hundreds of companies and institutes would have been difficult enough in a free and open society. The USSR system was all black holes.

Korolev, at OKB-1, had two official reporting obligations. His space projects, presumably the peaceful and scientific ones, required approval by a council in the Academy of Sciences headed by Mstislav Keldysh. Until his death Korolev would collaborate intensively with that gifted and powerful personality. His missile projects, however, had to go through an entity called simply Committee No. 2. Both programs, though, needed the blessing of joint decrees from the Central Committee of the Communist Party (CC of CPSU) as well as the Council of Ministers.

It was virtually always Ustinov, who from about 1964 or 1965 on was responsible for defense issues as Secretary of the Central Committee of the Communist Party, and Leonid Smirnov, chairman of the Military Industrial Commission, with whom Korolev had to deal. A graduate engineer, Ustinov was clever, politically astute, and conniving. Roald Sagdeev, in his memoir, calls him and Smirnov the "gray cardinals," who were not well-known "outside the military-industrial empire and party circles."[12] Smirnov's Military Industrial Commission controlled, said Sagdeev, "on a day-to-day basis hundreds of companies and enterprises belonging to different ministries," including Korolev's.[13] But Korolev knew how to end-run Ustinov, and would do so—going directly to Stalin, Khrushchev, or Brezhnev when it suited him.

The tactic did not always work, as in the 1959 instance, mentioned in an earlier chapter, when Korolev wrote to Brezhnev, then Secretary of the Central Committee of the Party, responsible for the defense complex, demanding, unsuccessfully, that the engines for the R-9 ICBM be developed by Kuznetsov rather than Glushko.

Korolev sometimes contrived to make his petitions face-to-face with the top political leaders by inviting them to inspect the current technology, sometimes making sure that they were well lubricated with spirits. Viktor Shakmatov, who served in the military at Baikonur, recalled such an occasion in a conversation he overheard between

Korolev and Shakmatov's boss, General Alexei Nesterenko, the first commander of the Baikonur cosmodrome. According to Shakmatov, Korolev told the general:

> A month ago, under the cover of a review of examples of rocket and space technology, I invited Nikita Sergeyevich Khrushchev to OKB. We showed him everything we could. Khrushchev was satisfied and the event ended with some elementary drinking. Besides Khrushchev, the leaders of various ministries, including the Ministry of Defense, were invited, and all of our chiefs. When everyone had decently drunk, Khrushchev went up to a tribune to make a speech. He complimented our work highly and told us that we were of great help to him in his effort to "destroy" imperialism.
>
> When he asked how he could help us I immediately put in front of him drafts of Decrees for the Council of Ministers to sign. He signed them without even examining them.

Korolev then scolded Nesterenko, who had been complaining about the absymal living conditions provided for his men. Nesterenko said:

> The Americans chose for themselves a range at Cape Canaveral. That is a resort . . . we are, practically, in a desert and moreover, there is nothing for people to live in . . . we got capital investments for three houses for officers, but with our increase in staff we need six. That means, we must leave our people in "zemlianki" [underground pits with a roof].

Korolev was unsympathetic, and blamed the military for not fighting their own political battles. "Sergei Pavlovich," Shakmatov remembered, "his face getting dark and solemn," responded angrily:

> In my speech of response [to Khrushchev] I said that the military was helping us very much and that without them we could not have done what we did. But, I said, the military had problems which need to be resolved at the governmental level.
>
> Khrushchev literally popped up: "What do you need, military comrades? Tell me! Let me sign!"

The generals in the audience, however, failed to speak up, and Korolev "spitting his words" at Nesterenko, according to Shakmatov, shouted:

> And those generals with [he used a vulgar expression that translates very roughly into "defective testicles"] shouted in a single voice: "Thank you for your trust, we are ready to perform any task of the Party, etc."
>
> Alexei Ivanovich, I got 75 millions in a single evening, yet your generals did not want to use their brains.[14]

Shakmatov's account of this incident is refuted by Sergei Khrushchev, who told me that his father could never have signed even a draft of a Council of Ministers decree under such circumstances, although he might have penned a note on whatever document was handed to him asking Brezhnev to look into the matter. In any case, it reflects one Korolev staffer's recollection—perhaps aggrandized over the years—of the Chief Designer's style toward the Soviet and military leaderships.

Exercising an effective downward strategy, toward his own design bureau and toward the hundreds of enterprises that furnished the research, testing, systems, subsystems, materials, and services he needed, was an equally formidable challenge. One of Korolev's deputies for scientific research, Pavel Kostyukevich, remembered the situation in early 1956, the year before the launch of the first ICBM and the first Sputnik.

> Everything was new: the material for coating the nose section, the welding technology, the manufacture of power assemblies of the rocket stages, the development of test-bench designs for final development of the rocket, the launch structures, the construction of plants for production of liquid oxygen, the manufacture of a powerful railroad transport for bringing the rocket to the launch site, and thousands of other large and small problems. . . .
>
> Many enterprises and plants, as well as sectorial scientific-research and technological institutes were involved in the development of the rocket. And for all the participant organizations it was necessary to have specific assignments, precise control and qualified treatment.[15]

At times Korolev decided that certain hardware, which in the United States would almost surely have been subcontracted, would have to be made in-house because of stringent deadlines. One such case involved a rig for launching a missile from a submarine—probably the R-11FM, which was the USSR's first submarine-launched tactical missile, using storable propellants. It was the spring of 1954, and Pavel Novozhilov, who had joined the design bureau four years earlier, recalled that the rig "had to ensure the holding of the rocket during the choppiness of the ocean, and its freeing at the moment of launch. All the work had to be done in six months. The task was new. We had no experience."[16]

Korolev came down hard on his team, insisting that the rig be developed in the bureau, even though it was, as Novozhilov recalled, "already loaded down with many different and complex orders." He pitched in himself with two working groups—one on the missile adaptation, one on the rig. "Korolev joked a lot" but was "demanding, and made clear, specific decisions. . . . We used the method of express planning, in which the plant technologists and the supply section workers worked together with the designers."[17]

Korolev's extraordinary pressure on his staff is exemplified even more strongly in the next phase of the operation. The hardware was installed on the submarine and the missiles were ready for loading into the hold when a strong wind came up. Operational regulations called for suspending the loading since the brakes on the crane could give way. "It wasn't in the character of the Chief to sit and literally wait for the weather to clear . . . " wrote Novozhilov. Reinforce the brakes, he demanded. The launch tests went on, successfully, in front of the fleet command.[18]

It is interesting to note that the U.S. Polaris Fleet Ballistic Missile system did not even begin to take shape until 1955, at least a year after the start of the R-11FM development. In fact, R-11FM had its first launch from a moving sea stand in May 1955, and then, on September 16, 1955, the missile was launched above water from a submarine. Korolev's bureau then transferred responsibility for the development to a newly organized design bureau headed by Viktor Makeev, and on December 23, 1958, the missile was launched from below water.[19]

The Polaris program lagged by about two years. The first U.S. concepts considered had been based on the Army's Jupiter missile and various solid-fueled versions of Navy missiles. But the Polaris concept prevailed because it promised smaller, less heavy warhead, propulsion, and inertial guidance technology suitable for launch from submerged submarines. The Polaris system development did not begin, however, until January 1957. The system became operational on November 15, 1960, when the *USS George Washington* (SSBN 598) left Charleston, South Carolina, on its first patrol with sixteen 1,200-nautical-mile Polaris A1 missiles.[20]

Upgrading suppliers for the components needed for such new weapon systems as submarine-launched missiles sometimes required Korolev to enroll support from Ustinov. For example, there was an urgent need for many new electromechanical devices and instruments for these missiles, as well as for other new spacecraft under development. In the spring of 1963 the Chief Designer organized an exhibit in OKB-1 of more than 100 displays of this equipment and invited Ustinov to examine them, after which the minister issued an order to the supplier companies to be more acutely responsive to Korolev's needs.[21]

Korolev literally "created the entire space empire himself—institutes, factories, labs—including organizations that didn't even know that they worked for him," said Yaroslav Golovanov. "They were given tasks to develop, say, instruments without knowing what the applications would be.

It's almost impossible to get some of the factual information, like how many engineers worked on *Voskhod.* It was probably hundreds of thousands."[22] Tarasenko estimates that about 800,000 people worked in the entire space sector at its peak.[23] It is difficult to compare that figure to the more than one million who were employed in the U.S. aerospace industry at its height of activity in the late 1960s, because the latter number included those involved in aeronautics.

It was usually Korolev's own bulldog tenacity, without help from the upper bureaucracy, that enabled him to keep the massive show on schedule. Gyorgi Pashkov, who was deputy chairman of the Military Industrial Commission on the Council of Ministers, remembered that

> Korolev had no qualms about going beyond the framework of his own direct functions as technical supervisor of the developments, and would declare war on the coordinators [of other enterprises] if they somewhere, somehow showed any indecisiveness or sluggishness. In the interests of the cause he would exacerbate the situation to its limit, and would always get his way.[24]

Sometimes Korolev lost patience with contractors whose products or services were not up to snuff and he would simply take over the work at OKB-1. An example was a vacuum spray technique for metals that had been developed at the Ukrainian Academy of Sciences but was turning out unsatisfactory results. As Mikhail Glazunov, a project planner, recalled, Korolev, "with a soft, Ukrainian accent . . . remarked bitterly and with annoyance . . . 'If things go on like this, we will have to build a blast furnace.' "[25]

Korolev had a special passion for ensuring the quality of any product whose failure might jeopardize the safety of the cosmonauts. Designer Vladimir Karin remembered an incident involving a faulty fuel system valve. Karin and his colleagues proposed a design change to improve the valve's operation but could not get the companies involved to sign off on the change. Korolev was furious. He phoned personally the directors of the companies and sent Karin to see them in his own Zil limousine to get the signatures, which proved still to be difficult, as Karin recalled: ". . . the red tape began: explaining the technical essence to each one, proving the correctness of the scheme's improvement. . . . "[26]

Korolev had snafus to deal with in his own bureau at times. Josef Gitelson recalled:

> Once we sent him a movie via classified post before one of our visits. But there were so many departments at Korolev's bureau, the movie was lost. They had looked all over for it but couldn't find it. So there we were in his office and there was no movie. He pushed a button and

shouted a message. "After a half hour we'll finish our talk and see the movie." Sure enough, in about twenty-five minutes the door of his office opened. "SP, the movie is ready."[27]

Bureaucracy in OKB-1 must have been minimal, because Korolev allowed his top people considerable autonomy in running their projects. "That was a pioneering management technique in those days. It shows that he wasn't an egoist," said Antonina Otrieshka, who was on Korolev's office staff for sixteen years. "He preferred dealing not with administrative chiefs but directly with the engineers. Everybody felt they had responsibility. He knew everyone's first name, and nobody wanted to leave the company."[28]

What Otrieshka remembers as autonomy doesn't sound like it when particular instances are considered. Lev Manenok, who started working at the design bureau in 1953, wrote that "Sergei Pavlovich often named commissions and work groups for some special one-time assignment . . . [and] often headed such groups personally." Manenok served once on a group that was assigned to calculate the strength of a new booster. Korolev had heard that the booster was turning out to be heavier than expected. After reviewing the calculations Korolev said, ". . . this will never do because it is too heavy . . . increase the critical tension [i.e., take a certain risk]." His technical assistant, Sergei Okhapkin, opposed the idea but Korolev insisted that another solution be found to lighten the weight. " 'Trust my experience. Ten percent of the weight of any construction can be reduced,' and he made a gesture with his left hand as if he was clearing off this excess weight. . . . "[29]

The Chief Designer could also be devious with his staff, using artifices that can hardly be termed straightforward. Vetrov writes, for example, that he would occasionally give the same assignment to several people, each of whom "was sure that he was the only man trusted with it."[30] Vassily Mishin remembers similar Machiavellianisms, quoting to me an aphorism of Sergei Pavlovich's, "A straight line is not necessarily the shortest way between two points."[31]

Indeed, Korolev intervened personally on many occasions when snags occurred. Gleb Maximov cited a 1958 intervention when "we could not find an organization for a long time which would undertake the manufacture and testing of the 'sodium comet' " whose vapors, illuminated by the Sun, were to signal the accuracy of the trajectories of the first rockets to the Moon. Korolev got on the phone and "literally in 5 or 10 minutes" signed up a contractor. "And you spent a month on this," he cracked, "looking at us with a clever grin."[32] This, at a time when OKB-1 was simultaneously developing not only lunar spacecraft but *Vostok,* Mars and Venus vehicles, the *Elektron* scientific satellite, the *Molniya* communications satellite, and the *Zenit* spy satellite.

Some of Korolev's actions were so autocratic that one wonders how his staff tolerated them. In 1964, when *Voskhod 2* was being developed, " . . . a curious thing happened, which became a legend," wrote designer Konstantin Shustin. "We were amazed when the telephone operator unexpectedly came into the room, disconnected all the telephones, stacked them in a large sack, and left." Korolev was angry that phone-calling was distracting his engineers from "discussing technical questions." The result, of course, was a paralysis in communications. Korolev relented after several days but ruled that there would be no use of the phones before 11:30 A.M.[33]

Since Korolev compared himself to Wernher von Braun, it is interesting to contrast their staff management techniques. Rather than meddle directly, a la Korolev, von Braun gave his people a lot of rein, depending on a simple practice to spot possible problems—one-page memos, sent to him every Friday afternoon by his top thirty-five engineers, managers, and scientists. "By Tuesday morning," remembered Ernst Stuhlinger, who was one of those thirty-five, "von Braun had read all of them, and had written his comments, suggestions and directives on the margins. Then each note was copied 35 times, and each of the writers received not only his own annotated note back, but also the annotated pages of his colleagues."[34]

Von Braun would never have considered a practice that Korolev sometimes used—bypassing middle management levels and going directly to his workers. Yuri Orieshkin remembered:

> Once there was a special need to deliver a spacecraft to the launch site to meet a difficult deadline. He and I went straight to the assembly shop. He invited four of the key old master workers into the smoking room, without the production bosses.
>
> "Would it be possible to make the spacecraft ready by [a certain date]?"
>
> He gave them some time to think it over and asked them to call him. One of them called and said, "We'll do it, but if you had ordered us we wouldn't have done it."
>
> They worked around the clock and met the deadline. They got bonuses later, of course.
>
> Sometimes he would deal directly with a young engineer who had been given some responsibility, rather than his chief. I remember once, for instance, when a young engineer had an interesting idea for the design of a component of the *Vostok.* It was in 1961. He inspected the young man's drawings, and asked him to give his arguments for the design.[35]

With his senior staff, however, Korolev was very demanding, sometimes harshly so. Vetrov told me that:

> I myself once got a "severe notation with a last warning" from him for issuing nonprecise data for the development of the guidance system on the second stage of the RT-l solid-propellant missile development. The second stage went out of control in a test firing at Kapustin Yar. All of our computer simulation told us that it should not have gone out of control. We had done automatic modeling on the ground using real hardware to control the aerodynamic surfaces. What I had not realized was that someone had modified the hardware and I had not checked to see if there had been a modification. This was in 1959, maybe 1960. When we analyzed the results and determined that the problem had been the aerodynamics, I was the culprit. I admitted that I hadn't checked to see if there had been modifications.
>
> It was typical of Korolev to get into all tiny details. There would be discussion, Korolev would follow carefully everything that was said, then he would speak bluntly to each guy and give him personal hell if he deserved it.[36]

After riling or castigating his team members, however, Korolev would usually calm down and lead them to the solution of difficult technical problems. One of the physicists who worked for him, Pavel Kostyukevich, offered a particularly sensitive description of the Chief Designer's personality traits in a memoir recalling the mid-'50s:

> Some considered him to be the soul of the collective, an attentive and smart manager who enjoyed the authority of the design bureau. Others believed that he was a wilful person, an undisputed authority not only in the design bureau but also in the entire institute, a harsh, demanding and abrupt man who essentially determined the work not only of the design bureau, but also the other NII subdivisions. A third group was simply afraid of him.
>
> . . . My first impression of Korolev in no way confirmed the stereotype which had been formed. . . . He seemed to me a calm and benevolent conversationalist with a soft, somewhat deep and pleasant voice, with a sense of humor, with a simple and trusting attitude, despite the difference in work positions. . . .
>
> As I subsequently understood, the divergence of opinion about Korolev was determined by the fact that he had a very complex nature: he could be categorical, demanding, abrupt, even coarse, and often was simple, friendly, tactful, ready and willing to render real aid. Also, he had a good sense of humor, sometimes mocking, sometimes good-natured. He knew how to concisely, clearly, and in an argumented

> manner present technical questions, how to specifically formulate a problem so that one wanted to solve it to the best of one's ability.[37]

Otrieshka recalled that "Korolev was very severe and sometimes lost his temper—but only with people who made mistakes. I thought of him as a god and tsar. He would often scream and yell and threaten to fire people but he never did. He would say to me, 'Why am I yelling at you? If you could hear how Beria yells at me!' "[38]

Orieshkin spoke warmly of another side of his legendary boss—his compassion for the lower-ranked people in the design bureau. I interviewed Orieshkin, then almost seventy, in his apartment in one of the many huge complexes of drab high-rise buildings in Moscow. Rail thin, he was dressed up for the occasion in a blue suit, white shirt, red polka-dot tie, and was wearing all his medals. His face was pale, his gray hair thinning, and during the interview he used an inhalator from time to time, maybe because of emphysema. It's clear that he was pleased to see me and to have a chance to talk about what must have been the highlight years of his lifetime. He said that

> Every morning between 8:00 and 8:30 A.M. Korolev had a "free line" on one of his office telephones for anyone to call him about a problem.
>
> Just ten days after I got the job he got a request from a woman draftsman for financial support. Her husband was in poor health. He asked me to follow up his message to the finance office to be sure that she was helped. He said that he personally felt guilty about this family because the husband had been injured on the job and he had forgotten about it.[39]

Such thoughtfulness by Korolev towards his staff was sometimes accompanied by generous bonuses, as if in contrition for his behavioral excesses. He had perks to offer that went far beyond what an American boss could command—not only bonuses, but apartment assignments and trips to fine health resorts. These inducements provided Korolev with the levers needed to turn out first-class work because, as Vetrov recalled:

> . . . he had to work with a system in which talented people were paid no more than lazy, untalented people. This tends to degrade the work output. So Korolev was careful to provide bonuses, and take special care of the good workers. In turn, they would be happy to work overtime.
>
> For example, in 1948 I got a bonus of 16,000 rubles, the value of almost two cars [you could get a car for about 10,000 rubles then]. My salary was 1,400 rubles a month, so my bonus was equal to my yearly salary. I got an "Order of Labor Red Banner" for the preparation of the documentation on the R-1 missile in 1956. I also got a "Sign of Honor"

> Order in 1957 for participation in the development of the R-2, R-5, and R-7. Only a few people got a bonus. . . . But our salary was only about the same as other engineers in other industries.[40]

There is no question that Korolev, in his dominance of the design bureau, had no counterpart in the American space program. The number of roles he performed is simply mind-boggling. Even in the operational phase of the missile and spacecraft launchings, as when he served as one of the capsule communicators for the Gagarin launch, mentioned earlier, he was omnipresent, especially in the early days. One of his former designers, Refat Appazov, remembered that in those times there was no Center for Flight Control. What is today a huge computer-controlled center, chockablock with graphic panels, in Kaliningrad—not unlike the control center at NASA Johnson Center in Houston—was in the late 1950s essentially the brain of Sergei Korolev himself. Appazov remembered how effectively that brain operated during one of the early launches.

> And then the hour of flight came. At the most crucial moments of the flight it became clear that there is a single person who has the best command of the arsenal of knowledge necessary for the implementing command of the flight. This was Sergei Pavlovich Korolev. He knew how to orient himself instantaneously in a complex situation, to involve the necessary people, to formulate fully specific tasks and questions, to unify all efforts, and to concentrate collective thought in the only correct direction. And he did this not only because he bore the greatest responsibility for the outcome of the flight. He really did see, feel, know, guess, and predict better than anyone else, and ultimately made his decisions.[41]

Apart from narrow flight control decisions, however, Korolev had much broader ones to face about how to keep his lunar exploration projects apace with those of the Americans. Those decisions were often constricted because he was without the advantage of the virtually open-ended budget that the United States had—a budget that enabled NASA and the aerospace industry to develop entire new technologies and infrastructure, supported by a microelectronics and computer capability that he could only dream about.

Chart A
Organizational Scheme of Government Rocket Programs of the Soviet Union, circa 1946

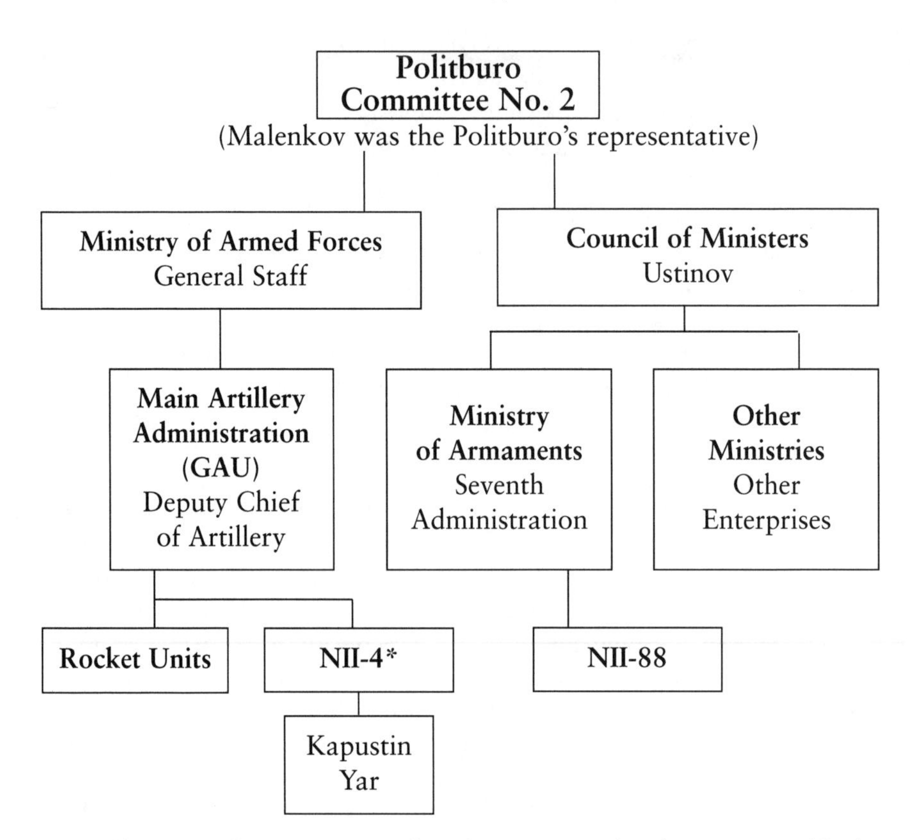

*Scientific Research Institute-4, within the Ministry of Defense, responsible for acceptance, development of testing procedures, technical requirements, etc.

Source: Tarasenko, *Military Aspects* . . . , p. 15

Chart B

Plan of Organization of Space Activities in USSR 1960s to 1980s

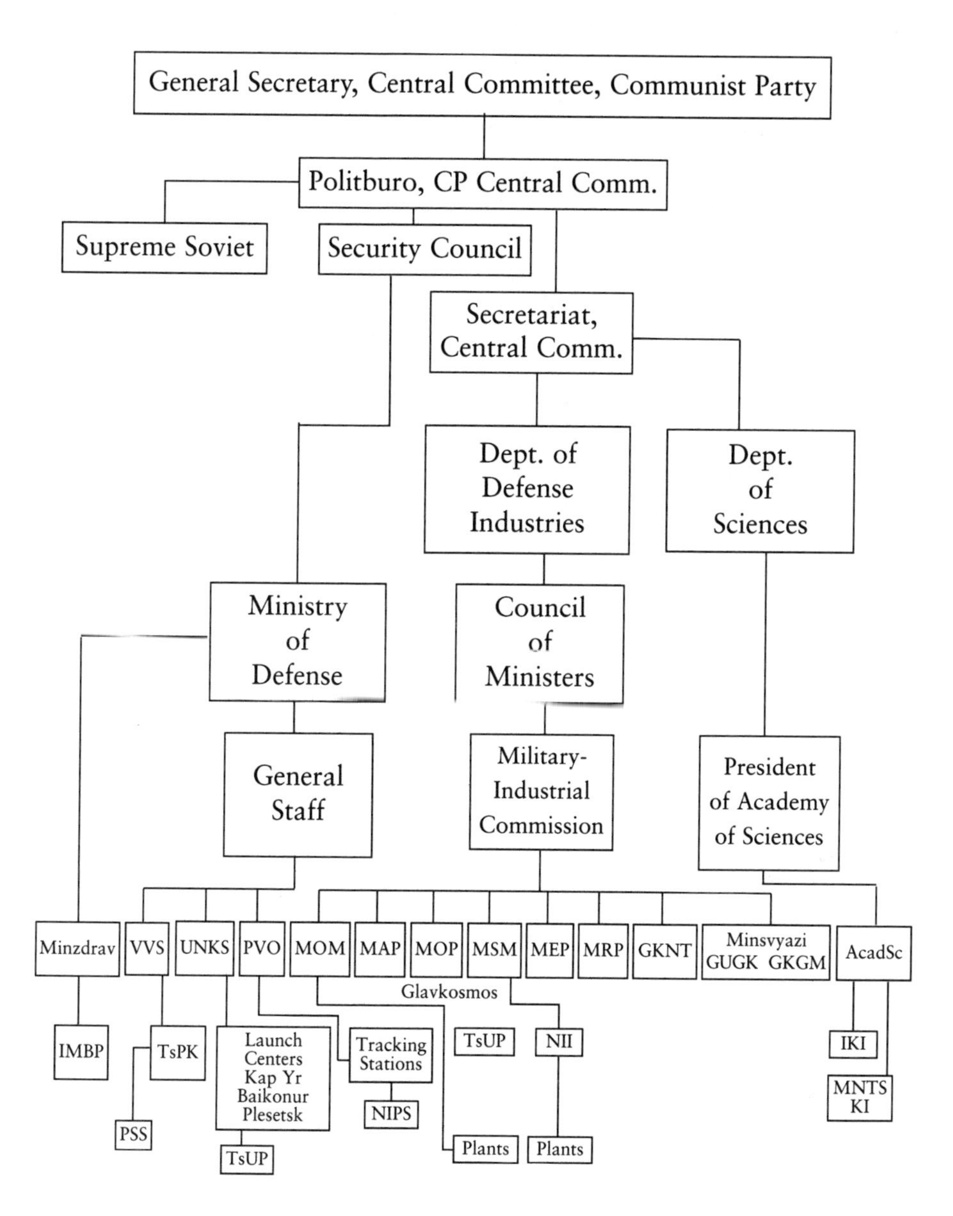

Source: Tarasenko, *Military Aspects . . .*, p. 19

GKGM: Government Committee for Hydrometeorology
GKNT: Govt Committee for Science and Technology
GUGK: Main Administration of Geodesy and Cartography

IMBP: Institute of Medical and Biological Problems
MAP: Ministry of Aviation Industry
MEP: Ministry of Electronic Industry
Minsvyazi: Ministry of Communications
Minzdrav: Ministry of Health
MNTS KI: Interdepartmental Scientific & Technical Council on Space Research
MOM: Ministry of General Machine Building
MOP: Ministry of Defense Industry
MRP: Ministry of Radio Industry
MSM: Ministry of Medium Machine Building
NIP: Ground Measuring and Tracking Stations
PSS: Search and Rescue Service
PVO: Air Defense
TsPK: Cosmonaut Training Center
TsUP: Center of Flight Control
UNKS: Administration of Chief of Space Means
VVS: Air Force

13

THE TECHNOLOGY: SIMPLE BUT RELIABLE

BY AND LARGE, Korolev put the simplest, most reliable technology that would do the job into a spacecraft or rocket and then stuck with it. Shining examples are the space launchers that emerged from the R-7 ICBM design. Those rockets, using upper stages and carrying various names—*Vostok, Molniya, Soyuz*—are still the workhorses for many missions. Over some forty years the rockets using R-7's first stage have had some 1,700 launchings, more than 98 percent of them successful, sending cosmonauts and satellites into orbit, and carrying payloads to the Moon, Venus, and Mars. Since each first stage has five four-chamber rocket packets, it means that more than 34,000 thrust chambers have been manufactured. No wonder the Russians were able to, and still can, routinely trundle the vehicles to Baikonur, and often with as little as two-week spacing, tilt them up, and fire them in rain, snow, or desert heat.

Achieving that record is itself a tribute to the Chief Designer, particularly since he was not able to draw on anywhere near the technology that was available to, or being developed by, the well-funded Americans. For example, the Korolev team suffered grievously from a lack of modern computer capability. "At first we thought of computers as bourgeois false science," Oleg Ivanovsky told me in an interview.[1] "We didn't have any."

Eventually, the Soviets did have computers, but they were primitive, and not everyone had access to them. One who did was Gyorgi Grechko, the cosmonaut who would eventually fly three times on *Soyuz* spacecraft to the *Salyut 4, 6,* and 7 space stations in 1975, 1978, and 1985. Before becoming a cosmonaut he had been an engineer at OKB-1, in the *Sputnik* years, and had direct experience with the computers available at the time. He recounted his experience:

> It was, I guess, 1956, or maybe 1957, we were working with what was then our biggest computer, called BESM [*Bolshaya Electronaya Shchotnaya Mashina*—Large Electronic Calculating Machine]. It was on Profsoyuznaya Street and was operated by the Academy of Sciences. It was built with vacuum tubes and was not very reliable. It sometimes took quite a bit of time to repair. It had been built, I guess, by Academician Ershov, now at the Academy of Sciences in Novosibirsk. There had been competition between several groups for the design of the computer. The computer had three "addresses." Example: two plus

two equals four. There was one cell for the first "two," a second cell for the other "two," and a third for the result. I suppose there are no such computers now but for programming, for use by engineers like us, it was very simple.

There was no computer language, no assembler language, just direct recording. I began to calculate the trajectory for *Sputnik* on electromechanical devices made in Germany by Mercedes and by Rheinmetal. It was very slow, manual, not enough accuracy. For the last part of the trajectory I should have had sines and cosines to the eighth decimal point. There were some tables but they were not very good. To establish one trajectory I needed to calculate many possibilities. Two girls working with us would use the electromechanical devices and tables of sines and cosines, working from 9 to 5. Then two new girls would come in and take over. I would work two shifts. I lived in the office and slept on the table. I would resume my work in the morning. I would compare their calculations. This was hard work because sometimes they would make the same mistakes and I didn't want them to talk to each other.

The computer we used occupied about ten rooms of the size of this hotel room [the hotel room was perhaps 15 by 20 by 9 ft. high]. I guess it was the most advanced computer in the Soviet Union at that time. Maybe the military had another one. At the time, Kurchatov's people [working on nuclear weapons] used it in the daytime and during the night Korolev's people. And for all the rest of Soviet science: maybe five minutes for the Institute of Theoretical Astronomy, maybe half an hour for the chemical industry. Because of all of the electric bulbs, the room was very hot. The cooling system was not strong enough to carry the heat away so that even in the winter time the ventilation to pull in air from outdoors was switched on for twenty-four hours. We worked in our coats. New workers would always try to switch the ventilator off but there was a big sign that said,

THE VENTILATOR IS A FRIEND OF THE LABOR.
LET IT WORK ALWAYS

. . . I was only an engineer. I had chiefs. The chief of my group was Refat [I guess it's Tatar] Fazylovich Appazov. The next higher boss was Svyatoslav Sergeevich Lavrov. The Korolev "lessons" about ballistics of rockets, although they are called the "Korolev lessons," were actually written by Lavrov and Appazov. The first program to calculate trajectories in OKB-l was made by Lavrov and one other person whose name I don't remember. I improved this program and used it to choose the trajectory. Somewhere in my house I have the paper tape of this trajectory. I also have a small piece of the *Sputnik* first stage booster, the R-7.

> The computer operating memory, RAM [Random Access Memory], was not large, I guess about 1,000 words, one word 16 bits. ROM [Read Only Memory] was 3,000 words, each word 16 bits.
>
> Above me in responsibility for calculating trajectories, first for Earth orbits and eventually for the Moon, Venus, and Mars, were Appazov, Lavrov, and then Korolev. Keldysh was a great scientist, while Korolev was a great technician, engineer. In Keldysh's institute was another computer and there was a counterpart to me who also calculated trajectories, which we would then compare. Sasha Platonov was my counterpart. . . .
>
> Keldysh's Institute of Applied Mathematics did scientific work on the basis of the laws of celestial mechanics, applied mechanics for satellites and Zonds. They created equations, and then we calculated the trajectories, using these equations, and we checked each other.[2]

The importance of the Keldysh Institute, to which Grechko refers, and of the scientific leadership of Mstislav Keldysh himself, to the Soviet space program, and to Korolev, was emphasized by many of the space veterans whom I interviewed.

"Before all launches," I was told by Efraim Akim, who worked at the Keldysh Institute in the early days of the space program, and is still there, "we had informal meetings hosted at NII-4 at first, then later at NII 88." At some point a computer complex was installed at NII-88, and it became the brain center. Akim described how the computers were used to make on-the-spot mission decisions:

> The business done was not supposed to be recorded. There were about twenty people who came to these meetings. Besides Korolev, Keldysh, Pashkov, Ustinov, there would be Gyorgi Tulyin, who was first deputy of the Ministry of Armament, Sergei Afanasiev, Chertok, Mishin, sometimes Marov and Okhotsimski, chief designers like Ryazanski (radio technology), Pilyugin (guidance and control), Barmin (ground systems), Viktor Kuznetsov (gyro devices). I was responsible for ballistics and navigation. These were remarkable meetings for me. There was a very creative environment. There were quick decisions, strong expressions from Korolev. . . .
>
> At NII-88, right next door to where we had these meetings was the room where we worked on the software and computers. All of the bosses would move back and forth between the meeting room and the computer room and we resolved many issues right on the spot. There were always problems during development of spacecraft and during the launch-readiness procedures that required decisions, and since the bosses of the military industry were present these decisions could be put into action right away.

An example . . . was during the preparation of the Moon launches. In 1959, when *Luna 1* was launched [on January 2], ballistics and navigation played a key role. The design concept depended on the ballistics. Every time a task was formulated we had to select trajectories, calculate the correction maneuvers, the delta V [velocity change], the launch vehicle capability required, the launch window. After a preliminary investigation we had to report the results to the Commission. When some system failed after launch, ballistics was at fault. Keldysh believed that all of the members of his team should understand all issues—celestial mechanics, physical processes, rocket engine operations.[3]

Grechko recounted an incredible anecdote that reveals not only the consequences of the unavailability of computers in those days, but the degree to which the Korolev design bureau invested significant responsibility in young engineers:

In 1958 we were getting ready to launch *Sputnik 3* from the cosmodrome. I don't like "Baikonur." It was a name invented for the Americans. Baikonur is hundreds of miles from the cosmodrome. It was for your CIA. I like the name Tyuratam, the railway station near the cosmodrome. I was there getting ready for the launch, which was funny because I was a young engineer with only three years experience, and I was the only one from my group at the cosmodrome. Appazov and Lavrov were back in Moscow. . . . The living conditions at the cosmodrome were awful. We lived in railway cars. The temperature at night was maybe 30°C [86°F]. Very hot. Our food was very bad. Not enough refrigeration. Sanitary conditions were awful, too.

Anyway . . . I received new instructions on how to fuel the tanks for the booster, signed by Appazov and Lavrov, and endorsed by Korolev. My signature was never on the cover page, sometimes on the last page. My instructions were to burn the old instructions and use these new ones. . . . The orbital apogee was supposed to be about 1500 kilometers I think. . . . I like to compare things. I learned my English by comparing Russian and English books. So I compared the old with the new instructions. What I was able to determine was that the new instructions called for a change in the quantity of fuel, but the change was in the wrong direction!

Because of the terrible living conditions nobody wanted to stay at the cosmodrome. They would come down just before the launch and leave right after. Only the engineers and Korolev would stay there. But, of course, Korolev had a house, and for us it was like a palace. If you see this house now, it will seem to be a very poor house, but they had hot water, and a restaurant. Korolev was maybe the only General Designer who went to the cosmodrome before the launch and stayed there until after the launch.

So only Korolev was there. . . . I suppose it was two or three days before the scheduled launch. After determining that the instructions were incorrect I went to Korolev and told him that it would be impossible to launch *Sputnik 3* with these instructions, which he had signed. Now in those years, as I told you, he liked to hear "No," but he saw the signatures of Appazov and Lavrov, with whom he had worked for many years. I suppose that 99 percent of the General Designers would have punished me, but Korolev didn't. There was no sign on his face that he disbelieved me, although I'm sure he did.

"What should we do?" he asked.

I told him that if we followed the new instructions we would have a crisis situation some hours before launch. Why? Because the kerosene tank would be full and cut off the fueling process before a proper amount of liquid oxygen was fueled. I could make the calculations which would specify the proper amounts of kerosene and liquid oxygen needed to launch the spacecraft to the proper orbit. He said, "OK, you should make the calculations."

I had no computer, no electromechanical devices, not even an arithmometer (a mechanical calculator), only paper and pencil. I calculated day and night new instructions for fueling the tank. These were very difficult hours for me. We had two staffs at the cosmodrome, one from OKB-1 and another from the military. The military men were mad because they considered the instructions signed by Appazov, Lavrov, and Korolev as if handed down by God. How could this young guy who put some figures on paper with a pencil overrule the bosses?

. . . What I didn't know was that Korolev went directly to the phone —a special phone to protect against your CIA— and he called Appazov and Lavrov and told them that their young engineer said it was impossible to fuel the tanks properly with the new instructions. They checked the instructions and told Korolev that this young engineer was an ambitious man, that he would be punished for disturbing the General Designer. Korolev said not one word to me about this conversation even though, having not believed me before, he must have been much more skeptical. But he didn't summon me to him. He came to my room, where I was doing the calculations. I was multiplying five- and six-digit figures by six-digit figures. It was hard work. My military colleagues, who were sitting next to me, were absolutely silent the whole time because they were sure that I would be punished because I was not right. But they were my colleagues so they felt very sorry for me. Korolev said that he would order me a telephone line to Moscow so that I could explain to my bosses the mistake in the instructions that they had made.

The train from the launch pad to our sleeping quarters was already gone but my colleagues stayed with me out of friendship, feeling certain that I would soon be in deep trouble.

So I got on the phone and told the bosses that this, this, and this were not right. Lavrov said, "Gyorgi Mikhailovich, what do you not understand?"

I told him.

"In an hour we will explain your mistake."

I sat at the phone with my military friends in silence. An hour went by, then two hours, then three, and they began to realize that I was right. Somebody broke out some "technical alcohol," and we began to celebrate. After more than three hours Lavrov called and said, "Gyorgi Mikhailovich, maybe we made a little mistake."

I told Korolev that they admitted that they had made this mistake and he said, "Go ahead and make your calculations for the new instructions."

He later told Appazov that he should "go to Baikonur and Grechko will be your boss for this launch." It was hell for Appazov and Lavrov, and hell for me. Appazov said, "Why didn't you tell us first?" I, a young engineer, had no right to call Moscow. I could only go to Korolev.

My reward from Korolev was a small piece of paper which authorized me to get an apartment. Until then I lived in a dormitory. But Korolev's secretary hid the paper because there were many powerful men ahead of me in line for an apartment. I didn't see the paper until years later.[4]

It was not only the computer limitations that hampered the Korolev team in the early years, but the electronics as well. Electronic components were, by and large, bulky and not very efficient. However, they usually worked, by dint of meticulous inspection, although they extracted a significant weight penalty. Boris Gubanov, one of the key designers from those days, recalls, however, that, "We had very good quality control. We selected one out of every hundred electronic components. Such a rejection rate would be unacceptable for civilian products due to the cost."[5]

As a consequence, systems reliability was not bad. The high rejection rates, of possibly inferior components, compensated for the lack of sophisticated quality-control procedures. The bottom line was that the failure of early Soviet missions was no worse than that of the Americans, who were working with much more modern technology and management systems.

There were, however, areas of very advanced technology that Korolev pursued as well. Pavel Vorobiev, for example, told me that when he came to OKB-1 in 1956 he sought out new areas and was told that he could work on nuclear propulsion. But after about three years, "I think it was 1959 or 1960," he said, the project was essentially stopped. "It was never cancelled but it was no longer an active program, although Kurchatov [the head of the nuclear weapons and nuclear energy programs] kept asking Korolev how we were doing."[6]

Crucially important factors in the success of Korolev's programs were the deep dedication and motivation of the workers in the manufacturing departments, who were sometimes given bonuses on the spot for high-quality work. And there was always a sense of urgency, no doubt communicated through the organization by the boss himself. In the case of the first *Sputnik*, for example, "the plant began manufacture of the body parts . . . prior to the preparation of a full set of blueprints," recalled Mikhail Kavyzin, one of the designers. He wrote:

> The casing of the body caused much worry and concern. The half-shell casings were made by stamping with many reductions. It took great skill on the part of the workers and technologists to get rid of the defects, the corrugations . . . typical for deep drawing. Much effort was required also for the preparation of the polished surface [to get the proper] . . . coefficients of absorption and reflection. It was impossible to allow even the slightest scratch.[7]

Korolev personally intervened at virtually every point in the manufacture. And while he was conservative, in that he stuck to proven technology on his flight vehicles, he was progressive in adopting innovative manufacturing techniques.

"How can we weld the sputnik by hand?" he asked exasperatedly. "We need automatic welding, and nothing but. The seam must be even, smooth, strong, and most importantly, it must ensure complete air tightness. . . . The old methods of checking air tightness are no good. Go to the specialized institute, get some vacuum specialists. . . . "[8]

NASA engineers and contractors, accustomed to impeccably pure clean rooms where spacecraft are handled like surgical instruments, will appreciate how difficult it must have been for Korolev to educate his entire team on the importance of a lint-free environment.

"In three days everything here must shine," Kavyzin recalled Korolev's admonishment while readying the sphere that became *Sputnik 1.*

"Hang white shades on the windows, dress everyone who works here in white coats and gloves. Paint the stand under the sputnik white and line the pedestal with velvet."

Those speeches, and Korolev's singular vigilance over every step of the process, had salutary effect, as Kavyzin remembered: "He assumed management not only from his office, but also to a significant degree at the work stations of the executors. . . . The building was transformed. We developed automatic welding. A new type of leak detector was developed to control air tightness."

The workers subjected the *Sputnik 1* engineering test model, which would itself fly later as *Sputnik 2,* to repeated dockings and separations. They made sure that the pneumatic locks operated reliably, that the nose deflector shield separated, and that the antennae came out. When *Sputnik 1* was flown to Baikonur, the pre-launch process was put into the hands of veterans of the ICBM launches.[9]

There are repeated examples of Korolev persuading bright young engineers and scientists to work at OKB-1 on practical manufacturing techniques. One such scientist was Alexei Lokotilov, who left the Academy of Sciences' Institute of Physical Chemistry, just after defending his dissertation in 1956, to head the chemical treatment and coatings plant. Uncertain about the decision he had made, Lokotilov was reassured by Korolev: "Young man, don't be upset. We are faced with serious problems. Soon we will perform work directly on space objects. So go ahead and work, think, and experiment a bit, but do so fruitfully, to the point."[10]

In his first year at OKB, Lokotilov—with Korolev's backing against the opposition of the chief metallurgist—came up with an innovation in chemical treatment of the toroidal hydrogen peroxide tanks for R-7, cutting the treatment time from forty-eight hours to twenty-four hours.

"Sergei Pavlovich devoted great attention to our shop, delved into all questions of technology, and energetically supported all the newest developments aimed at reducing the weight and increasing the strength and reliability of the assemblies," said Lokotilov, who claims that their practice of deep anodizing aluminum alloys for R-7 was performed for the first time in the world.

Asked to comment on that claim, Robert Schwinghamer, who headed materials and structures work on the *Saturn* launch vehicle at NASA Marshall Space Flight Center, said, "We didn't do anodizing [a process for producing a hard, brittle surface that protects against corrosion] because we were afraid that the anodizing process might break loose some of the aluminum substrate. The Russians might very well have been first. Their practice—I've been to Russia several times recently— was to try new things a couple of times and, if they worked, keep doing them. They didn't bother to accumulate statistical experience to justify a process the way we did."[11]

Sometimes the unavailability of materials prompted experimentation—with Korolev himself taking the lead. To make the huge bottoms of the fuel tanks, according to Lokotilov, " . . . we needed aluminum rolled sheet stock 3 meters in width, but domestic industry did not produce it. We decided to weld the tank bottoms out of two halves, and before stamping [it was necessary] to remove the excess metal on the weld seam flush with the primary metal." In that area, however, the strength was impaired, so "Sergei Pavlovich suggested the idea of making the weld seam of equal strength with the primary metal by means of chemical cutting (etching) of the latter." The result was a savings of 5 kilograms, or 8 to 10 percent, on each tank bottom.[12]

It is difficult to imagine one of the American space leaders, say Eberhard Rees at the NASA Marshall Space Flight Center in Alabama, exercising the close supervision over manufacturing that Korolev did. "It seems to me," remembered Gyorgi Korytov, deputy chief of the main assembly shop, "that he loved the general assembly shop like his own child. . . . Sergei Pavlovich could come early in the morning, and then again at midnight. He could drop in on the welders several times even on his day off."[13]

When things did not go right, Korolev could go into apoplexy.

"I will never forget the assembly of the first sputnik," wrote Korytov.

> That was in August 1957. Our shop was assembling the booster rocket. On that day we were leaving work very early, in our understanding—at 6:00 P.M. We had just come out of the shop doorway, when we saw the red Moskvich, from which Sergei Pavlovich and the head designer were getting out. . . . "Go on home, I don't need anyone," he said good naturedly. Having walked off a little, I nevertheless decided to return. I walked into the shop hall and saw: Sergei Pavlovich . . . conversing in a lively manner with the leading designer. Seeing me, he called me over. This was a totally different person. . . . In a sharp tone he called me by my last name, and not by my name and patronymic (I knew that was a bad sign) and asked: "Where are the containers?" I answered that they weren't ready, and would be delivered tomorrow morning. . . . Sergei Pavlovich grabbed me sharply by the arm. . . . We went out and saw the containers at the gates of our shop. . . . I broke into a cold sweat. He looked at me and asked quietly: "Are you a party member?"
>
> "Yes, you know that."
>
> I had seen him before many times in an agitated state, but I had never seen him like this. He suggested to the plant director and the party committee secretary, whom he hastily summoned to his office, to remove me and the chief of the containers shop from work and to exclude us from the party.

"These are not shop chiefs, but bad officials . . . ," he said. "They do not understand that the launch of the first artificial satellite is the start of the conquest of space. . . . "

As he had on other occasions, after blowing his top, Korolev, the next day, was calm again. "He patted me on the shoulder," said Korytov, "and smiled in a friendly manner: 'You really got it yesterday. I could have fired you!' "[14]

That sort of quixotic behavior must surely have been unnerving to his staff, as when he would even bypass a department chief to achieve a manufacturing improvement.

"In 1963," recalled Vladimir Zudanov, who worked in the experimental assembly plant, "we were making the craft for the flights to Venus and Mars." The spacecraft was seriously overweight, so Korolev announced, "I will pay a premium for one kilogram of reduced weight."

Irked by the lighthearted reaction of the chief of the design section he said, "You don't know how to earn money. Let's go to the mechanical shop, to the workers." He then drew a sketch for a milling machine operator and asked if he could duplicate it.

"Yes, I can," was the response. Korolev then ordered the shop chief to give the operator a bonus based on the weight of the filings from the material saved.[15]

"He was always at the shops," said Vakhtang Vakhtangov, chief engineer of the experimental plant. "He knew all the supervisors, section chiefs and very many of the foremen and workers . . . the last names, first names and patronymics of the best copper workers, the best welders and assemblers."[16]

One of Korolev's long-time deputies, Sergei Kryukov, emphasized the degree to which the Chief Designer relied on illustrations for carrying out complex designs. In this modern age of computer-aided design and manufacturing it seems nostalgic that Korolev, forty years ago, was issuing meticulous "notes" on what should be illustrated, and exactly how. Kryukov quoted from one document specifying what the Chief Designer wanted:

> Basic overall view (or several of them);
> Data tables (basic data, comparative data, etc.);
> Technological diagrams (may be skeletal), fitting and assemblies, etc.
> Skeletal diagrams of ground structures;

He even specified how big the drawings should be and how prominent the lines and the numbers! "All this should be large; all lines 4–5 mm thick; height of numbers 30 mm, etc."

He would specify the need for scale mock-ups: "mock-up of product to 1:100 scale (lightweight, plastic or paper, but high quality); mock-up of elements for transporting and welding, assembly; mock-up of launch or part of launch."

The notes would go to specific individuals, along with longhand remarks, phrased in, as Kryukov described it, "delicate form," such as "it would be very nice if S. O. Okhapkin and his associates take on this work." The notes might have been phrased delicately, but they specified cruel deadlines. Although the note that Kryukov cited as an example called for a "program of work . . . to last for many years" it demanded an "outline" in two days.[17]

While Korolev's attention to practical manufacturing techniques was scrupulous, he also paid careful heed to the need for fundamental research. He realized, in the early 1960s, for example, that such spacecraft as the first *Kosmos* satellites, the *Molniya* communications satellite, and the interplanetary craft, would be operating for long periods of time in an arena—cosmic space—which was quite different from Earth. How would their mechanisms respond to the space environment? There was much discussion of the subject by physicists and vacuum specialists, recalled Vladimir Syromiatnikov, one of the engineers who worked on those spacecraft. "Some . . . even predicted cold pressure welding of friction and contact surfaces. . . . " So Korolev went to Academician Anatoli Blagonravov at the Institute of Machine Science and asked for experiments in deep vacuum, using actual satellites and spacecraft. " . . . hundreds of large collectives of specialists took part" in these tests, said Syromiatnikov, who added: "In these years, I had occasion to work with many leading scientists and engineers in space technology. However, Korolev had the greatest influence on me."[18]

There were other influences, however, that deeply affected Korolev himself, and OKB-1 generally: the Communist Party apparatus, and the fear and secrecy that so thoroughly pervaded the working environment and daily lives of every staff member.

14

THE PARTY, THE PARANOIA

"EXTERNALLY KOROLEV was a strong Communist," Boris Chertok told me. "He had to be before his rehabilitation. 'Watch yourself,' Korolev would say, 'Remember the places we were before [the gulag and sharashka prisons].' "

Chertok, a Jew, must have been very good at watching himself. He is perhaps the quintessential survivor-pragmatist in Russian space annals. As this book is being written he is in his eighties, and still going to work regularly at RSC Energia.[1]

He had worked with Korolev from the days in Germany just after the war. In 1950 he was demoted, as were other Jews in OKB-1, during a period of particularly intense anti-Semitism in the USSR. It is not known how Korolev reacted to the demotions, although no one, in my many interviews, accused the Chief Designer, himself, of personal prejudice against Jews.

"Korolev was very able," said Chertok, "in using high-ranking officials to support his plans. Ustinov would complain that it was difficult to work with Korolev because he [Ustinov] liked working with people who were obedient and Korolev wasn't. All the other general designers followed orders."[2]

A fundamental step for Korolev was to become a member of the Party. As early as 1948 he had begun to study Party history and theory in an evening course on Marxism and Leninism at the design bureau. "Because he was so busy," recalled Bella Apartseva, who gave the course, "Sergei Pavlovich often had to miss the lessons, and the partkom asked me to help him in his studies . . . we had to study at different times: early in the morning or very late at night. But he always found time for study. He passed his exams and graduated. . . . "[3] It would not be until four years later, however, that Korolev would actually join the Party, at age 45, in 1952. As Gyorgi Vetrov put it:

> It was extremely unusual for a former prisoner to become a member of the Party. Tupolev never became a member. . . . I was there when he was accepted. In the procedure he was asked about his imprisonment: "What were you sentenced for?"
>
> He became sad when he heard that question and said, "My confession was annulled."
>
> This event took place at a meeting open only to Party members at the OKB. It was in February 1952. It was in the big hall used for routine work on drafting boards by designers. There were more than a

hundred people present, representing essentially all of the Party members at OKB. Maybe there was special interest because the Chief Designer was the person being inducted. Svyatoslav Sergeevich Lavrov made the only speech. Now he is director of the Institute of Theoretical Astronomy in St. Petersburg. Then he was deputy chief of the design department, working on ballistics, and had the job of Party leader in his OKB department. Lavrov's speech included a mild criticism of Korolev's tough management style. Criticism was a normal part of the procedure for accepting members of the Party and, in fact, the criticism was in order because Korolev demanded much of his deputies, insisting that they do their best under very strict conditions, and sometimes his demands were unrealistic. It was about a five-minute speech. Lavrov was one of Korolev's favorites, however, although they were not what you would call friends. But the criticism did not affect their relations.

About one third of the OKB engineers were members of the Communist Party. There was a queue of engineers requesting membership. Not everyone was accepted. I, myself, became a Party candidate in 1951 and a member one year later. My job at the time was as deputy to Lavrov in a special division on stability and flight dynamics.

The Party policy was to attract people who could do important things. Korolev had no choice but to join, because the Party could take actions which would be detrimental to one's career. For example one of Korolev's deputies—Konstantin Ivanovich Trunov—who had been arrested before the war and was in the Kazan sharaga with Korolev, was laid off as a result of the Party's initiative. The Party had the authority to fire people without Korolev's approval. So Korolev felt he had to join the Party to protect himself. He also had to have some successes in performing the tasks of the Party to get into their good graces. Korolev's qualifying success was at the end of 1951, when the 600-kilometer range R-2 was accepted for military operations.

Korolev knew the rules of the game. He demonstrated respect for the Central Committee of the CPSU [Communist Party of the Soviet Union]. At all of the CPSU meetings at OKB he would end the meeting with "Under the leadership of the Central Committee of the Communist Party we will fulfill our task."

He attended all of the important meetings, but, of course, he spent a lot of time at the test range.

Korolev would outline the tasks to be undertaken by his design bureau and then get these tasks adopted by the Party conference. Having control of the decisions made by the Party was very important and enabled him to use the Party influence in achieving his goals. The agendas for all of these meetings were prepared in advance so he could set them according to his own objectives.

> He had the budgetary authority [for his projects] as Chief Designer, but he had to obey the decisions of the Party. The OKB would draft proposals for decrees of both the CPSU and the Council of Ministers. Then the CPSU and the Council would accept the proposals and adopt the decrees to go ahead. There was a hierarchy: starting with a defense group in the Central Committee of the CPSU and proceeding up to the highest level of the Politburo. It is important to know that the decrees of the CPSU and Council of Ministers assigned personal responsibility through the Ministry of General Machine Building (MOM) to Korolev. So Korolev ended up having the personal responsibility.[4]

Tactically, however, Vsyevolod Avduyevsky told me, it proved most advantageous for Korolev, when requesting rubles for his projects from "the highest levels of the Communist Party, to get Mstislav Keldysh to cosign his letters."[5]

Sometimes members of Korolev's technical staff were obligated to take on substantive Party jobs, as did Lavrov. Another was Oleg Ivanovsky, who had worked with Korolev at OKB-l, and was in charge of the production, testing, and launch preparations for the first Sputniks. Ivanovsky, like many of Korolev's engineers, was a veteran of the Great Patriotic War.

> I'm a Muscovite. In 1940 I was recruited into the army as a guard on the western Polish frontier, near Peremyshl. An event that became a milestone in my life occurred on April 12, 1941. On that day I was ordered to leave the frontier service and go to Sergeant school. The war began on June 22, and nobody from that post survived the German invasion. I never thought that April 12 would be symbolic in my life in some other way, but of course that was the date when Yuri Gagarin became the first man in space, and I was one of those who helped design the spacecraft that put him there.
>
> I was badly injured towards the end of the war in Czechoslovakia. I made it to Moscow, though, for the victory parade. There [he points to a clothes rack] is my Red Army uniform right next to the hat I wore on the frontier. I served most of the war in the cavalry.
>
> I was discharged in 1946 and six months later, at the beginning of 1947, I went to work at the Korolev design bureau. I had been a master of horses in the cavalry so I had no qualifications for the work. I had graduated from high school, not university, so my first job was to push boxes around, weld wires, I had the lowest "black jobs." The first man to give me greater responsibility was Chertok. He was my department chief.
>
> Two years later I graduated from the Moscow Energetics Institute with an engineering diploma after being allowed to study full time. I

> went back to the OKB as an engineer, then became a senior engineer. Unfortunately, I was pressed into serving as Secretary of the Party for OKB-1, full time!
>
> I was a technician and simultaneously I was Secretary of the Party organization. It was there that I first met Korolev. It was an unpleasant situation because we had an argument. We had a tradition of announcing the results of a Socialist competition. As Secretary of the Party I wanted to uphold the honor of my department. Yangel had just become chief of the department and I decided to help him by supporting a good rating for its production record. Korolev disagreed with me on the rating and I protested. Korolev went to Yangel and said, "Who is this guy who objects to my recommendation?"
>
> It turned out to be nothing serious, but I think Korolev remembered me for standing up for my position. . . . This was 1955 or 1956 and I worked full time for the Party until the middle of 1957, when we started active work on *Sputnik*. There was then a Party reorganization and I took the opportunity to relinquish my Party duties and go back to engineering on *Sputnik*.[6]

Later Ivanovsky was assigned to the Kremlin to work on the Praesidium of the Council of Ministers.

> I was head of the space department, responsible for keeping track of the development of all space technology in the USSR. . . . My job was to prepare all of the decrees for the government about space, and the Communist Party decrees about implementation of the work of all space enterprises—who, what, when the spacecraft should be made, which Bureaus would get the jobs.[7]

Sergei Pavlovich was a member of a small cell in the Technical Information Department at OKB-1, Vladimir Shevalyov told me, and paid his Party dues faithfully, but he was not necessarily respectful of the Party's edicts, even though his close colleague, Nikolai Pilyugin, called him a "crystal clear Communist."[8] Yuri Orieshkin, who was Korolev's general assistant for the last six years of his life, said that

> Korolev sometimes complained that Stalin, Khrushchev, and Brezhnev had no idea of the problems of technology development and sometimes tried to force you to be driven by the opinions of people who didn't understand the complexities, but sometimes he was obliged to obey their orders.[9]

Apparently, however, Korolev did put in time studying the Party's ideology. His long-time technical secretary, Antonina Zlotnikova, reported that she would sometimes find him engrossed, in his office at night, in reading one of the books from his own library of the complete works of Lenin.[10]

Like many others he admired Stalin in spite of the treatment he had received from the Generalissimo's minions. His daughter, Natasha, told me that her father always "thought that Stalin had no knowledge of the arrests during the purges. He thought that Stalin's death was a tragedy. He had respected him, visited him twice, said that he was very clever in asking questions, seemed well-prepared for their discussions about ballistic missile development."[11]

Vassily Mishin remembered accompanying his boss to one of the meetings with Stalin: "It was in 1947 or 1948. I can remember that he walked around the room smoking his pipe, breaking *papyrosi* [cigarettes] and putting the tobacco in. There were about a hundred people there—marshals, ministers, officials of the Party. It was in the building of the Central Committee and the subject was liquid rockets. Stalin was a strong supporter of their development. The U.S. had naval bases in Europe, owned the A-bomb. They didn't really need the ICBM. They could reach the USSR with bombers."[12] For those reasons Korolev, with his plans for long-range ballistic missiles, became Stalin's fair-haired boy.

On one occasion, however, Korolev witnessed bizarre behavior by the Party head. Vladimir Shevalyov, the Chief Designer's English-language expert, recalled:

> Once, at lunch at his table in Baikonur, he [Korolev] told us about an invitation he had to the Kremlin. The members of the Politburo were there. Malenkov was flattering Stalin. Stalin asked Korolev what he thought about UFOs, that Korolev should study them, that he had some foreign books on the subject and he would like Korolev to take them for a few days and give him a report on what he thought.
>
> "I went to my office and gave the books to the nice girls [the translators], caressed their buttocks, and reported back to Stalin that there were no UFOs and Stalin agreed with me."[13]

There were ritual attitudes towards America that had to be adhered to at Party gatherings. Shevalyov said:

> You could admire the USA and its accomplishments privately, to your close friends or your family, but you had to criticize America at Party meetings. I never heard Korolev criticize America though. I remember when John Kennedy was shot. I heard the news on the Voice of America and I phoned Korolev at his home to tell him. His reaction was sorrow in spite of the fact that his phone was surely tapped.
>
> "You get used to it," he would say of the phone tapping.[14]

The Party had to be dealt with not only to get rubles for OKB-1 projects but for decisions on social matters—like housing for the workers. In the post-War Soviet Union, housing, of course, was as important as it was scarce. The assignment of apartments required the signatures of not only the Chief Designer but the head of a workers' committee and the Party Secretary. One of the Korolev bureau veterans of those years, Pavel Kostyukevich—a physicist who did his turn at Party chores—remembered how difficult it was for Korolev when he crossed the system dealing with such matters.

> In November, 1953, I was elected deputy secretary of the Institute's (NII-88) Party committee. . . . A resolution defining a new rocket provided for the construction of an additional residential building. Under the conditions of the acute housing shortage at the time, fair distribution of apartments in this building took on a vital character.
>
> Korolev, believing that the collective developing the rocket was only the design bureau which he headed up, prepared a list for assigning units in the building to design bureau workers, without coordinating this with anyone at the Institute, the Party committee, and the trade union committee. They felt that such distribution of housing was unfair and rescinded it, making the decision to grant housing to all the subsections participating in the development of the new rocket, [including] the scientific sections, and the experimental plant.
>
> A stormy expanded meeting of the Party committee was held on this subject. It stressed the importance which the distribution of housing had acquired, specifically for the plant workers by whose hand the rocket was being created in accordance with the design bureau documentation. Consideration was also given to the fact that the adopted decision forced Korolev to rescind a number of these "promissory notes" to the workers of the design bureau. To the credit of Sergei Pavlovich, he, though painfully, nevertheless did survive the cancellation of his decision and subsequently himself performed the work at the design bureau on reviewing the list.[15]

Often it was the Politburo, a group so influential that Hedrick Smith, in 1990, called it the "holy of holies, pinnacle of Soviet power,"[16] that harassed Korolev most. It was that body, for example, that continued to insist on his anonymity. In 1960, when Nikita Khrushchev came to the UN in New York on the trip that included his infamous shoe-pounding episode, "Korolev was eager to accompany him but the Politburo said no, he must maintain his anonymity. I remember," Shevalyov said, "that Korolev got a phone call from the Politburo telling him that a launch must be made, that Korolev should do his best to make Khrushchev a hero in the U.S." Under great pressure, he tried twice to send probes to Mars, but both launches, one on October 10 and one on October 14, failed.

Korolev was careful to keep the Party leaders briefed on his developments, aware that they needed patient education to appreciate what might have seemed far-fetched concepts. As early as the 20th Party Congress, in February 1956, the year prior to the launches of both the first ICBM and the first *Sputnik,* he hosted a two-day session at OKB-1 for the delegates. " . . . we set up examples of all the previously developed and mastered products," recalled Kostyukevich, " . . . there were individual posters and tables . . . on transport carts was the first mockup of the rocket [the R-7 ICBM]—the central block with the 'sides' [boosters], and the nose section. . . . "

The next day Korolev had Alexei Isaev carry out a demonstration firing of one of R-7's control rocket engines. "In short," wrote Kostyukevich, "the meeting participants left convinced of the State importance of . . . the task, ready to enthusiastically . . . do everything in their power for the success of the common cause. The on-site Party organizations took this work under their special control."[17]

Certainly one of the reasons that Korolev was so dutiful in his relations with the Party was because of its influence on the living conditions, and the social and cultural amenities, afforded his workers. They weren't much, to put it mildly. Hundreds of his staff members were jammed into tiny apartments as well as some "100 barracks [which were] on the balance sheet of the enterprise," recalled Party Secretary Yevgeni Tumovsky.[18] Korolev had to devote a lot of time, which might have been devoted to technology management, to getting Party support for "the construction of housing, of a Palace of Culture, a stadium, children's institutions and recreational facilities. . . . "[19]

Korolev had to deal with three or four Party meetings in his design bureau per month. He was so busy, of course, that he could not always attend, but Party officials would save up critical issues for when he could be there. It must have riled him to have to answer sometimes to alleged management weaknesses, but apparently there were times when serious infractions came to his attention, as when " . . . new equipment . . . was lying out in the open and not being used." Korolev "exploded," recalled Alexandra Pustovoytenko, a design bureau staffer, and said, "That isn't just mismanagement, that's barbarism." More often, though, the subjects were prosaic, albeit important to the health of the enterprise. OKB-1, and Korolev himself, on many occasions, worked through the Party with local city officials on the problem of supplying sufficient water (he got the bureau a vitally needed extra 50,000 cubic meters of water per day) and even food (the design bureau actually constructed a storage building for 1,500 tons of potatoes and 500 tons of cabbage).[20]

There remains a trove of information on the involvement of the Party in decisions about space and missile development. Despite my forays into the Party archives, I was unable to get to more than a few of the documents. Gyorgi Vetrov told me that he "spent a full year gaining access to the archives" and expects to reveal some of what he found in a book he hopes to publish. "Drafts of the decrees of the Politburo and the Party," said Vetrov, "are also in the archives of RSC Energia. There is a huge amount of information still to be examined."[21]

One pervasive characteristic of the Party, which affected all of Soviet society, but most of all the sensitive military programs, was a smothering paranoia. Born in tsarist times, carried into the early days of the Revolution, and brought to an all-time high in Stalin's purges of the 1930s, the fear of government, apparatchiks, neighbors, Westerners, and fellow workers threatened such paralysis in the performance of daily living that it became the norm for Soviet citizens to keep the lowest possible profiles. But this had devastating results.

The psychological toll was certainly heavy on engineers and scientists who ought to have been able to test theories, try unorthodox designs, and—above all—communicate freely with their colleagues, using rational avenues for protecting classified information, without worrying about their necks.

As Nikolai Sheremetyevsky, a designer in OKB-1, wrote,

> . . . much of the work we did with S. P. Korolev's organization was labelled secret for reasons that were totally incomprehensible to me. In the first place, we were very far behind the Americans there. We weren't on the Moon, hadn't performed such a circus trick like landing people there and then taking them back up off the surface. Those were all control operations that we couldn't even dream of. And for that reason, the "secrecy" we had really got in the way of our integrating into world science and technology.[22]

Even the names of the Soviet missile and space organizations were secret, although if they had been publicly announced it would have been difficult to tell their functions anyway, since they had such innocuous designations as "OKB-1" (Korolev's bureau) or "OKB-456" (Glushko's). These organizations, and many others throughout the defense field, were known as "mail boxes" since they could be communicated with only through a postal address, not a specific location.

Even within a mail box there was difficulty communicating. Vladimir Karrask, now a deputy general designer at the Khrunichev enterprise in Moscow, remarked at a 1991 conference that, " . . . this is the first time in my life that I've taken part in a round table. You probably know, don't you, that those who work in 'mailboxes' have no idea of what goes on in the neighbouring room."[23] In 1958 Alexander Kostrov, then one of Korolev's young designers, did not even know why a third stage was being built for the R-7 rocket.[24]

Americans used to seeing huge signs designating aerospace plants as "Lockheed-Martin Corporation," "Boeing Airplane Company," and "McDonnell-Douglas Astronautics" are nonplused when today, though the secrecy barrier has been dropped, they approach Russian facilities that show no exterior identities at all. In Korolev's day even the gateway entrances were supersecret. One of the space journalists, V. Golovachev of *Trud,* recalled his first visit to the Korolev design bureau in Kaliningrad in 1965, almost twenty years after the plant began producing space vehicles. Korolev himself had to give Golovachev instructions on how to find the place over the *vertushka,* the government communications line that, as he put it, "practically precluded any eavesdropping. . . . I got the feeling that, even on this telephone, Korolev was speaking with caution—for many decades he had been surrounded by an atmosphere of the strictest secrecy." Golovachev recalled that he was told by the Chief Designer

> When you reach our area, there'll be a fork in the road. Don't turn right, just continue straight ahead for about half a kilometer. There will be an iron gate on the right, in a brick wall. Drive up to it, and it will open automatically. After you pass through, the gate will close.[25]

In 1990, some twenty-five years after Golovachev's experience, I and several colleagues were among the first Americans to go through that same gate, after our taxi driver tried vainly for a half hour to find it. The gate is still known as Korolev's "mousetrap." Our hosts made much of the fact that we were certainly the first group of *inostranetsii* (foreigners) to have our picture taken on the front steps of the design bureau.

As already noted, the people who worked in such organizations as the OKB-1 were all strictly anonymous—not only Sergei Pavlovich Korolev himself, but the key people under him. A fortunate few were allowed to be identified when they were picked to represent the USSR at foreign meetings, but even then they were forced to participate in a kind of masquerade. For example, for some thirty years—from the mid-1950s until Mikhail Gorbachev declared a policy of glasnost in 1985—modestly sized delegations of Soviet space specialists attended technical meetings in foreign countries wearing meaningless identification badges. "USSR Academy of Sciences" was often used, although the badge-wearer was rarely an Academician. "Intercosmos," an organization dedicated to space cooperation between Communist-bloc countries, was another favorite. Rarely in these delegations were there engineers with competence in rocket and space technology. For the most part they would be cosmic ray physicists, astronomers, experts in space medicine, astrodynamics, gas dynamics. They would give papers, often of high technical quality, but on arcane subjects. Also in the delegation, besides interpreters, were KGB men, whose behavior betrayed their calling. They could be seen picking up the technology papers, company brochures, and whatever other technical literature might be available, especially if it was free.

Americans formed friendships with some of the technical people, who must have been delighted to have gotten through the KGB screening process that allowed them to qualify for attendance. It was virtually impossible, though, to get technological, as contrasted with scientific, information out of these wary people. One could sense their discomfort when questioned about space projects that the USSR was undertaking, or rumored to be undertaking. There would usually be one plenary Soviet session, featuring cosmonauts, at which a presentation of the latest space spectacular, with film footage, would be made before a packed audience of American, European, Chinese, and Japanese space professionals eager to get even a glimpse of what the Soviets were up to. But performance data were usually hard to come by, and getting copies of the films and charts was out of the question. Often one would see members of the audience trying to photograph off the screen the Soviet slides and charts, which were usually handwritten and difficult to decipher.

This communications stonewall was, of course, punishing to the Soviet professionals. Oleg Gazenko, the space medicine expert, told a group of visiting Americans in 1994 that, "Between 1950 and 1960 we weren't allowed to publish anything. We could not even use our own names on committees. Not much, therefore, is known abroad about our capabilities, although we have some seventy volumes of research reports."[26]

I formed a special friendship with a space physicist who accompanied the first Soviet space exhibit to the United States at the New York Coliseum in 1959, and we have maintained our friendship. I soon learned, however, that I would not be able to glean anything from him about upcoming projects. At one meeting in Germany in the 1970s, over a beer, I asked, naively, if it were true that the Soviets were working on a space shuttle. I got an evasive answer. Perhaps emboldened by a second beer I said, "Look, Yuri, I know that there are certain things you tell me that aren't the truth, and you know that I know they aren't the truth, so I guess I won't ask you any more such questions." That provoked a relieved smile.

Anyone with the vaguest German connection in the USSR was in automatic trouble. Former cosmonaut Gyorgi Grechko had difficulty enrolling in the Leningrad Technical Institute when he was a student because he had been captured by the Germans as a child. "On my curriculum vitae, underlined in red, was the fact that I had been taken by German troops when I was ten years old when they captured Chernigov in Ukraine. . . . Even though I was later given a certificate from the Chernigov Communist Party that I was innocent of any German connection I was asked about this."[27]

It was not just foreigners that the Soviets were afraid of. They were often just as leery of fellow citizens whose questions might call for answers that could get them in trouble. Sergei Khrushchev, son of Nikita, has unusual credentials justifying his reticence during his career as a missile designer with Vladimir Chelomei's bureau. His comment: "Many years of working in a rocket design bureau and in a scientific research institute taught me to fill my out box with crafty responses to complicated incoming requests."[28] He might have added that his father's experiences taught him the art of craftiness the hard way.

As a form of compensation for their anonymity, the top engineers and scientists in such crucial fields as space and atomic weapons were given remarkable perks. Yuli Khariton and Yuri Smirnov, star physicists in the atomic energy field, wrote in 1993 of a decree issued in 1949, two months after the explosion of the first Soviet atomic bomb, which provided that "several particularly distinguished participants in the research, led by Kurchatov, were named 'Hero of Socialist Labor,' awarded bonuses, and given Zis-110 or Pobeda cars, the title of the Stalin Prize Laureate of the First Degree, and dachas. Their children were to receive a free education in any educational establishment at state expense. The recipients and their wives were awarded free and unlimited transportation by rail, air, and water, anywhere in the Soviet Union, for as long as they lived." The last privilege, reported Khariton and Smirnov, was revoked by Khrushchev.[29]

It is not surprising that the paranoia that so deeply penetrated the Soviet psyche produced, as a byproduct, a rich lode of black humor. Khariton tells a story about Lavrenti Beria, who had Kremlin responsibility for both the nuclear and rocket programs. " . . . when deciding which prizes to award, [he] used a simple principle. . . . Those who in case of failure would have been shot were to receive the title of Hero; those who would have been given the maximum prison term were to be awarded the Order of Lenin, and so on down the list."[30]

Korolev himself, having to live with the national secrecy phobia, was nevertheless conscious of the need to document the achievements of his bureau for posterity. He saw to it that a large wing of OKB-1 was devoted to a museum of accomplishments, replete with full-scale hardware, performance charts, and photos of the main protagonists. Alas, no one but a select few officials were permitted entry. Today, however, that museum is one of the most interesting stops on the tours of visiting Western space professionals. Until recently, though, it excluded exhibits of projects still considered too sensitive. As mentioned in the previous chapter, for example, there was no evidence of the existence of the N-1 program for landing a cosmonaut on the Moon, or, more understandably, of the *Zenit* spy satellite.

Since the OKB-1 museum was inaccessible to the public, Korolev did his best to keep the Soviet citizens aware of the accomplishments of those space exploration triumphs that he was allowed to talk about through other means. Besides writing the "Professor K. Sergeev" columns, he had a special deputy responsible for putting on presentations in the Red Square. Vladimir Syromiatnikov, one of his engineers, has the opinion that his "role in the popularization of the achievements of space technology" by cultivating journalists, and producing special film and TV programs, "is inestimable."[31]

I interviewed Korolev's staff chief for filmmaking, Viktor Frumson, one Sunday after he had shown me one of his works at the Korolev home-museum. A rumpled, heavy-set man in his seventies, he said that he commuted two hours each way to Kaliningrad from his high-rise apartment off Leninsky Prospekt, where he has lived for thirty years.

It was a fairly typical rundown Moscow apartment building, approached by puddled dirt paths strewn with garbage. There were kids outside playing on broken swings. A man was repairing his sad-looking wreck of a car, with pieces of his engine on the ground. We went up to a high floor in a clackety elevator. The stairwells were filthy. We entered a messy apartment. He showed disappointment when, after setting out beautiful little glasses, Maxim Tarasenko, my research associate, and I turned down his offer of vodka. We drank a tasty apple juice instead as he recounted some of his experiences working as a "propagandist" for Korolev for some twenty years:

> I started working for OKB in 1947. Korolev was very conscious of the importance of films and exhibits for the public and for the government leaders. The Kosmos pavilion at VDNKh [Exhibition of Economic Achievement near the Kosmos Hotel in Moscow] was his idea. In 1957 I was put in charge of exhibits—I organized the ones at VDNKh, the Polytechnic Museum in Moscow, the Tsiolkovsky Museum in Kaluga, even floats for parades.

> In 1964 I began to do film work, beginning with *Voskhod,* on which Komarov, Feoktistov, Yegorov flew. I made more than 150 films, thanks to Korolev's strong commitment to film-making. . . .
>
> Before 1964 films were ordered by the military, by Marshal Nedelin. The first versions were classified, for historians. They are now being released. Then a version for the general public would be released. The public films would show an early single-stage launch vehicle instead of the real one.[32]

It is difficult, in retrospect, to understand why such ruses were exercised when surely the Soviets must have known that details of their launch vehicles had long been known to the West. But a mania for stemming unauthorized disclosure of even basic information prevailed. The inordinate fear of leaks extended even to doctoral theses by students. Vitaly Sevastyanov, the cosmonaut, wrote a theoretical paper on a two-man mini-shuttle. The Minister of Aircraft Industries phoned him one night and said, "Korolev has read your document and has written on it, 'To keep for eternity.' "

"My diploma work was classified and had to be destroyed."[33]

The *Pravda* and *Izvestia* reading public were studiously kept in the dark on the realities of Soviet space events, except for the occasional "Professor K. Sergeev" article.

Of course the hush-hush phobia inevitably led to speculation, often wild, about events that might or might not have occurred. A stinging article by one Leonard Nikishin in *Novoye Vremya* in 1991, entitled "Space Dramas: We Knew Nothing of Some, Others Did Not Really Occur," deals with some of them. For years there were rumors that numerous cosmonauts had died in space-related accidents. All have been disproven except the actual deaths—Valentin Bondarenko in a fire during a training test in 1961; Pyotr Dolgov in the test of a Vostok ejection seat in 1962; Vladimir Komarov in the *Soyuz 1* crash in 1967; and Gyorgi Dobrovolsky, Viktor Patsaev, and Vladislav Volkov, who were asphyxiated when *Soyuz 11* lost oxygen on return from the first mission to the *Salyut 1* space station in 1971. Nikishin rebuts the rumors but then questions: "But who to believe after decades of official lies and silence?"

He cites the inaccurate account of Komarov's demise as an example. " . . . they dreamed up something clumsy about 'tangled parachute lines,' although in actuality the ship's two parachute systems failed."

Pravda collaborated faithfully with the Party line on the secrecy of the manned Moon program, even to the point of suppressing a note written by Yuri Gagarin that implied such a mission was expected. His handwritten note, written on April 4, 1963, just after *Luna 4* missed the

Moon by more than 5,000 miles—and probably suppressed for that reason—was auctioned to an American buyer at Sotheby's in New York in December 1993. It read:

> The Soviet Union has made another giant step toward mastering space. The Luna 4 unmanned station has been sent to the Moon, which will enrich our knowledge of this nearest neighbor of our planet. We hope that, before long, man's foot will step on the Moon's surface.[34]

The Soviet leaders sometimes became victims of their own passion for secrecy. Nikita Khrushchev, for instance, was still getting assurances that the USSR would stay ahead of the United States in space at a time, in the mid-sixties, when—with the Gemini program getting underway—their incipient loss of leadership must have been obvious to the space professionals. According to Sergei Khrushchev:

> At the range he [Nikita, in 1964] had been shown the new, three-stage Voskhod, which would soon be launching a spacecraft into orbit, and had met its crew—Komarov, Feoktistov, and Yegorov. Father was bursting with pride for our country, which had overtaken the United States in space. The people around him 'yessed' him on everything, trying to support the illusion that the United States would continue to trail behind, and that the first land of socialism would become the world's leading technological power.[35]

A better knowledge of American developments would have been advantageous to the professionals as well. One of the highly respected engineers who was not privy to that knowledge was Efraim Akim.

"Between 1956 and 1961," he told me, "we had no information on what was going on technologically in the West. Even if we had read the information we would not have had time to dig through it. The pace was very intense. If we had kept it up we would have accomplished a great deal more in space exploration."[36] Sergei Khrushchev refutes this. "We got all the open Western publications regularly," he told me. "We also got the KGB reports on American weapons, like the full report on the Bomarc [Boeing's ramjet missile]."

Korolev himself also kept a close eye on American progress, through Shevalyov. "I was Korolev's English expert," he told me, "not as an interpreter but as a follower of English-language publications, mostly American." An engineering graduate of MVTU, Shevalyov worked at OKB-1 for ten years, from 1956 to 1966, first on ballistic missiles, then on the *Vostok* structure. He became quite useful to Korolev as a technical person who also knew English. He recounted those years:

I had studied English at MVTU and also with a private teacher, and so started doing translations on weekends for extra money. I would get 20–40 rubles for one printed page, 40,000 characters, 22 typed pages. I was asked to make many translations of US publications like *Aviation Week, Missiles & Rockets,* and the journals of the American Rocket Society and AIAA. There were five to seven translators in the department, most of them women specializing in German. I had no English speaking practice since it was strictly prohibited to speak to foreigners, but I could listen to the English-language radio—BBC, Voice of America.

Quite suddenly in the summer of 1960, my chief of the technical information department, Popov, had told me that Korolev wanted a person to keep him informed on foreign technology. I was summoned to Korolev's office. Mishin was there and he gave me an offer, telling me that I would have an office near Korolev. We made a three-year agreement, although I then had a technical job and was slated for a promotion. What's more, I had plans to go to the mountains on vacation. But Mishin said, "A wise man never goes to the mountains. At your age (I was twenty-seven) I normally took a girl to the beach and had a gay time."

A new building was put up in 1961 and we all moved into it. My first duties were to keep Korolev informed of the news from U.S. magazines and newspapers, popular magazines, but not the scientific journals. But then, since he had been misinformed about some U.S. achievements in space technology and had not been keeping track of the journals, I got the job of reading them. The Academy of Sciences gave him hard currency for the subscriptions.

These publications would come to us, and if a certain mark was on the cover—a hexagon stamp—it meant that it was not to be passed around. One U.S. publication, I remember, printed a chart of Soviet missile bases. That one had the hexagon mark. At the Lenin Library I was allowed to read the *New York Times* and other publications, but only to take notes from them, not to borrow or photocopy them.

When the first U.S. satellite was launched, the American newspapers were brought in and I was told to translate two full pages over night, that Korolev urgently wanted this information.

In spite of his value as an English translator with engineering credentials, poor Shevalyov found himself in hot water when he took it on himself to enrich the knowledge of Korolev and his staff about American space progress, using film footage from Moscow TV and other sources. He told me the sad tale:

On Moscow television they showed sometimes brief footage on U.S. spacecraft launches. I got a letter from Korolev, took it to the TV station and made copies of the footage. Then I put together the short segments

> into a one-hour film. I also went to NITSKD (the Scientific Research Center of Space Documentation, which had secret archives) and made from their footage a one-hour film on the Mercury launches, and then a two-hour film on the experimental launches of apes, the Shepard flight, and the Glenn flight.
>
> I was eager to show the one-hour film to Korolev so he said OK, but told me to invite the other department heads. About 800 people came to the screening in our big auditorium. I had translated the English commentary word for word. What's more, I used the same triumphal enthusiasm in my voice as the American narrator. After about ten minutes of this I was stopped and told to leave the auditorium. I asked my aide to continue the narration. Korolev had left the auditorium and was in his office, livid with rage. How could I present that film in such a way? Why did I use that tone of voice? Didn't I realize that these Americans were imperialist enemies? He was using the Party line words.
>
> I said, shall I stop? He said no, go on, but use a different tone of voice. From then on our relationship cooled.[37]

It was perhaps understandable that Korolev should be prickly about the "triumphal enthusiasm" of the "imperialist enemies," because the progress of the Apollo program, after the Glenn flight, became as rapid as the buildup of funds supporting it.

15

Racing Apollo: The Odds Were Enormous

It might just have been possible, but the odds against the Korolev team putting a cosmonaut on the Moon ahead of *Apollo 11* were enormous. The Sea of Tranquility, where Neil Armstrong and Buzz Aldrin were to land on July 20, 1969, was one of the possible areas the Russians had in mind for Alexei Leonov, the likely cosmonaut for the Soviet mission. Korolev's launch vehicle, the gigantic N-1, which towered almost as high as the U.S. *Saturn V*, might have done the job if it had been able to mature. Absent, however, was the very large investment in rubles that would have been required to bring the vehicle to an acceptable state of reliability. And so the timetable for the development was utterly unrealistic.

The fact is that in the 1960s neither Khrushchev, nor Brezhnev after him, was willing to give Korolev the money to subject the N-1 to anywhere near the developmental effort and the rigorous testing that the *Saturn V* got. The Kremlin's top ruble had to go either to thermonuclear warheads and their ballistic missile and submarine delivery systems or to space systems with specific military missions.

So the Soviet manned lunar landing program never got the priority it needed. For instance, although it is difficult to believe, the N-1's 30-engine first stage was never static-fired as a unit. The reason was to save time and money. To make matters worse, competitive approaches to the manned lunar landing mission were solicited from Korolev's rivals, consuming precious rubles, manpower, and design expertise. What was kept secret in the USSR would have been exposed as a national scandal in the United States. But in a paranoically secret society it was possible not only to allow the scrimping on tests and the costly duplication, but even to pretend in the end that there had been no manned Moon program at all.

Someday historians may gain access to the minutes of the Kremlin-level meetings that must have heard the stormy arguments on the Moon initiative. Of equal interest will be the bookkeeping ledgers that show how many rubles were allocated to which developments. So far, however, archival officials have stonewalled efforts to access the most sensitive documents. Fortunately, personal memoirs, and interviews with many of Korolev's colleagues who are now willing to speak openly, have produced candid accounts of what happened, even if all of the detailed documentation may not be yet available.

As recounted in an earlier chapter, Gyorgi Vetrov has documented an April 30, 1955, meeting at which Korolev raised the idea of lunar exploration in a meeting with government officials. But it was in April 1956, more than a year before the *Sputnik* launch, that he made the first serious proposal to mount a Moon mission. "The real task," he said, in a speech to the Soviet Academy of Sciences, "is to develop a rocket to fly to the Moon and back from the Moon. This task is most easily solved by starting from a satellite, but it can also be solved by starting from Earth. Somewhat more difficult will be returning to Earth the equipment that will be on the satellite or the rocket that goes to the Moon. But it must not be believed that the proposals I am making are extremely remote."[1]

The depth of Korolev's thinking about how he would systematically build towards a manned mission to Earth's only natural satellite is manifested in a 1958 report from a book titled *The Creative Legacy of Sergei Pavlovich Korolev,* which was eventually published in 1980. The book reprints some of Korolev's most significant writings during the period from 1932 until he died.

An introduction eulogizes Korolev in the stilted Cold War language of the time, " . . . this remarkable representative of our science and technology, Communist, member of the new socialist formation. . . . "[2] Korolev's Communism may have been more pragmatic than passionate, but remarkable he certainly was, as indicated by the boldness and foresight of his proposed program for exploring the Moon and beyond, including manned flights to Mars and Venus. It is all meticulously outlined in the report referred to above, entitled "Most Promising Works on the Development of Outer Space," coauthored by Korolev with another great Soviet space visionary, Mikhail Tikhonravov.[3]

Required by 1958–60, says the report, are a three-stage version of the launch vehicle that put up *Sputnik,* plus a more advanced launch vehicle, and, apace, the development of the "interplanetary technology," that is, the instrumentation, guidance and control, computer trajectory analysis, checkout facilities, and other advances that would underpin the whole effort.

In the 1958–61 period an objective, the report continues, would be the design of a 10- to 20-kilogram solar-powered "research station" for landing on the Moon; in 1959–61 a spacecraft for photographing the lunar surface; in 1960–64 a fourth stage for the launch vehicle that would enable a spacecraft to orbit the Moon; in 1958–60 a heat shield and reentry system for the first manned orbiting spacecraft; in 1959–61 robotic spacecraft with radio and TV equipment capable of flights to Mars and Venus; in 1962–66 the development of spacecraft rendezvous techniques; in 1963–64 a new launch vehicle that could put a 15- to 20-ton space station into orbit and permit the possibility of interplanetary flight; in 1961–65 the design of a two- or three-man space vehicle for prolonged stay in space, and the development of an ion engine-powered spacecraft for manned flight around the Moon; in 1963–66 the creation of a robotic space vehicle for flights to Mars and Venus; beginning in 1962 the design of a space station in orbit for studying the effect of weightlessness on plants and humans and for studying radiation effects on animal and vegetable organisms.

"After the realization of those planned works," the report says, " . . . can be set the following tasks: manned flight to Mars and Venus, manned flight to the Moon with landing and return to Earth; construction of a permanent station-colony on the Moon (beginning with research in 1960)."

The whole document is breathtaking in its audacity. It goes on to call for studies, starting in 1959–60, of new chemical fuels and nuclear rockets good for carrying "great weight" for satellites and "extraterrestrial stations"; in 1958–61 on the solution of rendezvous problems; in 1959–63 on space station assembly, including the use of carrier rocket "housings" as "finished sections of the station"; in 1960–65 on closed life-support systems and the design of pressure suits for space operations; in 1958–62 on the development of space power systems for space stations and "interorbital apparatus"; in 1959–65 on the prospects for new developments in long-distance radio communications.

There, in one report, was the grand plan, the whole kit and kaboodle, even going beyond a manned lunar landing to sending cosmonauts to explore Mars and Venus using nuclear rockets. Alas, however, the final sentence of the report is as wistful as the plans are bold, "The present material has not been discussed or coordinated with the basic developers. . . . " Indeed, this was a wish list, and while Korolev almost certainly did discuss his wishes with the "basic developers" he failed during the subsequent years to get approval from the Communist hierarchy for much that was on the list.

Still, he had a big head start on the United States. During the same year that the report was issued, 1958, the newly formed NASA was still struggling to get its space effort underway. It had taken over a modest complex of aeronautical research laboratories—the National Advisory Committee for Aeronautics (NACA)—whose budget in 1958 was only about $100 million, and whose total employment was about 8,000.[4,5]

It had been bad enough to be four months behind the Soviets in putting up a satellite, but the fact is that in 1958 the Americans were about four *years* behind in rocket engine capability alone.[6] The four-chamber engines designed by Valentin Glushko, the RD-107 and RD-108, which had launched the R-7 intercontinental ballistic missile and the *Sputniks,* were much more powerful than the engines that put up the beleaguered *Vanguard* as well as those that finally orbited *Explorer,* the first successful American satellite, on February 1, 1958.

NASA's manned program had not gotten underway officially until the authorization of Project Mercury on October 5, 1958, just five days after the new agency came into being. The Mercury project, which would put John Glenn in orbit less than four years later, had been virtually ginned up during the previous summer by a group of ten engineers working for less than three months.[7] The Eisenhower administration was not as enthusiastic about the prospects for the Mercury program, however, as its protagonists were. T. Keith Glennan, Eisenhower's appointee to head the new NASA, writes in his memoirs that, "The philosophy of the [Mercury] project was to use known technologies, extending the art as little as necessary, and relying on the unproven Atlas. As one looks back it is clear that we did not know much about what we were doing."[8]

Eisenhower was always cool towards manned space programs. Glennan reports from a private meeting with the president, on October 13, 1960, that, "I told him something of the costs that appear to be involved in Project Apollo, the follow on to Project Mercury. He expressed himself once more as having little interest in the manned aspects of space research."[9]

That would not be the attitude of Eisenhower's young successor, John F. Kennedy, although it must be said that the Russians' accomplishments during his first few months in office had a great deal to do with his newfound zeal for space. On April 12, 1961, only three months after Kennedy's inauguration, the Korolev team had lofted Yuri Gagarin on the R-7 launcher and sent him around the planet in the first manned flight of the *Vostok,* to world acclaim.

Kennedy's actions from then on were rapid fire. On April 14 he called a meeting of his close advisors with James Webb, the NASA administrator who had just succeeded Glennan, and his deputy, Hugh Dryden, to explore ways to catch and pass the Russians.[10] His reasons were compelling: "that the United States intended to maintain its position of world leadership, its position of eminence in commerce, in science, in foreign policy, and in whatever else might develop from space exploration."[11]

On April 19 he called in Vice President Lyndon Johnson, whom he had appointed chairman of the newly formed National Space Council, and asked for recommendations on an accelerated space effort.[12] In retrospect, Kennedy's decisive actions on the space program seem more pragmatic than visionary when one realizes that on these very days the CIA-abetted invasion by Cuban exiles of the Bay of Pigs—an action he had countenanced—turned into a fiasco of failure. On April 20, the day after he realized the Cuban mission was a disaster, he directed Johnson "to be in charge of an overall survey of where we stand in space. . . . Do we have a chance of beating the Soviets by putting a laboratory in space, or by a trip around the moon, or by a rocket to land on the moon, or by a rocket to go to the moon and back with a man. Is there any other space program which promises results in which we could win? . . . How much additional would it cost?"[13]

Speculation continues on whether the decision to go to the Moon might have been a way to reassert the president's leadership at a time of bitter defeat. "I think the president felt some pressure to get something else in the foreground," his science advisor Jerome Wiesner, is quoted as saying. "I'm sure it wasn't his primary motivation. I think the Bay of Pigs put him in a mood to run harder than he might have."[14] In any case, over the next few weeks JFK continued to push Johnson to grill the space experts on which project might best tip the balance towards the United States.

Wernher von Braun, one of those explicitly questioned, wrote Johnson a long memo that included the statement that the United States had "an excellent chance of beating the Soviets to the first landing of a crew on the moon," pointing out that the Russians would need to increase their rocket engine capability by a factor of ten to accomplish the same feat.[15] Von Braun knew, of course, that several years of work on the huge F-1 engine had already taken place. On May 5 Alan Shepard had become the first American in space, having flown a 15-minute suborbital flight in a *Mercury* capsule. Then, on May 25, President Kennedy, towards the end of a speech to the Congress on "Urgent National Needs," said,

> Recognizing the head start obtained by the Soviets with their large rocket engines, which gives them months of lead time, and recognizing the likelihood that they will exploit this lead for some time to come in still more impressive successes, we nevertheless are required to make new efforts on our own. . . . I believe that this Nation should commit itself to achieving the goal, before this decade is out, of landing a man on the moon and returning him safely to earth.[16]

The contest was on in earnest. Although starting the race way behind, the Americans had a lot going for them besides the F-1 development. Their principal competitor, Korolev, had no Kennedy, no Johnson, or, for that matter no James Webb, the savvy NASA administrator, to fight his political battles. What's more, Korolev, as deeply dedicated as he was to human exploration of the Moon, already had administrative and developmental responsibilities which, in the U.S., were shared by several NASA centers plus numerous industrial firms. Besides developing ballistic missiles and running the cosmonaut program, Korolev was working on third and fourth stages for the R-7; designing and testing the USSR's first communications satellite, *Molniya,* and its first reconnaissance satellite, *Zenit;* and developing robotic probes for Mars and Venus investigations.

But the Russian lead was formidable. Following the *Sputnik* success, Korolev's robotic program of Moon reconnaissance spacecraft, beginning in 1959, had greatly impressed the world. Eventually the U.S. robotic programs for Moon reconnaissance—Ranger, Surveyor, and Lunar Orbiter—which had also suffered numerous failures, produced much more comprehensive photos and scientific data, but the Russians had been incontrovertibly first in this phase of lunar exploration. While all of these programs, both American and Soviet, professed scientific motives—the acquisition of knowledge of the lunar surface and its environment—their findings were crucial to the eventual objective of landing a human on the Moon.

Nonetheless, the Moon mission had been high on the agenda of the thinkers and planners in both countries for some time. Von Braun, for example, although hardly the only advanced thinker in the U.S. space program, had a drawerful of conceptual designs for manned missions not only to the Moon, but to Mars and beyond, and in the early 1950s got national attention for his ideas in a series of articles in *Collier's* magazine.[17] But those ideas were germinating in the USSR as well, and they were known to those in the United States who, like von Braun, were paying attention. In a postscript to the first of his *Collier's* articles, in 1952—five years before *Sputnik*—the former Peenemunde head acknowledged that the "Soviets claim a head start." However, he stated, "the advantage in the competition to conquer space probably rests with us—if we move quickly."[18]

The man chiefly responsible for the Soviet head start, Sergei Pavlovich Korolev, had dreamed of lunar travel even before the 1958 paper referred to earlier. The Korolev design bureau's archives show that the dreams became working plans for a heavy-lift launcher capable of a manned lunar mission at least as early as September 14, 1956, more than a year before the *Sputnik 1* launch.

On July 15, 1957, almost three months before *Sputnik 1,* the concept for launch of a manned lunar landing crew—aimed at lifting off as much as 1,000 metric tons—received its first serious attention in the

Council of Chief Designers, as well as at the government level. But the upshot of the discussion was that the idea was premature, no doubt to Korolev's dissatisfaction. There was to be no movement on the project for the next two years. In fact, a negative action occurred in December 1959, when a decision was made to continue to downplay space developments that did not improve the Soviet Union's military capabilities. A very revealing article that refers to the Soviet military leadership's negative attitude toward the manned Moon concept during this period was written in 1994 by two veterans of the Korolev bureau who finally felt free to speak candidly. The writers were historian Gyorgi Vetrov and Sergei Kryukov, who had a key role on the N-1 development team.[19] The Ministry of Defense was always cool on space exploration, wrote Vetrov, feeling that "it was a direct threat to the defensive capability of the country."

But only about six months after the December 1959 decree, say the authors, an abrupt change of heart took place, very possibly a reaction to the aggressiveness during 1958 and 1959 of then Senator Lyndon Johnson, chairman of the Aeronautical and Space Sciences Committee, in pushing the neophyte NASA organization to move ahead. Johnson had personally grabbed the political leadership role of the space program shortly after the *Sputnik* launch, and he was not above using some rather overblown rhetoric on the subject of catching up with the Russians, as typified by this statement in 1958:

> Control of space means control of the world. . . . From space the masters of infinity would have the power to control the earth's weather, to cause drought and flood, to change the tides and raise the levels of the sea, to divert the gulf stream and change temperate climates to frigid. . . . There is something more important than the ultimate weapon. That is the ultimate position—the position of total control over earth that lies somewhere in outer space . . . and if there is an ultimate position, then our national goal and the goal of all free men *must* be to win and hold that position.[20]

With that kind of political boostership, NASA got carte blanche with its budget requests. In January 1959, NASA had given a contract to North American Aviation to develop the F-1 engine and in February established a working group on lunar exploration. In June NASA authorized the Army Ordnance Missile Command (the von Braun team in Huntsville, Alabama, which would soon join NASA, itself, en masse) to study the use of the Saturn launch vehicle for lunar missions. In August a NASA panel recommended that work be started immediately on "a program that would lead to a second generation three-man capsule with a potential for near-lunar return velocities."[21]

Realizing belatedly that the hard-won global prestige gained by the *Sputnik* and *Luna* successes was at risk, the USSR Council of Ministers reversed its 1959 decree sharply, issuing a new decree on June 23, 1960. It called for the "creation of powerful launch vehicles, satellites, spaceships and the control of space in the years 1960–67."[22] It was an omnibus decree, specifying the need for "a powerful complex rocket system with a launch vehicle liftoff weight of 1,000–2,000 metric tons, providing injection into Earth orbit for an interplanetary ship having a mass of 60–80 metric tons, and also a powerful liquid rocket engine (LRE) with high performance values, atomic rocket engines, and liquid hydrogen, ion and plasma engines. Other developments were planned by the decree, but probably the most important [was] the creation of the new N-1 rocket having chemical energy engines."

Interestingly, the moon was not the original target for the N-1, according to Vyacheslav Filin, who had just been hired by Korolev out of Moscow Aviation Institute. Filin says that N-1's first purpose was to put a 70-ton unmanned payload into a 200-kilometer orbit on the way to Mars. Only when the American Apollo program emerged were the big booster's priorities shifted to a manned lunar landing.[23]

Some of the concepts specified in the 1960 decree, such as the nuclear and hydrogen rockets, and the ion and plasma engines, were not pursued at the time. Also, the Ministry of Defense, although directed to prepare, with the cooperation of industry, a document about the potential use of spacecraft and powerful launch vehicles for military purposes, never responded at all. On January 15, 1961, Korolev wrote Marshal K. S. Moskalenko, says Vetrov, asking that he speed up the proposals but "the posture of the Ministry of Defense remained practically unchanged, which led to funding cuts, unfulfillment of preset plans and eventually became one of the primary reasons for the cancellation of work on the N-1."[24]

Nonetheless, some very ambitious studies ensued under Korolev, including the prospective use of the N-1 launcher for the Mars mission mentioned by Filin. Gleb Maximov, a member of Korolev's design team, for example, told me that he participated in the Mars mission study in the early 1960s. It called for N-1 to launch a three-man vehicle that would have orbited Mars.[25] Even earlier than that, in the late 1950s, I was told by Valery Kubasov, a young member of Korolev's advanced design team who later became a cosmonaut, "I was assigned by Tikhonravov to the design of a spacecraft for a Mars mission, before NASA worked on that." Korolev was building a formidable design team, said Kubasov proudly, and he wanted "engineers who demonstrated toughness, and who worked around the clock on solving problems. He wanted people who had the same passion, vision, and enthusiasm that he had himself. How else can you explain the fact that he had us working on designing manned Mars spacecraft even before the first cosmonaut flew?"[26]

The Moon mission, however, became paramount to Korolev in the early 1960s, even though it was not possible to admit this. Consequently, when *Soyuz* ("Union") was under development—although it was destined from the outset to be a one-man lunar transport—it was characterized by Korolev officially as part of an "orbital rocket complex."[27]

Theoretical work on what became *Soyuz* had begun as early as 1957—the same year *Sputnik* was launched—emphatic evidence of Korolev's efforts to get a lunar exploration program going right away. One of the participants in the conceptualization from the beginning was Vsyevolod Avduyevsky, a distinguished leader of Russian space technology research, who spent many years with the Scientific Research Institute for Thermal Processes (NIITP).

"In 1957 when the *Soyuz* spacecraft was being developed, I participated in the thermal exchange analysis," Avduyevsky told me. "There were different lifting forces on *Soyuz,* and a three-dimensional boundary layer. We developed a thermal shell for the front area, 2 meters across, to examine the turbulent boundary layer over a rough surface." That research led to the incorporation of a bell-shaped heat shield on *Soyuz*—a la *Mercury* and *Gemini*—differentiating it from the spherical *Vostok* and *Voskhod.*"[28]

Viktor Minenko, one of the OKB-1 designers, who is still active at RSC Energia, told me that by 1961 there were several different departments involving about 40 people in the *Soyuz* design.[29] At the time he had been doing heat transfer and aerodynamic studies for various missiles and space vehicles. "We used to read carefully the U.S. literature by the leading aerodynamicists—Ferri, Chapman, Van Driest, Lees,[30] and the top Russians—Sibulkin, Koropkin," he said and then recounted an anecdote about Korolev's direct intervention in the *Soyuz* development. It offers an interesting illustration of how the local Communist Party sometimes got into technical subjects:

> My first encounter with Korolev took place in 1961 at a meeting of the 11th department of the Party. Although it was unusual to devote a Party meeting to technical subjects this was an exception. We were in the midst of a competition between the 11th and the 9th departments on design approaches to *Soyuz.*
>
> I remember that Tikhonravov, Feoktistov and Ryazanov of the 9th department were there, and Maximov. The Party leader was against

inviting Korolev but he came. It was a 3-hour meeting, after work. I was not a Party member at the time but this was an open Party meeting. Nina Grechko (wife of Gyorgi Grechko, later a cosmonaut, but then an engineer also at OKB-l) was there too. So were Vladimir Timchenko and Tyulin.

I was ordered to speak on the configuration of *Soyuz* which we were proposing. We knew about the configuration of the U.S. *Mercury* capsule but we chose not to go that way. Our first criterion was stability. Feoktistov was in favor of a spherical design. After all of the people spoke, Korolev took the floor and criticized all of the designs.

"You are too far from a realistic design. Get back to Earth and look at the details," he said.

The inside insulation we had proposed was made of metal. He was against this. I was the leader of our team and he criticized me directly.

"Look at the mass budget. You should have some kind of margin."

I was an ambitious young man and I said it's not a problem to change the insulation materials. He didn't think I was a real designer, but more a mathematical modeler.

"You've lost a year on this project," he said. "I'll watch you closely now."

And he ordered us to get cracking. He wanted us to have a wood mock-up ready in two or three months. Bobkov was made leader of a special team using people from both the 9th and 11th departments, to do a new *Soyuz* design.

Although such pressure from Korolev for progress on *Soyuz* seems very explicit, Feoktistov told me that for some time the *Vostok* flights had actually diverted the Chief Designer's attention from the *Soyuz* design project. "He had wanted to prolong the use of *Vostok* and so it was difficult to work on *Soyuz* without his strong support. From my point of view it was risky to keep using *Vostok*. We had been lucky to avoid accidents."[31]

Soyuz had a quite sophisticated design. "The *Soyuz* appeared after *Gemini*," said Vassily Mishin, "but keep in mind, our program for the use of *Soyuz* was more complex than the American program. It contemplated the docking of two manned craft and the transfer of cosmonauts in orbit from one craft to the other. But the Gemini program envisioned the rendezvous and docking with automatic 'targets' only."[32] Actually, a more logical comparison for *Soyuz* would be to the *Apollo* command and service module, since both were meant to be lunar transports.

Soyuz, in the Korolev tradition, was an evolutionary rather than revolutionary step beyond *Vostok* and *Voskhod*. Its descent vehicle was the same *Vostok-Voskhod* spherical design, albeit with the new heat shield at one end. But there was an attached bell-shaped orbital module where the cosmonauts ate, slept, did experiments, watched television and movies, and kept their personal hygiene supplies. The NASA engineers had for some time considered a two-compartment vehicle for *Gemini,* too.

"This started out to be a very attractive idea," said Maxime Faget, one of NASA's principal spacecraft designers, "but as we went through the studies it became clear that less and less things were going on in that mission module, and everything that was vital for one reason or another . . . also was vital during entry so you either did it twice, once in the mission module and again in the command module or you did it once in the command module. . . . There were no systems and no particular activity that anyone really wanted to carry out in the mission module other than to stretch out and perhaps get a little sleep." But the idea was considered by Faget and his colleagues "for a long while," and since "we made no secret of these considerations," he wonders if it might have influenced the *Soyuz* design.[33] There has, in fact, been speculation that Soyuz was influenced by a losing design for the *Apollo* spacecraft submitted in 1960 by the General Electric Company.[34]

In any case, Korolev went with the two-compartment design for *Soyuz* and the Russians have stuck with it, making modifications over the thirty years during which more than 100 spacecraft have carried cosmonauts into orbit.

Also emerging from the preliminary design studies of this period were not just the N-1, but N-2 and N-3 for heavier payloads as well, although the latter two would eventually be merged into N-1. The money allocated to N-1, however, was hardly bountiful. As Kryukov writes,

> . . . the imposed deadlines were harsh, but the budget allocated was relatively small. . . . It was necessary to maximally utilize the existing technical base, try to make the rocket complex as simple and reliable as possible, provide the capability to use it for an extended period of time, and allow for future modernization. Different designs were discussed, fierce arguments took place, and finally it was decided that most problems, both military and scientific, could be solved with a rocket having a useful payload of 75 metric tons and a liftoff mass of 2,200 metric tons.[35]

Following Korolev's "simple and reliable" philosophy, N-1 was designed in three sections. An innovative idea was to integrate the sections in an assembly building at the Baikonur launch site. Propellant tanks would be large spherical structures that would actually be built at the launch site. Estimating the cost of such an advanced technology project such as this, especially during the preliminary design stage, is a dicey business. In any case, the OKB-1 estimators proved every bit as overly optimistic as U.S. space designers. Their calculation of the cost of manufacturing and launching ten rockets was 457 million rubles. The actual cost was 4 billion.[36] The "grave error" was not totally unintentional, however. As with U.S. contractors, it became a tactic to make a low initial estimate on the realization that once the project was underway it would involve such a substantial commitment that it would not be stopped. "It's not so easy to end a program or construction project on which one has already spent millions," writes Kryukov. "For this reason, individuals on both high and low levels eagerly played 'financial hiding games.'"

But deciding even the basic mission of the N-1 took several years. "Everybody thinks that the N-1 was designed to carry man to the Moon, but that's not true," says Vetrov, who told me that Korolev conceived of the big launcher as having multiple functions—capable of making military observations in near-Earth space, putting up a global communications satellite, gathering meteorological data, and carrying out robotic and eventually manned missions to the Moon, Mars, and Venus. "He was even trying to come up with a super reconnaissance military mission, and a kind of SDI."[37] Contrast that to *Saturn V*, which had one mission only, carrying astronauts to the Moon.

Clearly, Korolev had in mind bootlegging a manned lunar landing on the military programs. "As always, military problems stood first . . . necessary money could only be obtained for strengthening national security. . . . Therefore . . . the lunar program had to be realized 'by the way.' . . . This decision can hardly be judged reasonable," writes Kryukov. "The United States of America, seeking to take from us the role of space leader, set for itself a special goal: to be the first to land on the Moon. The Soviet Union should have either answered this challenge or decisively turned away from it and gone its own way. We did neither one nor the other, and thus did not concentrate our efforts on the solution of the 'lunar' problem."[38]

One blatant example of this lack of concentration occurred in 1961 when one of Korolev's bitter rivals, Vladimir Chelomei, was given a contract to develop a rocket and a spacecraft for a different lunar mission than the manned landing—a circumnavigation of the Moon by a solo cosmonaut—calling for hardware totally unrelated to the N-1 program.

It seems that Chelomei had had the political acumen to hire Khrushchev's son, Sergei, as a deputy for his guidance laboratory in 1959.[39] Nikolai Pilyugin, not long before his death, according to Vetrov, told a reporter that Chelomei's influence was so great, because of his Khrushchev connection, that if he had wanted the Bolshoi theater for his enterprise he could have gotten it.[40]

The hiring certainly improved his chances for favored contracts, in any case, although Sergei was a competent engineer and Chelomei, it must be said, had very impressive credentials as a missile designer.

Interestingly, Chelomei, like Korolev, Glushko, and Yangel, had a Ukrainian beginning. He was brought up in a very cultured environment in a family of teachers in a small town in northern Ukraine.

"Chelomei was the only true scientist among the chief designers, although his designs were successful in production," I was told by Sergei Khrushchev.[41] After some training in automotive technology, he went to Kiev Technological Institute, which Korolev had attended only eight years earlier. But their interests were different, even at Kiev Tech. As Yaroslav Golovanov puts it, "Korolev wanted to build and do, Chelomei—calculate and analyze." Chelomei would always compare himself to other Chief Designers by describing them as "constructors" while he was a "scientist." One of his professors said he had never met so gifted a student. He wrote a book on vector calculus at age twenty-two, and became, at twenty-six, the youngest winner of one of the fifty Stalin stipends offered to candidates for the Doctorate.[42] In 1941, he did his doctoral dissertation on rocket engines at the Central Institute of Aviation Motors (TsIAM) in Moscow.

In 1944 Chelomei went to the Kremlin and told Gyorgi Malenkov he could build a weapon like the V-1, which the Germans had used against England. Malenkov bought the idea and Chelomei spent the next twenty years building cruise missiles.

Apparently he was a difficult person to work for. "Chelomei didn't care about people, and so his relations with people were difficult," said Sergei Khrushchev. "If someone made a technical mistake, even a friend, he would fire him. He was a complicated guy like all geniuses, maybe he was not the biggest of geniuses, but he was on a level with von Braun. It's impossible to compare him to Korolev, who was a manager, or Yangel, who was an engineer."

A dislike of Ustinov was a characteristic that Korolev and Chelomei had in common. But their differences were profound, not only in their varying approaches to technology development, management, and human relations, but also in their work habits and personal demeanor. Korolev cared little for fine trappings or fancy clothes. Chelomei, however, "was very stylish," according to Sergei. "He looked like an artist, dressed well, wore natty ties. His office had elegant furniture, clean carpets. He spent two months designing his own desk. He was a good teacher, liked to talk, and he knew how to talk and lecture. He was very cultured, liked having associations with painters, writers. He had everything in his head. He could talk for two days about his ideas. He was thinking about new cruise missiles and laser weapons, like SDI, as early as 1962."

It is interesting to have such confirmation of what many Pentagon insiders long knew, that the Soviets—who later protested the Reagan SDI initiative so vigorously—had pursued their own SDI concepts for years. "I sat in on the session of the Council of Chief Designers which discussed [an SDI approach] called the Taran [Battering Ram] antiballistic missile system," Sergei Khrushchev told me. "I remember that Chelomei went to the blackboard and filled it full of figures which seemed to me to be silly but then 2 or 3 years later I realized that he had been talking about the future."[43] Taran was abandoned, however, when calculations showed that it was unacceptably expensive. "The Taran concept was a folly," says Maxim Tarasenko. "Chelomei offered his UR-100 (SS-11 in NATO) as an antiballistic missile but this was technically unrealistic since it would not have the thrust-to-weight ratio and maneuvering capability required."[44]

By the early 1960s, however, Chelomei was devoting almost half of his bureau's effort to spacecraft development. While his political clout turned out not to be great enough to usurp from Korolev the lead role in manned space exploration—an action he brashly proposed to Pyotr Dementiev, Minister of Aircraft Production—it nonetheless gave him license to grab for the circumlunar mission.

His launch vehicle would be the UR-500, which, in its inception was supposed to be a two-stage ballistic missile.[45] A three-stage UR-500K could place twenty metric tons into Earth's orbit, and could theoretically carry Chelomei's conceptual LK-1 spacecraft around the Moon with one cosmonaut. The first stage would use six new RD-253 Glushko engines fueled by the propellants he was determined to work with—nitrogen tetroxide and UDMH.

Korolev was, of course, deeply upset that the Chelomei-Glushko circumlunar mission was being offered with hardware unrelated to the lunar landing mission. Mishin recalls that he "made repeated attempts to consolidate both . . . programs or to at least use the developments of one for the other as much as possible. The first attempt was made in 1961, when he proposed using the N-1 (the first version, but with a 75 ton payload mass) sending two cosmonauts around the Moon. . . . "[46]

On February 20, 1962, Marine Colonel John Glenn became the first American astronaut to orbit the Earth, circling the globe three times in a Mercury capsule lifted off by an Atlas rocket.

In that same month, Sergei Khrushchev witnessed a meeting in Pitsunda, his father's dacha on the Black Sea, which was crucial to the future of the manned space program. It was a two-day session, hosted by the Premier for military generals and industry leaders. It says something about Soviet priorities in those times that the agenda was devoted primarily to briefings on the missile buildup. However, Korolev and Chelomei were given chances, towards the end of the second day, to advance their space flight proposals. The result: Korolev got a go-ahead for his N-1 program, Chelomei for his UR-500.

Korolev's determination to keep Chelomei out of the manned space flight business is shown by the fact that on March 10 he advanced another possible scenario for a manned circumlunar flight. It would use the *Vostok* launch vehicle that had carried Gagarin and Titov into orbit. *Vostok* would place three separate rocket stages into Earth orbit for assembly. A fourth launch would orbit L-1, the antecedent of the *Soyuz*. L-1 would carry its crew of three cosmonauts to a docking with the complex of three stages in Earth orbit. The assembled vehicle would then fly around the Moon and return to Earth. The concept was not pursued, however.[47]

In the meantime, NASA was riproaring ahead with its manned lunar landing plans. As mentioned earlier, one of the key needs for the project was a big first-stage engine, and the United States was well on its way to leapfrogging the Russians dramatically with the F-1, which would develop 680 tons of thrust in one chamber. In 1958 North American Aviation's Rocketdyne Division had begun the F-1's preliminary design, under Air Force sponsorship. Then, on January 15, 1959, NASA gave Rocketdyne a contract to continue to develop the engine, since it was a logical candidate for powering the *Nova* launch vehicle. *Nova* had been envisioned by NASA to carry astronauts on a direct ascent to the Moon. But the Kennedy directive to accomplish the lunar landing by the end of the decade essentially meant the end of *Nova,* which, it was felt, would have taken too long to mature.[48] On June 22, 1962, NASA decided that a multi-stage vehicle using lunar orbit rendezvous—*Saturn V/Apollo*—would be the optimum mode for the earliest success of the mission, and *Saturn V* 's first-stage engines would be the F-1's originally meant for *Nova*.

In July 1962, Korolev's preliminary design of the N-1 was blessed by a commission headed by Mstislav Keldysh. Korolev could not resist writing this longhand note in the margin of the design presented to the

Keldysh commission: "We were thinking about this in 1956–57."[49] Getting Keldysh's support was crucial to Korolev, since Keldysh was a prestigious scientist who later became president of the Soviet Academy of Sciences. He is generally credited with the lead theoretical role in the USSR's space program and he had direct control over the program's computer facilities.

A few months later, on September 24, a government decree "defined the scope and timing of the work, which provided for the beginning of flight testing in 1965," wrote Sergei Kryukov.[50]

The Korolev agenda, circa 1962, had on it not only the N-1 system, but also the first orbital flights of the cosmonauts; a series of unmanned probes to Venus and the Moon, most of which failed; the *Zenit* reconnaissance satellite, which also had its failures; and other programs.

One task worked on intensively at the time, an important milestone on the way to manned lunar flight, was the docking of two vehicles in space, a delicate maneuver requiring great skill and an intricate design.

"In the summer of 1962," remembered Vladimir Syromiatnikov, "we began work on the creation of docking installations simultaneously at several planning and design subdivisions. . . . The designs turned out to be cumbersome, complex to control, and contained many separate mechanisms. . . . Only by the spring of 1963 did the outlines of the future design for the docking assembly become clear: a moving pintle on an active spacecraft and an acceptor cone on the passive one."[51]

Korolev put his signature on the theoretical drawing of *Soyuz* on March 7, 1963.[52] At that time, though, the design approach was apparently still fluid. The imaginative Chief Designer, for example, even considered, in great detail, the possibility of using a rotor—a propeller—to replace the parachute in the spacecraft descent system. This was in mid-1963, just after the Bykovsky-Tereshkova flights in *Vostoks 5* and *6*.[53]

"We're tired of flying on rags. Sooner or later they can let us down," is the way he put it, in unwitting metaphor, to Igor Erlikh, who had been designing helicopters for the Yakovlev design bureau.[54] Korolev worked on Erlikh for months to give up his top job at Yakovlev and come over to OKB-1, even appealing to national loyalty. "He simply saw his duty," wrote Erlikh, "as doing everything he could 'so that they [the Americans] would not beat us,' so that we—he himself, I, our people, and our country—were first among all others in conquering space."[55]

"What will I do," asked Erlikh, "when the rotor landing is finished? Look for work at the helicopter plant, where my place will already be taken?"

"The rotor landing is not a one-time job. It has great long-term promise," countered Korolev, with what turned out to be unjustified optimism, since the idea never got anywhere.[56] In fact, the use of rotors had been pressed upon Max Faget and his colleagues at NASA when they were designing Gemini. "It was a lousy idea," he told me in a 1995 telephone conversation. "Where do you stow it? What about the weight penalty? Reliability?"[57] The idea was dropped by Korolev as well.

In the same period, Korolev faced a new problem. Ustinov, still fearing that too many eggs were in the fabled Chief Designer's basket, threatened to move the responsibility for *Soyuz* development and manufacture to another enterprise. As one of his long-time designers, Lev Vilnitskii, put it, " . . . obviously, the transfer . . . would have been a severe blow for the collective and its famous leader." The pugnacious Korolev fought back. He ordered his staff to organize for Party officials an exhibit of already-developed instruments, subassemblies—thermal regulators, pumps, ventilators, docking elements—and whole prototypes. When Korolev came, at about 10:00 o'clock in the evening, to pass on the display, he seemed at first defeated. "Our many years of work," said Korolev, "may go for naught." Vilnitskii says that "It was painful to see this willful, fearless man suppressed by such circumstances. But he was able to control his feelings and concluded his conversation on an upbeat note: we will fight and defend our brainchild." He won. *Soyuz* stayed at OKB-1.

In spite of an already enormous workload, Korolev continued to conceptualize ways of capturing the circumlunar mission for his bureau from Chelomei. In May 1963, he came up with one more approach, dropping the *Vostok* assembly spacecraft. One rocket stage, unfueled, would be placed in Earth orbit and fueled by tankers launched separately. A newly designed *Soyuz* spacecraft would be launched to dock with the rocket stage, then a burn of the engine would send the assembly around the Moon. This scenario proved too complex to pursue, however, and was abandoned.[58]

A few months later, Chelomei, who had been devoting almost half of his bureau's effort to spacecraft development, enhanced his reputation with the debut of his inaugural spacecraft, which made quite a sensation, both in the USA and USSR, impressing Korolev himself enough to have a look at it in Baikonur. Vladimir Poliatchenko, its lead designer, recalled:

> The first spacecraft design was the *Polyot* unmanned maneuvering spacecraft. The launch of *Polyot-1* on November 1, 1963 surprised the world. It surprised even Korolev. Many thought it was an anti-satellite weapon. [In fact, it was a prototype for ASAT. There was only one other *Polyot* flight, *Polyot 2* on April 12, 1964. Full-configuration

> ASAT spacecraft were tested in 1968, and the program, which Sergei Khrushchev described as "huge," ended in 1970, although Poliatchenko was not in a position to confirm all of this[59]]. *Polyot* was the first spacecraft to demonstrate the possibility of wide maneuvering. Korolev visited us in Baikonur. He was very interested in the construction details, and the engine, which was designed by Chelomei, although Isaev and Tumansky developed the combustion chamber. The Chelomei bureau made the fuel tanks, tubes, etc., and did the integration.[60]

Another spacecraft design by Chelomei, who was convinced by that time that space was where the action would be, was for a rocket plane similar to the American *Dynasoar*. The concept, however, never got anywhere in either the USA or USSR. *Dynasoar* was cancelled by the U.S. Air Force in 1963 and the Soviet program never developed beyond an atmospheric test flight.[61]

In mid-1963, perhaps due to the high publicity given to *Saturn/Apollo* in the West, the Soviet hierarchy decided to escalate the N-1 program's priority. It had been sold as primarily a military project. Now the military objectives, never clearly identified anyway, were dropped. As Korolev wrote on July 27, 1963, "The realization of performing a manned expedition to the surface of the Moon must be considered to be a main goal of a program to explore and master the Moon."[62]

In late 1964 Korolev dropped his original scheme of using multiple launches and Earth-orbit assembly, deciding instead to have a single launch of the N-1/L-3 system, with lunar orbit rendezvous a la *Apollo*. The plan called for increasing the payload mass to be lifted by N-1 from 75 to 92 tons, allowing a single cosmonaut to land on the Moon while a second cosmonaut would stay in lunar orbit. Both would return to Earth in a recovery section of the lunar orbiter. Intensive work was devoted to "approaches that would ensure the insertion of such a payload without radical revision of the published technical documentation, the design of the rocket units, or the special-purpose manufacturing equipment.[63] This was a huge order. Launch mass would increase from 2,200 to 2,700 tons. Six NK-33 engines would be added, making the total thirty in the first stage. A "flexible" system of controlling engine thrust would be adopted, thereby upping the thrust of all engines in the first three stages by 2 percent.

Meanwhile, development of the F-1 engine, using RP-1 (essentially kerosene) fuel and liquid oxygen, was well underway in the USA. Five F-1s would power the first stage of *Saturn V*. Also under development was the J-2 liquid hydrogen/liquid oxygen engine for the *Saturn V*'s upper stages. These engine designs were very bold, advancing the state of the technology significantly, and not everyone was confident that they could be pulled off. The Russians were among the skeptics.

"It was then the prevailing opinion in the USSR," said Boris Gubanov, the lead designer of the N-1/L-3 system's Block E, or lunar return vehicle, being developed at Yangel's bureau in Ukraine, "that the USA would never get such a powerful development [*Saturn V/Apollo*] going. The engines were too large. The launch vehicle was too big and the use of liquid hydrogen too complex. It was a major mistake for Korolev to underestimate the USA. It wasn't until 1964–65 that this mistake was realized . . . the USSR has no such engine [as the F-1], even up to now.[64] Korolev should have developed a bigger engine. There were thirty engines in the first stage of N-1. Control of thirty engines and sensing of malfunctions was too difficult. *Saturn*'s first stage had only five engines."[65]

In fact, it was the inability to sense malfunctions in the Kuznetsov engines that would prove to be one of N-1's biggest problems. This was not the N-1's only weakness, however. One of the cosmonauts who might have flown on a Moon mission, Gyorgi Grechko, points to another problem that plagued the Soviet space program, and still does to some extent. " . . . we had very bad electronics. Even the big booster, the N-1, could not lift its payload because the electronics were so bulky . . . its goal was to launch 125 tons but it could only launch 75 tons." Apollo needed only a 100-ton payload to reach an orbit coplanar with the Moon's, because launching into Earth orbit from Cape Kennedy's 28° latitude required less energy than from Baikonur's 45° latitude. "To design to 75 tons," said Grechko, "was impossible."

Grechko related this information with obvious sadness in a 1993 interview. Much beefier than he was in his cosmonaut days, Grechko was nevertheless still in fine health. With a pleasant, always beaming round face, and a brush haircut, he was nattily dressed for the interview in brown suit, blue dress shirt with spread collar, and tie. He loosened his tie and unbuttoned his collar for the interview. His biography gives him two Hero of the Soviet Union medals. One of the engineer-cosmonauts, Grechko graduated from Leningrad Institute of Mechanics in 1955, joined the Communist Party in 1960, went to work at the Korolev design bureau in 1956 and in 1966 qualified for the cosmonaut detachment. His regret over the troubles with the lunar missions was especially keen because he had been assigned in mid-1967 to the Zond group, which trained for a circumlunar flight. He was scheduled to fly as the instrumentation engineer on the second mission if the program had not been cancelled. Subsequently, in 1975, 1977, and 1985, he flew *Soyuz* missions to *Salyut* 6 and 7.[66]

Korolev's technical problems with N-1 were certainly no less troublesome than those posed by Soviet politics. As late as 1965, Chelomei and Yangel were still proposing alternates to N-1/L-3, no doubt to Korolev's vexation. *Apollo/Saturn V* was already well along in hardware development and the Russians were still entertaining other scenarios than N-1/L-3. Chelomei proposed a system based on the UR-500 called UR-700. The other possibility was the R-56, conceived by Yangel.[67]

Both the Yangel and Chelomei designs proposed a powerful new Glushko engine, the RD-270. If it could have been developed, the Chelomei/Glushko approach, the UR-700, might have made a sounder design for the Moon mission than N-1. The UR-700 would have used in its first stage nine RD-270 engines, fueled with nitrogen tetroxide and UDMH,[68] rather than the thirty NK-33 Kuznetsov oxygen-kerosene engines required for the first stage of N-1. UR-700's second, third, and fourth stages would have been the UR-500, later designated the *Proton*, which became a very reliable launcher now being marketed successfully worldwide. On the other hand, UR-700 would have taken much longer to mature than even the N-1. Korolev knew that the big RD-270 engine, for instance, would have required massive ground test facilities that didn't exist. The race with the United States would be lost.

"UR-700 was a pure bluff," in the opinion of Vetrov. "It was designed for a single-chamber Glushko engine of 600 tons thrust [RD-270]. The design approach was inherently faulty. Glushko was just trying to interfere and he was influential enough to do so."

Neither UR-700 nor R-56 received serious consideration, however, and Korolev's N-1/L-3 system kept its assignment. One yield of the UR-700 concept, however, was the construction of a prototype RD-270 engine. An engineering model of that engine is reverenced today in the museum of Glushko's design bureau, now called NPO Energomash, for having contributed significantly to the technology of its descendants. But how many precious man-hours and rubles had been devoted to its conception? It's as if the Aerojet, or one of the other American firms that had lost the Apollo competition to Rocketdyne, went ahead anyway—using government money—to design its own candidate Moon rocket engine while the huge F-1 was being developed for the *Saturn V*. U.S. citizens may deplore the excesses of American investigative journalism, but one wonders what might have happened in Russia if the journalists covering the space program—and there were some very able ones who knew what was going on—had been permitted to explode this scandal.

Certainly it was one of the most blatant instances of the negative effect of Soviet secrecy. Korolev's N-1, which needed all the help it could get, was not only an underdesigned, minimally tested, and undeclared program, but one harassed by possible competitors long after it should have had the government's exclusive focus. Korolev had received approval to go ahead with a detailed plan for the test launchings on August 3, 1964,[69] But only fourteen months later, on October 20, 1965, there would be an order issued by Sergei Afanasiev of MOM directing his subcontractors—such as Pilyugin for guidance systems and Barmin for launch facilities—to begin concept definition (*avanprojekt,* which comes before *eskiznyi projekt,* or preliminary design) for a new rocket (UR-700) "using backlog and facilities from the N-1." "After such an order," commented Maxim Tarasenko, the Russian space historian, "one can expect that people were less enthusiastic about doing their best for N-1, reasonably guessing that this program may soon be terminated and replaced by UR-700."[70]

Having to cut schedule and budgetary corners, Korolev aimed virtually every aspect of the design of the enormous N-1 launch vehicle at optimum producibility. Whereas *Saturn V*'s fuel tanks were load-bearing structures, and therefore quite rugged, the N-1's fuel tanks were less massive because they were spherical, and they were suspended so as to be subject only to fuel pressurization loads. Also, the spherical shape meant that scaling up the tanks in the various stages was relatively straightforward.

Korolev's sensitivity to comparison of N-1 with *Saturn V,* whose design details were fully public, is revealed by a May 25, 1965, memo to the Ministry of General Machine Building. Whistling past the gravestones, he tried to make a case for N-1's capabilities, while conceding its disadvantages. Even without the benefit of *Saturn V*'s powerful hydrogen upper stages, he stated, its performance differed "insignificantly" from that of N-1's. This, he claimed, was in spite of the fact that N-1's weight was 20 to 40 percent greater than *Saturn V* because of (1) the latter's use of "new very strong aluminum alloys"; (2) the "total weight of control telemetric equipment of *Saturn V* is one-fifth that of N-1"; and (3) "the necessity to divide the N-1 into separate units transported by railway dictated an increased number of joints which certainly made the units heavier than the blocks of *Saturn V,* transported by water."

Those claims seem belied by subsequent statements in the same memo, in which an obviously beleaguered Korolev recommends that Chelomei's rival launcher, the UR-500, be withdrawn "from further development and production," so that funds could be directed to an N-2 booster that would have liquid-oxygen/liquid-hydrogen upper stages, like *Saturn,* and develop "new very strong aluminum alloys similar to American alloys" as well as "more reliable control and telemetry equipment having one fifth to one seventh the weight."[71]

The memo did not carry the day. No additional funds were allotted for the hydrogen development, the new alloys, or the telemetry, and the UR-500 project continued.

Faced with the continuing need for frugality, there were numerous steps taken to minimize costs. All the while money was being poured into the *Saturn-Apollo* program. One can especially appreciate the expense gulf between N-1 and *Saturn V* by comparing the complexity and cost of the transportation of the vehicles alone. Ordinary rail, in the case of N-1, contrasts to the *Saturn V*'s system of transporting the first and second stages by barges to Cape Canaveral. The first stage started out through a specially built canal from the Michoud, Louisiana, assembly facility. The second stage sailed all the way from California, through the Panama Canal. The third stage was flown to the Cape in a unique cargo plane.

Actually, the Russians had originally studied various modes of transport to Baikonur, including by water and by air. Air transport would have required a plane capable of a 250-ton payload and building such a giant was considered unrealistic. Back they went to the original decision to transport the vehicle in sections by rail. However, the sections were first sent to the plant in Kuibyshev, where they were assembled with the engines. Then they were disassembled, put on rails again, and finally welded together in Baikonur.

Another telling comparison was the construction of facilities for engine testing. Besides Rocketdyne's static facilities in Santa Susana, California, *Saturn V* underwent comprehensive testing, with all five F-1 engines firing, at the NASA Marshall Space Flight Center in Huntsville, Alabama, under the careful scrutiny of the von Braun team.

Korolev's enforced emphasis on time and cost savings, in the end, proved to be one of N-1's fatal weaknesses. Building no large static testing stands at all, and therefore never test-firing the thirty first-stage engines as a system, was perhaps the most telling mistake.

The government decree of September 24, 1962, had called for flight testing beginning in 1965. Trying to meet such an unrealistic timetable placed enormous pressure on the design team. "Many scientific, design and manufacturing problems had to be resolved," wrote Kryukov:

> Among these, for example, were the welding of large-scale thin-walled vessels, development of electric, pneumatic and hydraulic connections between Blocks, the introduction of new construction

> materials. . . . Especially noteworthy was the critical and labor-intensive process of developing the "armature" (pipelines, umbilicals, valves) for the N-1. The large dimensions, high precision, impossibility of duplication demanded particular skill and a creative approach from the manufacturers. . . . Sergei Pavlovich Korolev gave to his "child" all his energy, all his "battering abilities."
>
> He went to Kuibyshev, Syzran, Leningrad, Baikonur. . . . It was necessary to hurry, to persuade, to prove. . . . It was particularly difficult to make changes to already approved drawings. However, it was even worse with financing. In 1964 OKB-1 was allocated 23 million rubles for all the work even though the need was for 45 million, Kuznetsov's OKB got 20 with a need of 50, Kuibyshev Sovnarkhoz got 9 while needing 23 million . . . a delay from the set deadlines began to appear. A decree on August 3, 1964 moved the date for the beginning of flight testing to 1966. But since the main goal had now become the landing of an expedition on the Moon, the necessity arose to very rapidly develop the lunar complex L-3.[72]

As with Korolev, NASA, during the same period of 1963–64, was confronted with technical problems that threatened to delay the program. But NASA never lacked for budgetary support from the Congress, even though Apollo had its individual critics. In contrast, Khrushchev's support for the Moon program was always ambivalent, and he held the Chelomei alternative over Korolev's head, although surely it was ridiculous to consider that alternative at such a late date.

"I witnessed a meeting between Korolev and Khrushchev in Khrushchev's dacha," recalled Yuri Orieshkin, one of Korolev's administrative deputies. "It was 1964. We flew down to Pitsunda, where the dacha was, between Gagra and Sukhumi on the Black Sea. I witnessed only the early part of the conversation. It was during the time when Korolev and Chelomei were 'at war' over the concept for the manned Moon program. Korolev wanted to protect his N-1 concept. Chelomei was promoting an alternative launch system. Korolev and Khrushchev went eye to eye and Khrushchev was persuaded to support N-1. Korolev's and Khrushchev's words to each other, though, were cold and unfriendly. Korolev felt that he was being interfered with in his job. When he was angry he lowered his brow and took a posture like a bull."[73]

When Khrushchev was dethroned in October 1964, Korolev had to deal with still another Soviet leader whose support for N-1/L-3 was equally ambivalent, Leonid Brezhnev. Orieshkin describes a Korolev-Brezhnev confrontation in March 1965.

> Korolev was phoning Brezhnev from the cosmodrome in Tyuratam . . . just before the launch of *Voskhod 2,* during which Alexei Leonov was to make the first walk in space [crucial to the N-1/L-3 program, since a key EVA—extravehicular activity—would be required in that mission to enable the cosmonaut, perhaps Leonov himself, to move in his space suit from the Moon orbiting spacecraft to the lunar descent vehicle]. There had been a malfunction in the previous unmanned test flight. An airlock had failed to be jettisoned in orbit. Brezhnev, who was apparently apprehensive that NASA might beat the USSR to the first EVA, said that it didn't sound necessary to delay the Leonov flight. "There was nothing dangerous. The explosive bolts are OK." Korolev said: "I have the responsibility not only for the technology but for the people and the whole project. Until I complete the program of testing I don't want to proceed. If somebody believes that it is possible to launch people after failures that have not been corrected, let him come and take my place." Brezhnev agreed with Korolev's insistence on further testing.[74]

Brezhnev need not have been apprehensive, I was told by Christopher Kraft, NASA's director of flight operations at the time, because Ed White's EVA was not even planned until Leonov performed his. Then, said Kraft, "We decided to have White do an EVA. We devised an umbilical cord and used a standard air gun to give him motion outside the spacecraft."[75] White's EVA took place from *Gemini 4* in June 1965, three months after Leonov's.

Both Khrushchev and Brezhnev, while eager to capitalize on the prestige value of a possible victory in the Moon race with the United States, always recognized the mission's difficulty, and they were not anxious to get into a spending contest with the Americans. So their public pronouncements were always carefully hedged, so that they could (and often did) deny the program's existence.

By contrast, once John Kennedy had told the U.S. Congress in 1961 of the goal of landing a man on the Moon "in this decade," the Apollo program had been transformed from a NASA concept into a national program of very highest urgency and visibility.[76] Kennedy's determination to prosecute the space program fully was manifest to the day of his death. His undelivered speech to the Dallas Citizens Council on the day of his assassination, November 22, 1963, included this: " . . . we have regained the initiative in the exploration of outer space—making an annual effort greater than the combined total of all space activities undertaken during the Fifties—launching more than 130 vehicles into earth orbit—putting into operation valuable weather and communications satellites—and making it clear to all that the United States of America has no intention of finishing second in space."[77]

Kennedy, more than two years earlier, on June 3, 1961, had broached to Khrushchev, at a luncheon meeting in Vienna, the subject of a possible joint cosmonaut-astronaut lunar mission. Khrushchev was at first reluctant to consider the idea, then responded positively, but finally turned it down unless it were combined with talks on disarmament.

Two months before his death, on September 20, 1963, Kennedy brought up the subject again, now based on his confidence that U.S. space leadership had been regained. It was a step that can be interpreted more as political gamesmanship than noble gesture. Appearing before the United Nations General Assembly he proposed that the USA and USSR go to the Moon together, confounding his own NASA Apollo team. "Space offers no problems of sovereignty. . . . Why, therefore, should man's first flight to the moon be a matter of international competition? Why should the United States and the Soviet Union, in preparing for such expeditions, become involved in immense duplications of research, construction, and expenditure? Surely we should explore whether scientists and astronauts of our two countries—indeed of all the world—cannot work together in the conquest of space, sending some day in this decade to the moon not the representatives of a single nation, but the representatives of all of our countries."[78]

Kennedy knew very well that the Soviets would be unlikely to respond positively to such a bold offer. A CIA report of November 11, 1963, would point out that the Soviets were in economic difficulties exacerbated by their choice of national priorities. The report reads that " . . . the military and space programs have had first call on the scarce resources of high-quality manpower and materials," and that, "In consequence, improvements in living standards have slackened, and general economic growth has fallen off from the high rate achieved during most of the 1950's."[79] The politically astute NASA administrator, James Webb, reported Kennedy advisor McGeorge Bundy, "is quite open to an exploration of possible cooperation with the Soviets and thinks that they might wish to use our big rocket, and offer in exchange the advanced technology which they are likely to get in the near future."[80] But Khrushchev was not about to take up the offer. In his memoirs he states that, "Had we decided to cooperate with the Americans in space research, we would have had to reveal to them the design of the booster for the Semyorka."[81] In retrospect, revealing details on the R-7 would have been much less deleterious to the Soviets than to have to show how unready they were for a serious lunar assault.

The Apollo team was greatly relieved. "I don't think one would ever take a Russian rocket and an American spacecraft and try to put them together," Robert Gilruth, head of NASA Johnson Space Center, recalls having said at the time to a reporter.[82]

During the next two years, the American program's leadership became very obvious to the Soviets. Boris Chertok told me that the pressures to try to catch up produced inevitable tensions in the N-1/L-3 team:

> I had a conflict with Korolev myself on the approach. I had responsibility for the lunar module design and we had weight problems. Our booster was too small to handle the weight needed to take three cosmonauts in the lunar module so we proposed two. But it would have been difficult to put even one up because the N-1 wasn't powerful enough. We needed 120 tons to Earth orbit. The N-1 could only lift 95 tons [the six engines later added to the original 24 had improved the lift capability from 75 tons, but 120 to 125 tons was the need]. Korolev was very upset about this, but he had hoped to improve the booster and reduce the weight of the module. He realized though, that it would be difficult to beat the Americans who were pouring huge amounts of money into their Apollo project, with the strong backing of President Kennedy. Some of the Soviet Marshals were opposed to the race, like Malinovsky, Moskalenko, Krylov [head of strategic rocket forces]. Marshal Nedelin's death had been a big blow to the program because he had supported the Moon project.[83]

While work on the N-1/L-3 program was going on intensively at OKB-l and the Kuznetsov engine facility in Samara, it was also occupying the time and skills of numerous government and industry organizations. Oleg Gazenko, former head of the USSR's premier space medicine center, the Institute of Medical and Biological Problems in Moscow, recalls that "Our work on the lunar program began in the mid-1960s. Understanding the lunar mission involved experiments that were very difficult and dangerous. For example, if decompression were to occur—such as by failure of the apparatus or by a micrometeoroid penetration of the spacecraft—would the cosmonauts be able to circle the Moon in space suits and be able to return to Earth? We put a man—not a cosmonaut, but a technician—in a space suit into a vacuum chamber. He was in there for ten or twelve days. We had to deal with thermal regulation, food, urine disposal. We concluded that it would work."[84] Dr. Gazenko's Institute had the major responsibility for passing on the physical and psychological adaptability of all Soviet candidates for space missions, and would have had a key role in choosing the cosmonauts for the actual lunar mission. The mock-ups of habitation modules used to test cosmonaut endurance on simulated Moon and Mars missions are still intact at IMBP.

Virtually all of the organizations involved in preparations for the manned Moon mission worked against mercilessly short deadlines and with marginal financial support.

Hindsight criticism of some of Korolev's N-1 decisions during this period was heard for years to come. For instance, Vyacheslav Kovtunenko—later chief designer at NPO Lavochkin, the major maker of Russia's robotic scientific spacecraft as well as the military's early warning satellites—was a young engineer at NPO Yuzhnoye, the Yangel design bureau in Dnepropetrovsk, Ukraine, during the N-1 development. Korolev came to Yangel and asked him to join the program by taking responsibility for the lunar landing craft. That was a "mistake," says Kovtunenko, and "we suggested that Yuzhnoye do the launcher—Yangel's bureau had very substantial experience building ballistic missile launchers—and Korolev work on the L-3. But Korolev was not enthusiastic about this and the idea came to nothing. The mistake of the N-1/L-3 program was that they worked a lot on the launcher but not enough on the space segment. So it turned out that the space part was much heavier than it should have been and the launch vehicle was not big enough, it had to be beefed up. I told Bushuyev, 'What are you doing? You don't even have the early drawings of the spacecraft and the launch vehicle is already done' . . . they knew how to build launchers and so they worked on the launchers."[85]

Early in 1965, it became clear that the N-1 program was in deep trouble. Ustinov, said Kovtunenko, called a "huge meeting to try to identify programs that could be postponed so there could be concentration on N-1. Korolev and other chief designers made their presentations during an all-day session. Afterwards, Korolev said to me, 'Well, if you think your spacecraft are that good I won't object if you continue to build them. I wouldn't want N-1 to replace your spacecraft.'" Presumably the meeting was unable to come to grips with the essential weaknesses of the N-1 system, although Kovtunenko's view is that if Korolev had not died he would have successfully brought the N-1, even with its weaknesses, to fruition eventually. Most space veterans agree with him, but some have their doubts.

During this period Korolev had to defend his choice of liquid oxygen and kerosene as fuels for his N-1 engines against Glushko's argument in favor of nitrogen tetroxide and UDMH. In a strongly worded letter to Sergei Afanasiev written in late 1965 Korolev cited earlier conclusions of an "expert commission" that the use of liquid oxygen and kerosene as fuels was far superior, in both energy content and economics, to nitrogen tetroxide and UDMH.[86] Moreover, he pointed out, oxygen engines had a very attractive growth potential, especially when combined with hydrogen and engine and nuclear engine stages. He went on to argue that N-1 launch vehicle improvements could be made in existing production facilities with minor modifications.

He then attacked Glushko's engine concepts in a handwritten insert that he apparently decided not to send to Afanasiev.[87] The handwritten comments include this devastating criticism, "One cannot but mention that for a number of years OKB-456 [Glushko's design bureau] . . . ceased to work effectively on development of realistic engines which could be used for practical purposes. OKB is perfectly isolated from the demands of life and spends its 'activity' in unneeded developments, spending tremendous sums of money for that. All this is at a time when there is such an acute need for good engines." Korolev then fired a shot at one of his subcontractors who, he charged, "performs its duties exceptionally poorly" on an N-1 engine component while carrying out an order from MOM on a new project that takes advantage of N-1 technology.

One consequence of the realization that N-1 was sure to be a loser to *Apollo/Saturn V* was that a decision was made to jack up the priority for a circumlunar flight by cosmonauts.[88] Vassily Mishin, who would become Chief Designer after Korolev's death, reports that there were several meetings on the circumlunar mission late in 1965, involving "long and heated discussions."[89]

Chelomei's OKB-52 had been assigned responsibility to develop the LK-1 spacecraft for that mission in the same August 3, 1964, decree that gave Korolev authorization to proceed with the N-1/L-3 lunar landing system. He had, in fact, proposed uprating the LK-1 spacecraft to carry two cosmonauts around the Moon instead of one, based on improved performance of his UR-500 launcher, which, on July 16, 1965, had carried into orbit the heaviest Soviet payload up to that time, the 12,200 kilogram *Proton, a* "physics lab."[90] But he was yet to make any real progress in the LK-1's development. Up to that time, it must be remembered, Chelomei's experience in designing manned systems had been primarily with boost-glide vehicles which, however, had not yet flown cosmonauts.

The UR-500/LK-1 system would have consisted of the booster, plus "an instrument-equipment unit, and a return module similar to the American *Gemini* spacecraft [located in the forward part of the spacecraft, it was enclosed by a fairing]. . . . With a conical shape, the return module had a lift-drag ratio which enabled it to perform controlled descent at planet escape velocity in the Earth's atmosphere, with acceptable g-loads and with a landing in a designated area of the Soviet Union. Electrical power would be supplied to the spacecraft systems by solar panels, which would open after the spacecraft entered a translunar trajectory."[91]

Korolev, and several other chief designers including Pilyugin, had been skeptical that the LK-1 would be able to accomplish such a difficult mission. According to Mishin, Korolev felt that Chelomei's design for the circumlunar mission was "technically useless."[92] Instead, he saw a solution to the mission which, at long last, could be carried out in his own design bureau, using elements of his planned lunar landing system. Two cosmonauts would fly around the Moon in the L-1 spacecraft, made up of a lighter-weight version of the *Soyuz* spacecraft plus the Block D upper stage being developed for the N-1/L-3. The new design would use the Chelomei UR-500K launcher, but Block D would boost the vehicle from Earth orbit to the Moon.

So why not use the UR-500K/Block D combination to launch a circumlunar mission that would gain great prestige for the USSR, one-up *Apollo,* and celebrate the fiftieth anniversary of the Revolution in one fell swoop? The new approach, now called UR-500K/L-1, was approved by the Smirnov Commission on December 15, with Korolev in the driver's seat, although the pressure on him of adding this new program to an already chockfull load was very great.[93]

Although it seems incredible, the already overcommitted Korolev was even considering yet another strategy for manned orbital operations during this period. Gleb Lozino-Lozinsky and his team at the Molniya design bureau in Moscow had been working for some years on the concept of a two-stage aerospace plane called *Spiral.* It could carry one man, change inclination, perform orbital maneuvering on reconnaissance missions, and could be recovered. Lozino-Lozinsky wanted to use Korolev's R-7 to boost the craft into orbit for testing. Lev Voinov, deputy chief designer under Lozino-Lozinsky, remembered his discussions with Korolev about the concept:

> He proposed different variants of the placement of the orbital plane on the rocket. He favored placing the vehicle in such a way as to not require a redesign of the rocket. We wanted a vertical launch. Korolev favored an unusual idea—a tether between the rocket and the space plane. In theory this was possible. But there were uncertainties. A tethered plane would be in the wake of the rocket exhaust. The hardware would have to be protected by a reusable heat shield. We did not accept Korolev's proposal.[94]

One of Korolev's motivations in getting involved in the project, it seems, was "so he could get a big order for R-7's to make them cheaper. He pushed this idea hard," said Voinov. "There were five launches scheduled . . . [including] "orbiting the earth in 1967. . . . Then Korolev was taken to the hospital and didn't return."[95]

Andrei Sakharov recalled, in his memoirs, seeing Korolev around this time at a meeting of the Academy of Sciences.

> I had just learned from a foreign broadcast that the Americans had used a gigantic Saturn rocket to boost a nineteen-ton space station into orbit, a long stride on their way to the moon. I couldn't resist asking Korolev if he had heard the news (knowing, of course, that we had nothing to match the American rocket). Korolev smiled, put his arm around my shoulder, and, using the familiar form of address, said to me, "Don't worry, we'll have our day yet. . . . "[96]

Not long after, on January 5, 1966, Korolev entered the hospital for what was expected to be routine surgery for removal of polyps from his rectum. He would observe his fifty-ninth birthday on January 12 in the hospital. On January 14, he would be dead.

16

The End of Anonymity: Burial in the Kremlin Wall

At the end of 1965 the pressure from Korolev's crushing agenda of space projects was mounting just as his relationships with Ustinov and with Sergei Alexandrovich Afanasiev, the head of the recently formed Ministry of General Machine Building, which funded Korolev's bureau, were getting worse and worse. Scheduling an operation, therefore, was hardly opportune. But the Chief Designer was drained physically, and perhaps a hospital stay, even including minor surgery, would be therapeutic.

Yaroslav Golovanov has written an absorbing and detailed account of the Kremlin hospital experience. He reported that Nina Ivanovna, Korolev's wife, had described his disposition before the operation as "extremely harassed. Everything exasperated him, even his slippers were in the wrong place . . . returning from conferences he was exhausted. . . . 'I can't work like this any longer,'" she recalled his saying.[1]

One of Korolev's OKB-1 colleagues, Vahtang Vachnadze, said that about a month before entering the hospital Korolev said, "I'll just reach sixty, and that's all. I'll not stay here a day longer. I'll go out and plant flowers." Words like those, said Vachnadze, were "not from his vocabulary."[2]

Prior to the hospital stay, in his letters from Baikonur to Nina Ivanovna, his despondency had been pronounced: "I am somehow unusually deeply tired . . . in the days of our troubles, it is especially heavy and hard, sometimes the little heart aches a bit and I . . . receive in large . . . doses Validol."[3] For some years Korolev had not been a well person. Even though he appeared physically strong—stocky, husky looking, square-jawed—he suffered numerous ailments. But the heart problems plagued him mostly.

Lydia Samoshina, who ran the medical clinic at OKB-1, recalled that one day she noticed that Sergei Pavlovich did not look well.

"I checked his pulse and asked if he had any heart medication and whether he needed nitroglycerine, and added, 'You smell of Validol.' 'Oh, these country doctors!' he responded, and quietly asked me not to tell anyone about this."[4]

Any kind of physical activity made him fatigued. Physicians had diagnosed a heart arrhythmia. Sergei Pavlovich consulted with Academician Vladimir Nikitovich Vinogradov and a treatment was prescribed that had no effect. He had a heart incident in February 1964 and spent ten days in the hospital. Six days later he was back in the hospital with acute cholecystitis—a painful inflammation of the gallbladder.[5] Later that same year, on November 3, 1964, his deputy, M. K. Tikhonravov, put a note in his diary: "In the evening we met with SP. . . . Low pressure. Things bad with the heart."[6]

Two years earlier, in the summer of 1962, just after the flights of Andrian Nikolaev and Pavel Popovich in *Vostoks 3* and *4*—the two spacecraft approached to within 4.9 kilometers of each other—he had suffered from intestinal bleeding. He had an "awful night attack of gastric and intestinal pains, and an ambulance had taken him to a hospital."[7]

The intense strain of his schedule kept mounting. In mid-1965, after a succession of flights—the *Voskhod 2* EVA of Alexei Leonov in March; the launch of the first Soviet communications satellite, *Molniya 1A*, in April; the failures or partial failures of three *Lunas,* one in April, one in May, and another in June—Korolev showed up at a Party meeting. After listening to some speeches, he said to Vladimir Lamkin, a Party worker, "Vladimir Ilyich, what do you think if I go home and lie down? I'm not feeling well. I can barely sit up." Lamkin "looked at his face. It was covered with beads of sweat. . . . I escorted him to his car . . . and thought to myself: In such a condition he could have left the membership meeting without asking permission. . . . Yet he could not."[8]

Then, in December 1965, the same month in which his design bureau was assigned the manned circumlunar flight on top of the N-1/L-3 lunar landing program, Korolev had a medical analysis that showed that he had a bleeding polyp in the straight intestine. Surgery for removal of the polyp was prescribed for January. Korolev moved his appointments to the latter part of that month.[9]

What seems astounding is that he masked his sickness from some of his closest office workers. Antonina Zlotnikova recalled that "I never remember him to be sick. He knew how to take care of himself so well! Even on January 4 he worked late. That is what he always said: 'I will die right here, at this desk!' "[10]

The very next day, on January 5, he entered one of the divisions of the so-called Kremlin hospital, a special clinic for top State and Party officials. Nina Ivanovna came to see him every day and talked with the doctors. There seemed to be no cause for undue concern, but to add to his vexations he was getting deaf, possibly due to the monstrous noise of the Baikonur rocket engines. He was given a hearing aid but neglected using it.[11]

On January 11 the Minister of Health, Academician Boris Vasilevich Petrovsky, who would perform the polyps operation, made a histological analysis—that is, he excised a small piece of polyp from the gastro-intestinal tract. The excision caused strong bleeding that was difficult to stop.[12]

While Korolev was in the hospital undergoing tests, his deputy, Vassily Mishin, took the brunt of more hassling from Afanasiev. On January 7, Afanasiev had reprimanded Mishin so severely about alleged failings of OKB-1 that Mishin sat down and wrote out a resignation. However, one of his colleagues saw what he was doing and phoned Korolev at the hospital. Korolev immediately phoned Mishin and said, "What are you doing?" "Writing the report," said Mishin. "It is hard enough to work with you, but with him there is no way." "Tear up the report," said Korolev, "ministers come and ministers go, but we stay in our own business."[13]

More than twenty five years after her father's death, on a bitterly cold December evening in her Moscow apartment in 1991, Natasha Sergeevna Koroleva recalled for me the circumstances of the operation. She had been reluctant at first to grant an interview because she was still mourning the passing of her mother, Korolev's first wife, Xenia Vincentini, who had died only about a month earlier. But she was intent on seeing to it that her father got the historical credit he deserved, not just in Russia, where he is now revered, but in the West, which hardly knows of his existence. Natasha is a warm, attractive woman. Her dark hair was graying graciously at age fifty-six. She is an eminent lung surgeon and professor at the Moscow Medical Academy.

The atmosphere for the interview was not somber at all, but warmly familial, with Natasha's entire family present: her husband, Russian Army Major General Gyorgi Shevchenko,[14] and her children—Andrei, twenty-nine, a lung surgeon like his mother; Sergei Pavlovich, twenty-four, namesake of his famous grandfather, engineer at the Korolev design bureau; and Maria, eighteen, a student at Moscow State University.

One of the rooms in Natasha's comfortable and spacious Moscow apartment is entirely devoted to her father's memorabilia. There are many framed photos on the wall and in albums: Korolev with other design bureau chiefs, Korolev with cosmonauts, Korolev at Baikonur for a launch. There are mementos of Soviet and American launches left by visiting cosmonauts and astronauts. Glassed cabinets hold numerous souvenirs. There is even a tin cup from Kolyma days. And in one of the many albums there are several pages of photos of the day when Sergei Pavlovich Korolev's remains were carried to the Kremlin wall by Leonid Brezhnev and his cohorts. Brezhnev, not a boat-rocker, had kept the Chief Designer anonymous, as had Stalin and Khrushchev, but reportedly it was his sensitivity that led to the decision to reveal

Korolev's achievements to the world, and to give him an elaborate interment. Natasha's large brown eyes were saddened by the subject of the interview, but she nevertheless recalled the story:

> Petrovsky thought it would be a simple operation, so simple that he scheduled another operation afterwards. He started at 8 A.M. [on Friday morning, January 14], used a rectoscope to remove the polyps endoscopically. My father hemorrhaged on the operating table, bleeding so severely that it couldn't be stopped. Petrovsky cut the abdomen to stop the bleeding, and found a cancerous tumor, which had not been visible before. He began to remove parts of the rectum to take out the tumor. This took a long time. My father had an anesthetic mask on for eight hours. They should have put some kind of tube into his lungs [intubation] but his jaws had been broken in prison so they couldn't use the tube. His heart was not in good condition and Petrovsky knew this. He completed the operation, but my father never revived. Another doctor, A. Vishnevsky, was called in from a resort, but nothing could be done. Petrovsky was a good surgeon so it's not fair to criticize him. He is now eighty-three years old and is Director of the Center of Surgery.[15]

Golovanov cites an account of the biopsy analysis from a book written years later by Petrovsky himself. The operation performed was a laparotomy (dissection of the peritoneal cavity). It revealed a large tumor that had grown into the intestine and the pelvis wall. Petrovsky used an electric knife to do a biopsy, which confirmed the existence of a very malignant angiosarcoma. Korolev would not have lived more than a few months, even if he had not been operated on.[16]

The shocking and abrupt death of this great figure in Soviet space annals has provoked, over the years, many second guessers of Petrovsky's surgical abilities. Absent the truth, a not unusual condition in Soviet society in those times, rumors that the operation had been poorly performed began to spread. This is, in fact, a case study of the consequences of the failure to disseminate the facts in the cover-up society that was the USSR. For example, James Oberg, in his excellent 1981 book on the Soviet space program, concluded that "Korolev was the victim of a botched operation, of medical malpractice brought on by the defects of Soviet science bureaucracy."[17]

As perceived through the maze of misinformation with which historians had to deal at that time, and on the basis of the skimpy facts available, such speculation is understandable. Golovanov dealt with the accusation that Petrovsky might have been unqualified for the operation by asking the Academician directly, getting this response: "It is difficult for me to speak on my professional qualifications. I can just say that in the '60's I operated a lot. I became Minister in 1965, five months before the operation on Sergei Pavlovich. And during all these five months I operated too. Operations on the straight intestine were made by me both before and after that, so, I had the experience. . . . "

In any case, in the Soviet system it is not likely that the responsibility of operating on so esteemed a figure as Korolev would have been given to a lesser person than the Minister of Health himself. As Golovanov puts it, "Here, the Soviet 'Table of Ranks' prevails." Korolev was, after all, a member of the presidium of the Academy of Sciences, twice Hero of Socialist Labor, and the "most secret Chief Designer" and, therefore, "only the knife of a 'leib-physician' [from a German expression, used in Russia since the time of Peter the Great, to denote a doctor who works directly for the tsar or some other high-level person][18] deserves your stomach."[19]

But even the Minister of Health admitted to Golovanov that he had not been prepared for the special situation that would obtain. "On that hard day in the Kremlin hospital," Petrovsky told Golovanov, "neither the Chief Surgeon, V. S. Mayat, nor his deputies were present. Absent also was consultant A. A. Vishnevsky." Why was this, asked Golovanov, since Friday, January 14, was not a holiday? No response.

There were other complications. Korolev had an unusually short neck, and he had apparently not revealed to Petrovsky that, as Natasha Koroleva indicated above, his jaw had been broken in prison, so that he could not open his mouth wide. Those two factors meant that an intubation tube could not be inserted. "I faced this phenomenon fairly frequently," Petrovsky told Golovanov, "while operating on people who went through the horrors of the repressions of the '30's. I have no doubt that during the interrogations in 1938 they broke Korolev's jaw. This circumstance forced us to make a tracheotomy—a cut in the throat to insert a tube."

Perhaps the most serious difficulty was with the application of anesthesia. Petrovsky did not want to use a face mask, possibly because it would have been hard to get a good seal because of the jaw fracture, and so a general anesthestic was used, but how Korolev's weak heart would respond to this was not known. No electrocardiogram had been taken, certainly a serious procedural omission on a patient with Korolev's heart condition. At the start of the operation, nitrous oxide, which does not relax the muscles, was used as the anesthetic. Furthermore, there was not enough anesthetic to last for the eight hours that, it developed, the operation would require. "There were no large oxygen pressure vessels . . . so small ones [were used] . . . which lasted twenty minutes."[20]

All of these complications, in Golovanov's words, "enforced the already high tension of all of the people in the operating hall."

Muscle relaxants were injected, but when that is done the patient becomes unconscious and autonomic (involuntary) breathing occurs. A breathing mask was applied. The tracheotomy was held off. It was felt that it was not needed, as Korolev seemed to be doing well. But then, as the mask was taken on and off, breathing became more difficult and the tracheotomy was performed.

"Certainly," writes Golovanov, "there was still nothing fearsome in a tracheotomy itself. But it was not expected in advance and, as is known to anybody, every unexpectancy . . . introduces some nervousness. . . . "

The main surprise was the sight of the tumor. It was "very big, like two fists," Golovanov quotes one Valentina Grek, a nurse who was in attendance. Petrovsky sent an urgent message to find Vishnevsky, the eminent cancer specialist, who eventually arrived. No doubt Vishnevsky, once he learned the situation, had responded quickly, because he and Korolev had known each other well. At Korolev's initiative, members of his team of space instrument-makers had briefed Vishnevsky on apparatus that might find application outside of the space field. From that briefing, wrote Yevgeni Volchkov, an OKB-1 designer, "by the end of 1961 we gave the surgeons the first experimental prototypes of a polarograph, a small artificial heart, and a number of instruments."[21]

The two surgeons, says Golovanov, may not have liked each other, but respected each other's experience. The operation was completed and the surgeons were satisfied that it had been successful. "Borya! Sasha! All has come out well!"

Alas, some thirty minutes later, Korolev's pulse stopped. Both Petrovsky and Vishnevsky, who had left the operating room, were rushed back and tried desperately to get the heart beating again with adrenalin injections, but to no avail. Korolev was gone.

A subsequent examination showed that his heart had been much weaker than expected. "It is not understandable how he walked with this heart," said a pathologist.[22]

The news of Korolev's death was received quickly, and in disbelief, by the space professionals who had been working with him. General Kamanin wrote in his diary on January 17, "It has been three days since he died, and I still don't want to believe that he is no longer among the living." He recalled that on January 14, "About seven o'clock in the evening the phone rang, and General Kuznetsov told me the terrible news: Sergei Pavlovich had died. . . . Like an avalanche, this terrible misfortune came down upon us rapidly and unexpectedly. The country has lost one of its most outstanding sons, and our cosmonautics had been orphaned. Korolev was the main author and organizer of all our space successes. His personal contributions to cosmonautics, the

Soviet people, and all of humanity are limitless. He could have done much more, but he left us when his talent was in bloom."[23] Those words came from the same diary which, on December 22, had revealed how difficult it had been to deal with Korolev in the days prior to the operation: "He spread himself too thin and tried to keep everything under his control; this explained his continual conflicts with Glushko, Pilyugin, Voronin, Kosberg and other chief designers."

Now, finally, it was time to tell the Soviet people who the Chief Designer was. Insiders, of course, already knew.[24] But the glory and plaudits that had gone to the cosmonauts had never reached him in his lifetime. Now they would. On January 16, *Pravda* ran the obituary, with a photo of Sergei Pavlovich wearing his medals, on pages 3 and 4.

There is perhaps nothing more solemn than a Russian funeral that ends up in Red Square. Imagine then, the funeral of a person whose eminence should have been proclaimed long before, having been chiefly responsible for the Soviet Union's monumental achievements in space for more than twenty-five years, but whose very existence was being declared, in most dramatic fashion, to a space-adoring Soviet citizenry, only at his death.

An article in the publication of the USSR Academy of Sciences describes the scene, while neatly indicating the pecking order for all of the organizations to which the space community was beholden. Not named—in keeping with the secrecy tradition—were the technical and design organizations that did the work:

> In the Hall of Columns of the House of Unions on 17 January the workers and employees, scientists and engineers, writers, military leaders, teachers and cosmonauts said farewell to a major scientist, the Academician Sergei Pavlovich Korolev. He was warmly loved and respected by the many, many people who worked with him. S. P. Korolev lived giving all of his creative energy to the concern of the people and to progress.
>
> Covered in flowers on a high pedestal was the coffin with the body of S. P. Korolev. Over the top was a red cover. The large crystal chandeliers were draped in crepe. Around the pedestal were many wreaths. On the ribbons one could frequently see the words: "To the outstanding Soviet scientist, twice Hero of Socialist Labor, Lenin Prize winner, the Academician Sergei Pavlovich Korolev."
>
> In the center were two large wreaths from the Central Committee of the CPSU and from the USSR Council of Ministers, wreaths from the Presidium of the USSR Academy of Sciences, from the conquerors of the universe—the Soviet spacemen, from the Moscow Municipal Committee of the CPSU, from the Moscow Soviet, from the USSR Ministry of Defense, from the rocket troops and from the Air Force.

> On the white marble columns were black and red buntings. At the foot of the columns were wreaths from the collectives of the scientific-research institutes, the enterprises of Moscow and other cities of the nation, from the design bureaus, from the scientific collectives and from the social organizations.[25]

It's not difficult to envision, and imagine the sound of, the playing of the dirge music, from the same account: "Within the arches of the hall the sounds of funeral music could be heard. Works by Tchaikovsky, Beethoven and Chopin were performed."[26]

Korolev's body, if it had not already been cremated, might very well have rolled over when he noted from the hereafter two of his chief harassers in the "honor guard" of "members of the governmental commission for organizing the funeral," namely S. A. Afanasiev and L. V. Smirnov.[27] Also in that group, however, was one of Korolev's long-time champions, and himself a major contributor to the Soviet space program, Academician Mstislav Keldysh.

In any case, "Each two or three minutes the honor guard changed," reports the writer of the article, who then begins to editorialize, injecting a ritual tribute to the Party:

> Now standing at the coffin were the comrades of S. P. Korolev at work. The slightly hoarse voice of the scientist still rang in their ears. Everyone remembered his words, the words of a patriot of our motherland and remarkable communist. "What we have achieved in conquering space," said S. P. Korolev, "is [in] the service of the entire people, the service of our Party, the Party of Lenin."[28]

Next up for duty in the honor guard were some of Korolev's favorites, the cosmonauts. There were Yuri Gagarin, who would be killed in an airplane crash two years later; Gherman Titov; Andrian Nikolaev; Pavel Popovich; Valery Bykovsky; Valentina Nikolayeva-Tereshkova; Vladimir Komarov, who was to die in *Soyuz l* the following year; Konstantin Feoktistov; Boris Yegorov, the *Voskhod 2* doctor, who died of illness in 1995; Pavel Belyaev, who died in 1970 after an operation; and Alexei Leonov. "Their faces were sad. Sergei Pavlovich Korolev was always close to them and for them he was an example of high service to the motherland, Party and science."[29]

Then came the academicians: N. N. Semenov, M. D. Millionshchikov, Boris Petrov, N. M. Sisakian, Pyotr Kapitsa, Vladimir Kotelnikov, A. Y. Ishlinsky, Leonid Sedov, and P. N. Fedoseyev, and then an umbrella category, surely mentioned so as not to leave anyone out: "marshals of the Soviet Union, ministers, leaders of the Party, Soviet, trade union and Komsomol organizations."[30]

Obeisance to the Party's big cheeses was next. Their decisions and opportunism, of course, had often bedeviled S.P., but there they were, "standing in guard around the coffin . . . leaders of the Communist Party and Soviet government [who] came to pay their respects . . . L. I. Brezhnev, G. I. Voronov, K. T. Mazurov, A. I. Mikoyan, N. V. Podgorny, D. S. Polyansky, M. A. Suslov, A. N. Shelepin, V. V. Grishin, P. N. Demichev, Sh. R. Rashidov, D. F. Ustinov, Y. V. Andropov, I. V. Kapitonov, F. D. Kulakov, B. N. Ponomarev, and A. P. Rudakov."[31]

Finally came "The last minutes of farewell. The relatives and family came up to the coffin. . . . " At noon the House of Unions was closed to outsiders. The funeral procession, carrying Korolev's ashes in an urn, "moved slowly towards Red Square, where thousands of workers from the capital had assembled."[32] In April 1961, Korolev had been shunted to a rear position in that same square when Yuri Gagarin was feted after his epochal flight. Now he would be eulogized emotionally in a succession of speeches, while the Party leaders who had kept him in obscurity looked down from the rostrum of Lenin's tomb.

Smirnov, deputy chairman of the Council of Ministers, led off with a ringing tribute to Korolev's "organizing abilities and inexhaustible energy, he knew how to organize the work of numerous collectives of scientists, designers, engineers and workers for the carrying out of great and complex tasks: for he was deeply aware that in our era of intensive progress in science and technology, it is only the labor of collectives which are armed with modern instruments and apparatus which can reach success. . . . Farewell, good friend and comrade! . . . I declare the funeral devoted to the memory of Sergei Pavlovich Korolev to be open."[33]

Keldysh recalled Korolev as "a young scientist in his thirties, full of burning faith in Tsiolkovsky's ideas on interplanetary flight and man's conquest of space" and his capability of "transforming the most complex scientific-theoretical ideas into reality."[34]

N. G. Yegorychev, first secretary of the Moscow City Committee of the Communist Party, had the obligation to report that "Sergei Pavlovich could always find time for active social and Party activity," even though those closest to him say that his attention to Party matters was, at best, indifferent.[35]

The last speaker quite obviously felt the loss of Korolev most deeply. It was Yuri Gagarin, who had been one of Korolev's favorites among the cosmonauts. The world's first human in space put into words his view of Korolev's place in history. "The name of Sergei Pavlovich is linked with a whole epoch in the history of mankind—the first flights of the artificial earth satellites, the first flights to the Moon and to the planets, the first flights by human beings in space, and the first emergence of a human being into free space."[36]

The Academy of Sciences article concluded: "The funeral is over. An urn containing the ashes of the deceased is carried to the Kremlin wall, and under the thunder of a farewell salute is placed in a niche."[37] The niche is covered with a black marble plaque on which appears the following:

Sergei Pavlovich

Korolev

$\frac{30}{19\ \text{XII}\ 06}$[38] $\frac{14}{19\ \text{I}\ 66}$

Those who may wish to visit Korolev's tomb these days will not find it easy. There is limited access to the long wall of past Soviet leaders and heroes that includes, besides Korolev, cosmonauts Gagarin and Komarov; and Vladislav Volkov, Gyorgi Dobrovolsky, and Viktor Patsaev, who died in the *Soyuz 11* accident in 1971, unless one is willing to give one of the guards a financial inducement.

17

After Korolev: Demise of the Circumlunar and Lunar Landing Missions

THE LEGACY THAT Korolev left for Vassily Mishin was awesome. Having far less charisma, not to speak of political clout, than his revered predecessor, and with inadequate financing and infrastructure, Mishin was nonetheless expected to take on the great leader's brutal workload and timetable for beating the Americans.

Korolev's chief deputy and colleague since the Germany days, Mishin was forty-nine years old when he got his assignment. There were three major—one might say herculean—tasks that Korolev had started, Mishin wrote in a 1990 memoir, poignantly titled "Why Didn't We Fly to the Moon?":

1. Final development, ground testing, and execution of docking of two manned spacecraft, with the transfer of cosmonauts through open space from one craft to the other (the Soyuz program).
2. Final development, ground testing, and execution of a circumlunar flight by two cosmonauts, with their return to Earth in a recovery capsule at escape velocity (the UR-500K/L-1 program).
3. Final development, ground testing, and execution of the landing of a single cosmonaut on the Moon, his return to the orbiter waiting for him in lunar orbit with the other cosmonaut, and their return to Earth in a recovery capsule at escape velocity (the N-1/L-3 program).[1]

Those tasks were overlaid on other substantive jobs in OKB-1, such as the *Molniya* communications satellite, which was still in its early stages. Two early *Molniya* failures had been followed by two successes in 1965, but the program still had its problems, as is made clear by the fact that its first launch under Mishin, on March 27, 1966, would result in still another failure.

Sagging spirits could not have been improved by the performance of Neil Armstrong and David Scott, only eleven days earlier, on March 16, in *Gemini 8*. The two astronauts accomplished the first manned docking in space, linking their spacecraft with an Agena target vehicle. That flight, at the same time, demonstrated some of the hazards of the space flight contest. The spacecraft's attitude control thrusters became clogged, spewing fuel into space and causing the spacecraft to tumble crazily—revolving once a second at one point. Armstrong solved the problem by shutting off the attitude control system and activating reentry control, which meant coming to Earth right away, two full days earlier than planned, after six and a half orbits.

The post-flight report by Chris Kraft, who directed flight operations for the Gemini and Apollo programs, depicts how serious the crisis faced by Armstrong and Scott was. "The spin rate was up as high as 550 degrees per second. That's about the rate at which you begin to lose consciousness or the capability to operate." Of Armstrong's use of the retrorockets to stop the spin, he wrote, "That was truly a fantastic recovery by a human being under such circumstances and really proved why we have test pilots in those ships."[2]

Throughout 1966 the last of the Gemini missions—*Gemini 9* through *12*—saw the USA going well up the manned space flight learning curve, not without some real problems, but in the end with very substantive achievement.

Gene Cernan, flying with Tom Stafford in June, did a record-breaking two-hour EVA on *Gemini 9,* but Cernan was totally exhausted at the end of the stint. Wearing a backpack a la Buck Rogers he was supposed to carry out some manual tasks but they proved very difficult. He was supposed to make some simple electrical connections but " . . . he found that every move was more time-consuming than he had counted on. When he seemed to have a job under control, he floated helplessly from the spacecraft. There was no way he could maintain a solid body position. There were a few footholds and handholds there, but they were woefully inadequate. He needed positions that would allow him to use leverage."[3] It seemed to be a jinxed flight, as the attempts at rendezvous with an Agena target vehicle were frustrated as well. A first Agena target never reached orbit when the Atlas malfunctioned. Then *Gemini 9* slipped its launch window, missing a chance to link with a second Agena.

However, in *Geminis 10, 11,* and *12,* the art of rendezvous was perfected. *Gemini 10,* with John Young and Mike Collins, rendezvoused with two different Agenas, one of them still in a parking orbit from *Gemini 8*. *Gemini 11,* with Pete Conrad and Dick Gordon, docked with an Agena on its first orbit. Gordon, doing a forty-four-minute space walk, connected the Agena and *Gemini* with a 30-meter tether. *Gemini 12*'s crew was Jim Lovell of *Apollo 13* fame and Buzz Aldrin, who would be the second man on the Moon in *Apollo 11*. They docked with an Agena and then Aldrin, who did his Ph.D. thesis at Massachusetts Institute of Technology on orbital rendezvous techniques, performed five and a half hours of space walking, dispelling NASA concerns about EVA difficulties. In their 1994 book on the *Apollo* program, Alan Shepard and Deke Slayton wrote that, "He [Aldrin] proved such a master at it that he seemed to be taking a leisurely stroll . . . [and] didn't show even a hint of heavy breathing, perspiration, or an increased heartbeat."[1]

As for the Russians, there were no manned missions at all in 1966 as the *Soyuz* spacecraft continued to be developed. A big morale boost, however, came from the superb lunar reconnaissance missions performed by the Lavochkin design bureau to which Korolev had passed responsibility for robotic exploration. Following the finding, by *Luna 9* in February 1966, that the Moon's surface was firm enough to support a spacecraft, came rapid-fire orbital reconnaissance with *Lunas 10–12*, in March, August, and October.

On November 16, buoyed no doubt by those successes, the Keldysh commission approved a new draft plan for N-1/L-3 drawn up by Mishin. Ten weeks later, on February 4, 1967, came another decree from Ustinov—*Postanovleniye* 115-46—which authorized Mishin to keep going with the program. It seems strange, however, that the latter decree, titled, "On the Progress of the Work on the Development of the UR-500K/L-1" failed to mention N-1/L-3 in its title. Was this a manifestation of a realization that the circumlunar mission was the USSR's best bet? In any case, the decree set unrealistic target dates for N-1/L-3's development—flight design tests by the third quarter of that same year, 1967, and execution of the lunar mission itself by the third quarter of 1969, which would have meant beating the Americans.[5]

But the Americans were not about to be beaten. Even as Mishin got his authorization to continue with N-1/L-3, the U.S. intelligence community was reporting convincing evidence that the Soviets were far behind in the manned lunar landing contest. A National Intelligence Estimate dated March 2, 1967, states:

> Several factors militate against their [the Soviets] being able to compete with the Apollo timetable as originally planned. The status and pace of construction of Area J [the proposed N-1 launch facility] indicate that the launch system will probably not be ready for test until about mid-1968. When a launch vehicle is available, we would expect to see a series of tests for man-rating the system extending over at least a year before a lunar landing would be attempted. In the meantime, the Soviets will need to check out a new spacecraft, to test reentries at lunar return velocities, and perhaps to develop a water recovery capability. Depending upon the technique selected, they may also need to test rendezvous and docking techniques and equipment.[6]

The report was right on. As shown earlier, the "docking of two spacecraft" and "transfer of cosmonauts" were part of Task One in the Korolev legacy. Both were essential steps in the lunar landing scenario.

Taking these steps, and going on to the landing of a cosmonaut on the Moon before the Americans, must have seemed the longest of long shots at the beginning of 1967. Still, the Mishin team would make a good run at them, and they might very well win the preliminary race—the orbiting of two cosmonauts around the Moon and their return to Earth—Task Two in the Korolev legacy. But that was a formidable challenge, too.

"The short time frames given for realization of the program," writes I. B. Afanasiev, "called for maximum use of the existing ground base, documentation, and materiel of the Soyuz program."[7] The designers redesigned the L-1 from the basic Soyuz design, lightening it so as to be able to carry two cosmonauts around the Moon. "They removed the living/working section [bytovoy otsek] . . . some of the reentry vehicle systems (including the reserve parachute system)." Also sacrificed were some of the solar battery panels and a backup maneuvering rocket system. More jet nozzles were added to improve Earth reentry control, and the thermal protection shield was modified to be able to provide protection during Earth atmospheric descent at the velocities expected for lunar return.

The first unmanned test of the UR-500K/L-1, designated *Kosmos 146*, took place on March 10, 1967, a little more than a year after Korolev's death. This was only six weeks after the catastrophic *Apollo 1* fire killed astronauts Gus Grissom, Edward White, and Roger Chaffee in a pre-launch test at Cape Canaveral. That fire made it inevitable that the U.S. moon landing effort would stand down for a lengthy period, giving the Soviets a chance to play catch-up.

Kosmos 146's success, therefore, generated optimism. It orbited Earth for eight days while Block D was ignited twice. The Soviets were looking good again.

Continuing at a fast pace, the second test, *Kosmos 154,* took place less than a month after the first one, on April 8, 1967, with the spacecraft orbiting Earth for two days. This time, though, there was trouble when Block D did not ignite due to a failure in the control system of the Proton launcher.

Two weeks later came a monumental catastrophe. On April 24, 1967, Task One in the Korolev legacy ended in the tragic death of Vladimir Komarov in *Soyuz 1,* the first manned flight of that vehicle. Komarov was supposed to perform the USSR's first manned docking, with Valery Bykovsky, Yevgeni Khrunov, and Alexei Yeliseev in *Soyuz 2,* but the mission was troubled from the start when a solar panel failed to deploy, greatly reducing the power available. With that failure *Soyuz 1* was not able to stay in orbit long enough for its rendezvous, so the

Soyuz 2 launch was scratched. After experiencing orientation difficulties Komarov's spacecraft, returning to Earth on its nineteenth orbit, got tangled in a fouled parachute that had not been packed properly, and crashed, killing the cosmonaut. There would no more launches of *Soyuz,* the spacecraft on which the lunar command module was to be based, until eighteen months later.

That death, incidentally, following Korolev's own passing, put an end to a not-well-known hope the Chief Designer had had that would seem to further characterize his desire to see the space flight adventure immortalized. Korolev had been dissatisfied with the rather inarticulate descriptions sent to Earth by his orbiting cosmonauts. Henry S. F. Cooper, a *New Yorker* writer, quoted Yaroslav Golovanov as saying that, "He wanted us to send up a poet—someone of Lermontov's calibre." That idea was nixed at a higher level so he then decided to include a journalist—Golovanov—and a TV newscaster. Both passed physicals but after Korolev's death, and then Komarov's, the plan was dropped.[8]

The next setback for the Soviets was with the circumlunar mission. On the third unmanned test of the UR-500K/L-1 system, on September 28, 1967, the 7K/L-1 spacecraft was supposed to go into a highly elliptical orbit. However, the first stage of the UR-500K Proton booster failed. Only five of its six first-stage engines worked, and the launcher was destroyed, although the emergency rescue system saved the reentry capsule. Following custom, the failed launch was never announced.[9]

One can appreciate the strenuous demands on the Mishin bureau during 1967 by noting that in a single month following the September 28 failure, Baikonur witnessed the successful launches of the sixth and seventh *Molniya* communications satellites and the unmanned docking of two *Soyuz* craft, designated *Kosmos 186* and *188,* on October 30—the world's first automatic docking. The good luck broke on November 23, when another 7K/L-1 failed to orbit after one of the four engines in the second stage of the Proton failed. The emergency rescue system again saved the reentry vehicle.[10]

Only two weeks earlier, on November 9, the Apollo program had moved back into gear, almost ten months after the fatal fire, with the launch of *Apollo 4*. This was a successful "all up" orbital test of an unmanned *Saturn V/Apollo,* including reentry of the command module, simulating a lunar return. The 36,656-kilogram payload orbited was the largest ever.

As for the Russians, as of November 1967, there had been no flight tests of N-1 and the dates for the tests were officially slipped a year by a new directive. But, says Mishin, " . . . the dates for the execution of the lunar mission were supposed to ensure our country's priority over the United States." This was pure posturing because "by then, it was already clear that the dates set by these directives were unrealistic. They were not backed up by funds, or production capacities, or resources."[11]

By 1968, a race that had begun to swing towards the Americans in 1965 found the Russians getting farther and farther behind. The year began with the *Apollo 5* mission which, on January 22, placed an unmanned lunar module into orbit. With no launchings, Soviet progress in lunar exploration was still going slowly. Although the upper-stage engines had been tested, only groups of four engines in the thirty-engine first stage had been fired statically.[12]

As per the mysterious Soviet name designation and numbering system, which continually baffled Westerners, the next circumlunar test, launched March 2, was designated *Zond* (probe) *4,* although absolutely unrelated to *Zonds 1, 2,* and *3*. The latter had been relatively small spacecraft, less than 1,000 kilograms. What's more, they had all failed on missions to Venus and Mars in 1964 and 1965. This new *Zond* was totally different. It was, in fact, the same 7K/L-1 vehicle as *Kosmos 146* and *154,* weighing more than 5,000 kilograms. Westerners therefore called it the "Heavy Zond." Alas, those first two Heavy Zonds got Kosmos designations because they failed to achieve lunar or planetary trajectories.

Zond 4 was the first officially announced Heavy Zond because it flew a lunar distance. The Moon, though, was on the other side of the Earth at the time. This was not inadvertent. The test was for lunar distance rather than for actual lunar approach. But the mission was frustrated as the spacecraft had to be destroyed on reentry when a motion sensor failed.

Apollo progress was not without glitches. On April 4, *Apollo 6,* unmanned, went up on *Saturn V.* First it experienced a severe case of pogo effect in the giant first-stage F-1 engines—the same problem that had afflicted Glushko in developing the R-7 engines. Although it lasted only ten seconds on *Apollo 6,* it would probably have caused an abort if astronauts had been aboard because of the violent shaking of the entire vehicle. The solution to the problem, however, illustrates the massive level of engineering attention and money put into the Apollo program. To eliminate the pogo problem, NASA had 125 engineers and 400 technicians work some 31,000 man-hours. The fix, finally, involved installation of a system of shock absorbers. *Apollo 6* had other alarming engine problems. Two of the five J-2 hydrogen engines in the second stage shut down altogether, an occurrence expected to be highly unlikely, and the single J-2 engine in the third stage failed to restart for a simulated translunar injection. All three J-2 failures, it turned out, were related to the same problem—ruptured fuel lines due to a weak bellows that had to be replaced by a stronger one. Actually, one of the second stage J-2s shut down automatically because its wires were crossed with the failed engine. Nevertheless, the spacecraft fulfilled its mission of orbiting the command and service module and recovery from the Pacific Ocean.[13]

Getting Soyuz back on track was a top priority for the Russians. This was achieved with another successful unmanned docking of two spacecraft "identical to Soyuz,"[14] *Kosmos 212* and *213,* on April 15, 1968. Then, however, Mishin experienced more problems. Two Heavy Zonds that were supposed to fly unmanned around the Moon failed. One, launched on April 23, did not reach orbit. A short circuit caused the emergency rescue system to trigger during the firing of the second stage. The other Zond, being readied for a July 21 launch, suffered launch vehicle damage on the pad during preparations on July 14 and never lifted off. The Block D oxidizer tank had cracked because it was overpressurized. It exploded, killing three men.[15]

Soyuz got back on track with an unmanned launch, called *Kosmos 238*, on August 28. Then, on October 26, *Soyuz 3,* carrying Gyorgi Beregevoi, came close but failed to dock with the unmanned *Soyuz 2.*

Both *Zond 5* and *Zond 6,* launched on September 15 and November 10, made it around the Moon unmanned. *Zond 5* took a good photo of Earth from 85,000 kilometers but on return "operator error" caused an attitude control failure that placed the spacecraft into a ballistic trajectory that would have generated so many g's it would have killed the cosmonauts.[16] It landed in the Indian Ocean, thousands of miles off course, and was pulled out of the sea by a Soviet rescue vessel.

Zond 6, the tenth L-1 spacecraft, flew by the Moon at a distance of 2,400 kilometers, photographed both the Earth and the Moon, and landed in Kazakhstan after a prescribed double-dip glide reentry—a maneuver designed to minimize heating of the descent module. However, a gasket failed and the reentry vehicle was depressurized. This would have led to loss of oxygen in the cabin—once more producing a condition that would have killed the cosmonauts. There was also a parachute systems failure so that the spacecraft crashed on landing, killing biological specimens onboard.[17] As Afanasiev puts it, "The hopes of the cosmonauts of the 'lunar detachment' to be able to soon make a flyby of the Moon were rapidly dissipating under the onslaught of the ever newer emergencies that were occurring with each new launch. The manned launches in the L-1 program were postponed. In ferreting out the causes of the emergencies and overcoming the difficulties, our program, with time, fell behind the American program."[18]

In between *Zonds 5* and *6*, on October 11, astronauts Walter Schirra, Walter Cunningham, and Donn Eisele took off on a *Saturn IB* and flew 163 orbits in a nearly perfect *Apollo 7* flight, the first manned test of the *Apollo* spacecraft.

But it was *Apollo 8* that finally quashed Soviet hopes to win the circumlunar race. On December 21, 1968, Frank Borman, James Lovell, and William Anders went all the way to the Moon, orbited it ten times, made photos of the lunar landscape and, most memorably, photographed a spectacular Earthrise over the Moon's rim. The latter picture, immortalized by thousands of posters, is credited by many as heralding the environmental conservation movement worldwide. *Apollo 8*'s success was also significant in that it was a direct step toward the larger objective, a manned lunar landing, while testing out the hardware and operations required by that landing.

The global impact of the *Apollo 8* mission was particularly powerful when the astronauts radioed to the world a Christmas message of peace, the astronauts quoting from the Bible. Score one for God, debit one for Communistic atheism. Chris Kraft says that *Apollo 8* was not meant merely to beat the Russians to the circumlunar achievement but rather as a planned step in the testing of the whole Apollo system. "We had experienced weight problems with the lunar landing module," he said, "and so it wasn't ready to be test flown. So we decided to move up the circumlunar flight."[19] While this was the operational reason for the circumlunar flight, there seems little doubt that the possible imminence of a Russian try, reported at the time by the Central Intelligence Agency, added some urgency to the U.S. mission.[20]

"The political sense of continuing the circumlunar program of the L-1 had been lost," said Afanasiev. Nevertheless there were, anticlimactically, three more Zond tries because " . . . immediately stopping a flywheel once it has been started is virtually impossible . . . the spacecraft were built, the launch vehicles were waiting. The schedule of flights had to be observed."[21]

But pursuing the circumlunar schedule in 1969 would produce some painful experiences, mixed with even more painful troubles with the lunar landing program. The latter began the year spectacularly, however, on January 14–15, with what Mishin calls "the world's first orbiting space station."[22] Although that description would seem an exaggeration since it was, actually, the first successful docking of two manned vehicles, the mission was, nevertheless, very impressive. *Soyuz 4*, carrying Vladimir Shatalov, docked with *Soyuz 5*, carrying three cosmonauts—Yevgeni

Khrunov, Alexei Yeliseev, and Boris Volynov. Both Khrunov and Yeliseev did an EVA to enter *Soyuz 4,* which returned to Earth after forty-eight orbits. Volynov came down alone in *Soyuz 5* after fifty orbits.

Then things went bad for the Mishin team. The first circumlunar attempt of the year was on January 20. Liftoff of the UR-500K was nominal but then the second and third stages both failed and the booster was destroyed, although the L-1 capsule was once again saved by the emergency rescue system. In the Soviet "tradition" this launch got no official name.

Only a month later, on February 21, came the long-awaited premier launch of the N-1, carrying a simplified L-1 spacecraft rather than the L-3 complex. Ambitiously, if imprudently, N-1—on its very first test launch—was to send L-1 unmanned by the Moon. But this was not to be. Afanasiev tells the story:

> During liftoff, in the interval between seconds 3 and 10, a spurious command issued by the KORD system [KOntrolia Raketnykh Dvigatelei, the system for controlling the rocket engines] shut off the A [first stage] unit's LFREs [Liquid Fuel Rocket Engine] Nos. 12 and 24, which were performing well. At 66 seconds, the elevated vibration caused by acoustical loads ruptured a line that feeds oxidizer to the gas generator of one of the LFREs; the leaking liquid oxygen started a fire in the aft section. The rocket could have continued the flight, because the fire was growing rather slowly, but at 70 seconds, a general command shut off all the LFREs of unit A. The engine developers feel that the logic of the command to shut down the entire system was wrong in that case. The emergency rescue system triggered; the reentry vehicle of the L-1 spacecraft, which was to perform a flyby of the Moon on that flight, landed several dozen kilometers from the launch site.[23]

Afanasiev points once again to one of the persistent weaknesses of the whole N-1 program as the main culprit, "The source of the problem was that the engines were not adequately tested. . . . "[24] It was, in fact, because of the inadequacy of engine testing, and the bald fact that lack of time and resources to develop bigger engines had dictated the use of a large number of small engines in the first stage, that the KORD system had been devised. Mishin described how it worked:

> The size of a single LPRE [Liquid Propellant Rocket Engine] in unit A was chosen on the basis of the condition of minimal expenditures for its development and manufacture. In order to enhance reliability, plans called for the backup of individual LPREs. For example, the first stage could perform flight with two pairs of opposing peripheral engines switched off; the second stage, with one pair, and the third stage, with one engine switched off.

> In order to cut off malfunctioning engines situated opposite each other, a special system was provided for monitoring their operation [KORD]. Unfortunately, that system was not able to react to the rapidly occurring processes [for example, those which precede the explosion in the turbopump assembly's oxygen pumps]. But such defects would have been eliminated during the final development testing of the individual LPREs and checked for during those engines' delivery tests.[25]

The last statement is surely wishful thinking, since the malfunctioning of engines continued to plague the N-1, as was the case with the second launch attempt, on July 3, 1969. The result was catastrophic. Mishin writes that a piece of metal somehow entered the oxidizer pump and exploded one of the thirty first-stage engines. The explosion damaged cables linked to adjacent engines, a fire erupted, and the rocket fell over, destroying the entire launch complex.[26] It is now known that it was a bolt that was ingested by the engine, leading—in the coverup fashion of the times—to a report that the engine was "sensitive to incoming foreign objects."

N-1's calamities, hushed up of course, were in contrast to the spectacularly successful launch on July 16, and landing on the Moon on July 20, of *Apollo 11,* only seventeen days after the N-1 explosion. Soviet acknowledgment of the landing by Armstrong and Aldrin, while Mike Collins orbited the Moon, was understandably restrained. *Pravda* reported the mission each day, starting with the July 16 launch, in small bulletins datelined Washington. The first bulletin was printed at the bottom of the world news column on an inside page, mixed with mundane items from Cairo, Brussels, and other cities. On July 22, on page 5, there was a six-column story datelined Washington—not Cape Canaveral or Houston—heralding the landing. The first four pages of the issue carried the full text of windy speeches by Brezhnev and the Polish Communist Party secretary, Wladyslaw Gomulka, who had been visiting Moscow.

Armstrong had stepped on the lunar surface at 9:56 P.M. Houston time on July 20, and since Moscow is nine hours ahead of Houston time, it was too late for the July 21 *Pravda*. On July 23 there was an update on the space walk, again on page 5. On page 6 an essay by Leonid Sedov, the international "front man" for the Soviet space program, congratulated the United States on the flight, although not without ritually citing all of the Soviet "firsts" in a preamble—first satellite, first manned flight, first walk in space, first impact on the Moon of a robotic spacecraft, first spacecraft to Venus, and first to fly by Mars.

Not announced to the world, however, was the fact that the Soviets had made one last effort to upstage *Apollo 11* with what was to have been the first robotic lunar soil return mission. It was *Luna 15,* which had been launched on July 13, and, if successful, would have brought the first Moon material to Earth just before the *Apollo 11* astronauts returned. An absorbing hour-by-hour tracking of *Luna 15* by the resourceful team of spacecraft trackers at Britain's Jodrell Bank is described in an excellent book on the Soviet space program by Brian Harvey. Alas, *Luna 15,* after fifty-two orbits, smashed into the Moon on July 21 and was destroyed. As Harvey puts it, "The crash site was the Sea of Crises and it marked the graveyard of the Russian ambitions to somehow beat the Americans to the Moon."[27] *Pravda* reported the mission on page 4 of its July 22 issue, with this head: "Flight of Automatic Station *Luna 15* Completed." There was no mention at all of the intent to return lunar soil.

With the *Luna 15* failure added to the loss of both manned races, the Soviet program was indeed at a low point. Nonetheless, the two manned programs were to continue for several years more. *Zond 7* became the very first 7K/L-1 to complete its full journey without mishap, albeit unmanned, circumnavigating the Moon at a distance of 1,230 kilometers and soft landing on August 14, 1969, about 50 kilometers from its designated landing spot south of the city of Kustanay in northeastern Kazakhstan.

"Theoretically," says Afanasiev, "it was that launch that could have been manned, but the leadership, after the triumph of *Apollo 8,* didn't want to give a 'go ahead' to a manned launch."[28]

The Soyuz program stayed in high gear with the launch of three spacecraft on consecutive days, October 11, 12, and 13, 1969. All told, there were seven cosmonauts in orbit—Gyorgi Shonin and Valery Kubasov in *Soyuz 6;* Anatoli Filipchenko, Vladislav Volkov, and Viktor Gorbatko in *Soyuz 7;* and the veterans Shatalov and Yeliseev in *Soyuz 8.* All three ships orbited eighty times as the crews did automatic welding, navigation experiments, and maneuvering.

But the mission was soon eclipsed by the second lunar landing with Pete Conrad and Alan Bean in *Apollo 12,* as Richard Gordon orbited, November 14–24.

The famous, and almost ill-fated, *Apollo 13* mission took place in April 1970. Jim Lovell, Fred Haise, and the late Jack Swigert converted the lunar module into a rescue vehicle and returned safely to Earth after a liquid oxygen tank in the command module exploded en route to the Moon.

Still in hopes of getting there themselves, the Russians sent Vitaly Sevastyanov and *Vostok 3* veteran Andrian Nikolayev on an 18-day

Earth orbital mission in *Soyuz 9* in June, breaking the record for the longest stay in space.

Further brightening of the Soviet outlook came from the outstanding successes achieved by the Lavochkin bureau with *Luna 16,* on September 12, 1970, which returned soil samples, and *Luna 17,* which performed magnificently by landing a robotic rover, *Lunokhod 1,* on November 17. It roamed the Moon's surface for eleven months, taking more than 20,000 photos.

But the lunar target remained unreachable for cosmonauts. One last gasp in the circumlunar program took place with the launch of *Zond 8,* on October 20, 1970. The L-1 went around the Moon successfully but had an attitude control glitch on return. The reentry was ballistic, and the spacecraft again landed in the Indian Ocean. It was the thirteenth L-1 spacecraft. "The fourteenth and fifteenth, equipped for circumlunar flight, went unused," Afanasiev points out wistfully.[29]

The UR-500K/L-1 program's demise was not only a calamity in itself. It had not flown a single cosmonaut, and its contributions to the N-1/L-3 knowledge base had been modest. In fact, its development had hindered the progress on N-1/L-3, which had its own serious problems.

Following the first two failures of N-1, there was an intensive effort to prepare the L-3 lunar orbiter and lander complex for manned flight. The lunar orbiter, a redesigned Soyuz, sat atop a newly designed lander. Mikhail Yangel, whose bureau designed the lander engine assembly, wanted unmanned orbital tests. The first, designated *Kosmos 379,* consisted of the lunar lander, minus legs, called *T2K*. It was launched on the Soyuz (SL-4) launch vehicle on November 24, 1970. *T2K* tested its main propulsion system after three and a half days in orbit—the time it would have taken to fly to the Moon—throttling the engine as it would have for lunar landing. The following day, matching the proposed stay of the cosmonaut, the landing platform was ejected and the main engine restarted as it would have for lunar ascent.[30]

On December 2, 1970, *Kosmos 382* was launched on a Proton. The payload was a prototype lunar orbiter. One of the objectives was to check out the multiple-start capability of the oxygen-kerosene Block D engine in simulated lunar duration flight. The other objective was to test thoroughly the storable propellant thrusters on the lunar orbiters in simulated lunar flight maneuvers.

The Americans started the new year with the highly successful *Apollo 14* flight of Alan Shepard, Edgar Mitchell, and Stuart Roosa, January 30 to February 9, 1971. It could not have amused Mishin to watch Shepard loft a golf ball across the lunar landscape.

The Russians also began well in 1971. *T2K* was successfully tested in orbit, unmanned, on the *Kosmos 398* mission on February 26.[31] Then came the launch of the world's first true space station, the 18.5-metric-ton *Salyut 1,* on April 19. Based on the Chelomei Almaz design, it was built by Mishin's bureau, and was lofted by the Chelomei-designed Proton launch vehicle. Four days later came a third success, albeit a partial one, as *Soyuz 10,* carrying Shatalov, Yeliseev, and Nikolai Rukavishnikov, docked with *Salyut 1,* but, as Mishin describes it, "we discovered some defects in the docking mechanism which prevented the cosmonauts from transferring to the space station."[32]

Then, however, came two heart-rending failures within a few days of each other in June. On June 6 *Soyuz 11,* which, with a new docking mechanism, carried Gyorgi Dobrovolsky, Vladislav Volkov, and Viktor Patsaev to a rendezvous with *Salyut 1.* The three cosmonauts docked successfully and after twenty-three days carrying out biomedical and other scientific experiments—another record stay in orbit—headed back to Earth. "Everything went according to plan," wrote Mishin, "until the moment during the descent when the instrument section of the spacecraft separated from the cabin. A pressure release valve opened prematurely, causing the cabin to depressurize abruptly killing the three cosmonauts."[33]

The deaths occurred on June 29. Two days earlier, on June 27, a third attempt had been made to launch N-1, from the second complex at Baikonur, the other one still a shambles after the 1969 explosion. This time the payload was mock-ups of the lunar orbiter and lander. Afanasiev recounts the event:

> Soon after the liftoff, unexpected gas-dynamic moments [eddies and countercurrents] at the base caused the rocket to roll. The rate and angle of roll grew steadily. At 39 seconds, the gyroscope-stabilized platform of the launcher control system hit its stops, and at 48 seconds, the large amount of torque started the destruction of Block B [second stage]. The emergency rescue system was only a mockup and, naturally, did not fire. At 51 seconds, when the angle of turn reached 200 degrees, the KORD system issued a command that cut off all engines of the first stage.[34]

The failures, so close together, must have traumatized the whole Soviet space program. But the Mishin team struggled on. Always in deep trouble trying to meet harsh deadlines with minimim budgets, Korolev's successor had only partial control over what monies he got. "Our country," he wrote, "could not afford to spend the kind of money that was spent by the United States for the Apollo/Saturn program."

> By 1 January, 1971, the total spending for the N-1/L-3 program (or more accurately, the amount written off for that program) was 2.9 billion

rubles. The largest financial "infusion" did not occur until 1970 (around R600 million). But even those funds, allocated directly to the ministries, were being spent at their direction, without controls. The monopolistic practices of the departments, about which everyone is talking today, were already in full flower at that time. There were serious shortcomings both in the organization and in the coordination of the work in that program.[35]

While Apollo surely outspent N-1/L-3 by a big factor, in terms of manpower effort N-1/L-3 was actually larger. Apollo, at peak, employed some 417,000 people in NASA and industry, whereas "There were 500,000 people working on N-1/L-3," says Mishin, "involving 26 ministries, 500 research institutes, plants, design bureaus."[36]

The estimate for the cost of Apollo was $24 billion in 1969 dollars, calculated to be more than $100 billion today, although even that number would likely change depending on the accounting system used. What Mishin's estimate of 2.9 billion rubles for N-1/L-3 could translate into is likewise open to question. One report, cited by the U.S. Senate in 1966, estimated that a realistic research-dollar-to-research-ruble conversion in that time was 3.5, which would mean that N-1/L-3 cost $10.1 billion.[37]

Mishin today makes no attempt to hide his contempt for the ministry leaders from whom he had to seek his money and direction, as well as the system under which he was forced to operate. "Overall supervision . . . was carried out," he writes, "by CPSU [Communist Party of the Soviet Union] Central Committee Secretary D. F. Ustinov through the USSR Council of Ministers' Military Industrial Commission [L. V. Smirnov, chairman], to which only the defense-related sectors of industry were accountable . . . [some] enterprises would fail to meet the deadlines for deliveries of components to the 'head' ministry [General Machine Building] and its 'head' OKB [ours], which were responsible for completing the work on the program within the designated time frames. We did not have any levers of influence on our own suppliers. In a word, the organization of the work on the N-1/L-3 program was typical of the 'period of stagnation' of our society."[38]

Depression must surely have set in when *Apollo 15* saw David Scott cavorting on the Moon's surface in a lunar rover in July, only a few weeks after the *Soyuz 11* and N-1 debacles.

A second test of an unmanned *T2K*, designated *Kosmos 434*, on August 12, lifted spirits somewhat. Then more than a year passed in the Soviet program, during which *Apollo 16*, in April 1972, landed John Young and Charles Duke on a plateau near the Moon's Descartes mountains, from which they drove a second lunar rover through large fields of boulders.

The fourth and final N-1 failure occurred on November 23, 1972. At launch time much hope rode with the vehicle, which had designed numerous fixes to eliminate prior problems, including the addition of vernier engines in the first and second stages to better control vehicle roll. The huge assembly rose majestically from the pad, this time carrying a full-scale lunar orbiter and a lander mock-up. Afanasiev:

> The start and liftoff went well. At 90 seconds, according to the cyclogram, six LFREs of the core propulsion system of unit A were cut off. Then, apparently because of large nonstationary loads caused by a hydraulic shock when the core propulsion system was shut off abruptly, lines for feeding fuel to the LFREs of the core propulsion system burst, and a fire broke out. The failure developed within a matter of one or two seconds. The rocket continued its flight and exploded at between 107 and 110 seconds, when the fire in the aft section reached a critical point. The emergency rescue system triggered soon after that.
>
> An investigation established that the abrupt shutdown of the LFREs led to fluctuations in the fluid column in the feeder lines. They ruptured, and a large amount of fuel spilled onto the shut-down, but still hot, engines and into the section with LFREs still operating.[39]

More engineering fixes for N-1's problems were devised and the Soviets continued their *Luna, Soyuz,* and *Salyut* missions in 1971 through 1973. *Luna 21,* built by Lavochkin and launched on January 8, 1973, was a particular highlight, carrying *Lunokhod 2,* which traversed the lunar surface for five months, taking 80,000 photos. A fifth N-1 launch was scheduled for August 1974, almost two years after the launch of *Apollo 17,* the last of six U.S. missions, which had landed twelve astronauts on the lunar surface. But before it could take place the program was cancelled by Leonid Brezhnev and Mishin was fired.

"N-1's cancellation was a cruel mistake," says Mishin. "We'd have gone to the Moon later, but better. . . . I think that the N-1/L-3M, using hydrogen upper stages, which would have been capable of putting 150 tons into LEO, was our best chance for a good vehicle for lunar exploration. There was a statement recommending N-1/L-3M, signed by all of the chief designers—myself, Glushko, Pilyugin, Barmin, etc."[40]

Since international prestige had been the program's main reason for existence, as far as the top political leadership was concerned, Mishin's anger at the cancellation seems naive. One even wonders why the pragmatic Brezhnev waited until 1974, two years after the fourth test failure, and more than five years after *Apollo 11* had won the race, to lower the boom. Whatever, it must have been delicious vengeance for the cancellation to have been carried out quickly by one of Korolev's bitter rivals, Valentin Glushko, who was appointed by Brezhnev to succeed Mishin.

Still stewing over his failure to get his engines on the Moon mission, Glushko was delighted to halt all N-1 work as soon as he was the boss. He "incinerated every notion of the N-1 with a hot iron," writes Sergei Kryukov.[41]

For many years visitors to the museum in the Korolev design bureau in Kaliningrad were unable to see even a spot of evidence that there ever was an N-1 program. There was, Kryukov wrote ruefully, "not . . . a single word about the N-1/L-3 complex, and not a single participant in the work is named. . . . Yet at the cosmodrome launch facility (in Baikonur), the following are still being used: the assembly and testing building of the N-1, even, the fueling facility and its launch pad, the command and telemetry complex, and many, many things developed for our lunar project. However, specialists coming there do not always know that this is the 'heritage' of a gigantic space program, which was tragically interrupted 20 years ago."

Sad as they were about the cancellation, however, most veterans of the N-1 era believe that the moon landing race with the Americans could not have been won with the budgets and the technology available. Their challenge was all the more unrealistic when one realizes the workload imposed on OKB-l as a result of what seem clearly to have been bad decisions by Ustinov and his lieutenants in the Military Industrial Commission.

Mishin quite naturally considers himself the program's scapegoat. But some veterans of N-1/L-3 hold Mishin, himself, at least partially responsible for the program's failings. Unable to exert Korolev's powerful political leverage, he failed to get the top-level support that the project deserved. It is also alleged that he was not an effective administrator. Nonetheless, his complaints about underfunding and mismanagement by the Soviet bureaucracy seem justified, especially when comparison is made to the Apollo program. The latter not only had generous funding but clear sailing through the U.S. Congress. Also, it force-fed the development of a systems management approach that is still a model for organizing large complex projects involving new technology. Apollo, too, had its administrative problems, notable among them allowing contractor North American Aviation to design the spacecraft with flammable materials in a pure oxygen atmosphere, thereby making possible the *Apollo 1* fire. But by and large Apollo became an exemplar of good management techniques, leading often to the adage, "If we can go to the Moon we ought to be able to . . . ". Apollo was also characterized by gutsy decisions aimed at minimizing the development time and mission complexity.

One such decision was to use a single launch vehicle with rendezvous of the returning lunar module and command module in the Moon's orbit, rather than stage multiple launches with Earth orbit rendezvous of the system components. Another astute decision, fostered by George Mueller, NASA's associate administrator for manned space flight, was to test the Saturn V/Apollo "all up," flying untested elements of the system all together rather than one by one. That decision was a big factor in meeting Kennedy's "before this decade is out" stricture.

Mishin's envy of the Apollo organizational and operational arrangement is made clear in his writings, "The (U.S.) program was not cloaked in secrecy, which facilitated the free exchange of data between all the interested organizations, and the flow of information was organized not only vertically (from the higher organizations to the lower, and vice versa) but also horizontally, i.e., between contractors. The free flow of information made it possible to track and monitor the course of the work."[42]

By contrast with the relative simplicity of the Apollo organizational reporting, there were in the USSR manned space program numerous government organizations involved, starting with the Party, and including several ministries, which managed numerous institutes and design bureaus that were responsible for the spacecraft and launchers, systems and components. The designers had to cope as best they could with this complex arrangement.

On the other hand, the Soviets claim that they would have been more conservative than the Americans in carrying out the mission operationally. "We believed," said Efraim Akim, "that spacecraft should mature completely in the automatic mode. Only then would we put a man aboard. I'm excited about what my U.S. colleagues did with Apollo, but we felt that using manual control and a single engine for the return spacecraft with no redundancy was a big risk. We didn't want to take that risk. Our approach was that a human mistake must not be the cause of death. That's why we try to mature the spacecraft to 99 percent reliability."[43] That practice was not always successful, witness the Komarov tragedy, and the deaths of Dobrovolsky, Volkov, and Patsaev in *Soyuz 11*.

Since the glasnost era, dozens of newspaper and magazine articles, and numerous books have been published in the open Russian literature rehashing the aborted manned Moon programs. Arguments were elaborated, sometimes heatedly, about funding inadequacies, personality conflicts, political interference, and technology choices.

One of the more contentious issues was the choice of the N-1 engines.

Here are a few of the opinions on the engine design recorded in interviews:

Yaroslav Golovanov: "The big mistake was not developing a large engine. Korolev kept Glushko from developing the F-1 type engine because of the time it would have taken to build the ground test facilities. His main problem was to solve the very complex problems of achieving the thrust and capacity required for the moon mission with the simplest technology."[44]

Efraim Akim: "Korolev would have worked out the N-1 problems eventually. I don't agree that a big rocket engine was what was needed."

Boris Gubanov: "Afanasiev [Sergei, Minister of General Machine Building] called a meeting after Korolev died in 1967. Mishin presented a graph showing that he promised to fulfill a manned lunar landing in April 1970. It was at that meeting that Pilyugin was the only person to state that this was an 'avantura,' an uncommon word to use in an official environment, meaning an unrealistic crazy dream. It was unrealistic with respect to time and with respect to engine performance. It was impossible to complete the full cycle of testing. Many systems which should have been tested on the ground were delayed until flight test. The rocket was untested. It was 'raw.' Test flights began in 1969. There were four flights. They all failed, against a background of U.S. successes, one landing after another. My opinion is that the decision to cancel the N-1 was correct. Many articles say that a fifth N-1 flight would have been successful, but nobody noted that Mishin himself admitted that it would have been successful only if the Kuznetsov engines had been fully tested, and that would have taken another five years . . . if Mishin says that the government was responsible for the problems in the N-1 development, he is being unfair. All of us worked in the same system. I believe the failure of the N-1 was Mishin's own fault. He turned out to be a poor organizer and designer who didn't defend the importance of testing. Korolev was responsible for the original scheme and Mishin was his right hand."[45]

In apparent disagreement with Gubanov and others about the alleged lack of sufficient attention given to engine testing, I. B. Afanasiev, who wrote a series of articles in 1993[46] reviewing the procedures followed between 1963 and 1967, points out that Kuznetsov's N-1 engine specialists even benefited from an exchange of experiences with their counterparts in the Glushko design bureau who had begun development of the RD-253 two years earlier (RD-253 was the very successful engine used on the Proton).

"Specialists in Kuznetsov's OKB," he wrote, "possessing a great experience in the field of gas turbine engines and a high culture of aviation manufacturing, using advanced technical and technological solutions, managed to make a perfect LRE [liquid rocket engine]. That was achieved first of all because of turbopump performances higher than those of the RD-253." He says that "the NK-15 and NK-15V,[47] originally designed for heavier operating regimes, were downrated for use on the N-1 in order to increase operating capabilities and reliability." Artificial damage of the combustion chamber walls was even introduced to test survivability, says Afanasiev. "Those tests confirmed the large margin of durability and workability of the engines, as well as their high reliability." Still, Glushko was critical of the engine, whose progress he watched "carefully." Just before serial production of the engine was authorized, one of the NK-15 prototypes began to smoke during a test-firing. Glushko, at a meeting of a joint commission of chief designers of liquid rocket engines (surely a uniquely Soviet invention), said, "You see for yourself that the engine is bad. It is not suited for work, much less for installation on such an important vehicle as the N-1." However, says Afanasiev, "it became clear to all members of the commission that the defect was purely technological. Soon it was eliminated and no similar malfunction ever occurred. The commission recommended the engine for serial production."

Mishin still feels pride about the N-1 and its engines: "N-1 was a good design, using the best kerosene and oxygen engines in the world, even now . . . the first closed cycle engines ever made." It must be said in support of that statement that the N-1 engine that would have been used on the fifth test, which never took place—the Kuznetsov NK-33—is thought of so highly today that it may power a future American commercial launch vehicle.[48] "Its thrust-to-weight ratio is higher than any lox-kerosene engine ever made," conceded Ivan Bekey, a NASA propulsion expert who visited the Kuznetsov plant in Samara in 1994. "It also has an impressive multiple-start capability. One of the engines had seventeen firings, and although they were not full duration, all they had to do was replace the seals. We saw the fully disassembled engine. The pre-burner assembly was beautiful."[49]

The problem in the late 1960s, however, was that there were thirty of those engines needed for the first stage of N-1 and they needed exhaustive testing as units and as an ensemble. "The Americans," said Mishin, "were able to test an entire assembled engine module on their test stands and install it on the launch vehicle and fly it without a takedown inspection. But we tested in pieces and did not even dare to think of firing all thirty motors in the first stage as a full assembly. Then the pieces were assembled, without guarantees, of course, that they were properly run in."[50] Sergei Khrushchev recalled that only two of each batch of six engines were actually tested; the other four were installed with no testing at all. "This system of statistical control did not prove appropriate," he wrote in 1994, "as the first flight tests demonstrated."

What's more, the designer, Kuznetsov, who had taken the job of developing the engine because Glushko had quit the project, had "no experience in creating such powerful machines and . . . no test facilities base for their development. . . . "[51]

Although cast in the role of the heavy by some, Glushko loved his new power and visibility at the helm of OKB-1, which he eventually renamed NPO Energia.

Glushko's original designs—looked at objectively in retrospect—were sound. However, "One of the technological mistakes of Glushko, which had an appreciable influence on subsequent history, was his long-held belief that hydrogen was not appropriate for rocket engines. The result was that the United States incorporated hydrogen engines in the flight to the Moon while we were still discussing it." That rather serious indictment came from Yuri Demyanko, chief of the space energetics division of the Scientific Research Institute of Thermal Processes (NIITP) in Moscow.[52]

From his new seat of influence, after putting an end to the N-1 program, following Brezhnev's orders, Glushko tried to get support in 1974 for a new manned lunar expedition, using, of course, his concept for a new powerful rocket called the Vulkan. By that time, however, the Soviet military wanted to put their rubles into building a shuttle, as the Americans did.

"It must have made Korolev spin in his grave when Glushko was made head of the Korolev design bureau," Mishin told me.[53]

For whatever reason, Glushko never had been forced into a role of anonymity as Korolev had been. In the middle of the Cold War, he had even been allowed to write a book (published in English in 1975) recounting his rocket engine accomplishments.[54] And he made sure that it would be his own account that would document space history by becoming chief editor of an encyclopedia titled *Kosmonavtika* in 1985, four years before his death.[55] Although his own biography in that volume shows uncharacteristic restraint, he could not resist "deselecting" some of his enemies, Mishin prominent among them.

He was very gracious, however, in eulogizing Korolev in a memorial book published in 1980. After praising SP's "unusual organizational abilities" he wrote a line that revealed the hard-won experience he had shared with his sometime collaborator and antagonist. "I would especially like to stress his goal-orientedness, and the persistence that characterized his activity."[56] Finally, he wrote, " . . . in the history of the development of our country's rockets, by the size of the contribution made . . . Sergei Pavlovich occupies first place after Tsiolkovsky."[57]

Glushko died on January 10, 1989, 23 years after Korolev's passing. Mikhail Gorbachev and other Party leaders signed his obituary.

As for whether the Russians would have beaten the Americans if Korolev had lived—the answer is a clear no. As with other heroes whose lives ended at the peak of their achievements, Korolev has been virtually canonized by a country which, after all, recognized no saints. But as substantial as his achievements were, he was not omnipotent. His human capabilities are put into perspective by one of his close advisors, Mark Gallai, the former test pilot who had responsibility for training the first cosmonauts: "I respect Korolev's abilities in developing complex projects. However, I'm not sure we would not have had failures. The design decisions on the Komarov spacecraft were all made during Korolev's time. The same with the N-1."[58]

It was not until 1989 that the existence of the N-1 program was finally acknowledged, seventeen years after the booster's fourth and final failure. The first detailed summary of the N-1/L-3 program appeared in *Izvestia* in August of that year.[59] It was probably, in fact, the pressure of Russian journalists which, in the year of the twentieth anniversary of the *Apollo 11* landing, resulted in the authorization to declassify the N-1 story. The first Westerners to get a glimpse of the actual hardware were three professors from the Massachusetts Institute of Technology who visited the museum of the Moscow Aviation Institute in November. Professors Jack Kerrebrock, Laurence Young, and Edward Crawley photographed elements of the vehicle and gave the photos to the *New York Times,* which ran a front page story on the revelation. Only then, well after the inception of the glasnost period, did memoirs begin to appear in the Russian literature by those who had been involved in the momentous program.

Perhaps the saddest testimony to the ludicrous nature of Soviet officialdom's secrecy phobia about the Moon program is revealed in a recollection by Roald Sagdeev. Sagdeev, the former director of the USSR Academy of Science's Space Research Institute, is now a professor at the University of Maryland. In his memoirs, Sagdeev tells of asking Mstislav Keldysh why the N-1 project wasn't closed sooner, since the Americans had already made several trips to the Moon when the cancellation was made.

"Young man," Keldysh told him, "you should understand simple wisdom. Nobody wanted to cancel it. It would have meant confessing before the government the failure of this expensive project, probably worth two billion rubles. The documents we have to sign now are very crafty. You will find out how smart we are. We are closing the N-1 project not as losers, but as ingenious people who have found a much cheaper and more intelligent way to explore the Moon."

"Mstislav Vsevolodovich, do you mean the successful series of small unmanned probes sent to the Moon to return with lunar sample?" asked Sagdeev.

"Yes, of course," said Keldysh.

"By that time," writes Sagdeev, "I was no longer a simpleton who would buy such an explanation. I immediately interjected, 'But these are two completely different genetic lines in the program. The people who design and send unmanned rockets are indeed the heroes, except they have nothing in common with the N-1 team.'

"Keldysh looked at me seriously and said, 'There is an opinion at the top that it would be better for all sides, including the future space program, not to create negative emotions.'"

Sagdeev writes, "I was completely stunned by such official hypocrisy."[60]

Hypocrisy was only one of the sins that did in the Soviet Moon program. Add first of all inadequate funding, then bureaucratic bungling, the vicious rivalries between some design bureaus and their chiefs, political infighting, duplication of effort, and perhaps the trait that most seriously hampered progress in the Communist system on many fronts—a paranoid secrecy that meant that virtually no part of the program was examined by the press or the citizenry.

Scenarios for Landing Cosmonauts on the Moon

In undertaking the mission, one difficulty the Russians faced was geographical. A launch from Baikonur, which is at 45 degrees latitude, requires more energy to reach the plane of the Moon's orbit than a launch from Cape Canaveral, which is at 28 degrees latitude.[61] From Baikonur a temporary orbit must be achieved and then a transfer must be made to an orbit coplanar with the Moon's orbit, whereas from Cape Canaveral there are windows for launching direct to the Moon.

Even after the 1964 decision continuing the L-3 development, and right up to the time of the cancellation of the program, there had been no final approval given to send a cosmonaut to the Moon. The decision was to be made after the completion of tests of the elements of the N-1 and L-3 system in unmanned flights all the way to the Moon.

By 1970 there were still several approaches being considered by the former Korolev design bureau, (called TsKBEM for Central Design Bureau of Experimental Machine Building, during this period). Korolev's own favorite, assuming the tests were successful, had been called simply N1/L-3. It involved a single N-1 launch and lunar orbit rendezvous. However, as I. B. Afanasiev points out, "One of the biggest difficulties . . . was acknowledged to be the lack of reliability in the docking of the lunar orbiter . . . and the lander . . . after it ascended from the lunar surface, a difficulty that stemmed from the poor capabilities of the electronic systems of the two spacecraft, the poor knowledge of navigation conditions near the Moon, and the impossibility of rendering the cosmonauts comprehensive support from Earth. . . . "[62]

Recognizing those difficulties, and aware that the U.S. success with *Apollo 11* in 1969 made it logical to design a mission that could put more than *Apollo*'s two people on the Moon and allow them to stay on the lunar surface for a longer period, Mishin planned a follow-on scenario, which also had been conceived during Korolev's time. Called N-1/L-3M, it would have involved two launches of the N-1 booster, with docking in lunar orbit, and the landing of three cosmonauts.

A third scenario, which got no name, would have required a single launch of a more advanced booster using engines fueled by liquid hydrogen and liquid oxygen in the upper stages. This strategy, too, had been developed during Korolev's time and, in fact, three different engine design bureaus—headed by A. M. Isaev, A. M. Lyulka, and N. D. Kuznetsov—had been assigned to preliminary work on the engine development.[63] Naturally, Glushko, who was quoted with relish by Mishin as having said in 1961 that "liquid oxygen is nowhere near the best oxidizer, and liquid hydrogen will never be of any practical use in rocket equipment," was not included among the designers.[64]

A fourth scenario, which plagued Korolev politically almost to the end of his life, but which never got serious support, was Chelomei's approach, which would have involved development of a new launcher, the UR-700. It would have delivered two cosmonauts directly to the lunar surface and returned them to Earth without the need for lunar rendezvous.

A fifth scenario, which would have used Mikhail Yangel's R-56 booster, never developed beyond the conceptual stage.

Detailed below are the N-1/L-3, the N-1/L-3M, and the UR-700/LK-700 scenarios:

The N-1/L-3: One Launch, Lunar Orbit Rendezvous, One Cosmonaut on the Moon

By 1970 the outline of this scenario had been rather fully developed. It would include (1) delivery to the Moon of two lunar rovers;

(2) delivery to the Moon of an unmanned spacecraft to back up the lunar lander plus two lunar rovers, equipped with TV cameras, which would have moved over to the lunar lander and its backup and checked them out; (3) delivery to the Moon of the lunar lander carrying one cosmonaut, leaving one cosmonaut in lunar orbit, with the cosmonaut on the lunar surface driving one of the rovers to the backup lunar lander if something was wrong with the first one.

The cosmonaut was to spend six hours on the lunar surface. The whole system on the launch pad was to include the four-stage N-1 11A52. The stages, translated into English from the first four letters of the Russian alphabet, are identified as A, B, V, and G. The Moon vehicle, L-3 plus Block D, was the fourth stage, carrying its own propulsion. The lunar orbital station (LOS) consisted of the lunar orbital complex (LOK), which carried the cosmonauts; the lunar landing complex (LPC), which itself included the lunar lander unit (LPU); plus Block D (D in the Russian alphabet follows A, B, V, G); Block E, plus the lunar cabin (LK); then Block I (I); the controls (PAO); the Earth reentry vehicle (SA); the cosmonauts, compartment (BO); and the docking rendezvous cone.

The two lunar rovers would have been launched on the Chelomei UR-500K (8K82K) launcher, about one week apart, two or three months before the cosmonauts. The rovers would have gone to the area on the Moon where the cosmonauts were due to land and send TV pictures back to Earth to identify an ideal spot for the landing.

About one to two months later, the backup lunar lander would have been launched and landed near the rovers.

No less than twenty to thirty days later, the main spacecraft with the two cosmonauts would have been launched, using the first three stages of the N-1 to put the spacecraft into LEO (low Earth orbit). The orbit would be so chosen as to enable the fourth stage to ignite twenty-five hours, or seventeen orbits, after the launch, while the spacecraft was at orbital perigee. If something were to go wrong, another try for ignition could be made two orbits later.

The fourth stage would take the lunar orbital spacecraft to lunar transfer orbit and separate—it wouldn't go to the Moon. Flight time to the Moon would have been 101 hours. Block D would make two corrections, one eight to ten hours after leaving Earth orbit, the second one ten to twenty-four hours before the start of deceleration to the Moon (the time to be determined by orbital parameters).

Then Block D would decelerate the spacecraft to a circumlunar orbit about 150 kilometers above the lunar surface. The spacecraft would be in circumlunar orbit for seventy-seven hours, or thirty-nine orbits, including the time when the cosmonaut descended to the lunar surface, stayed for six hours, and returned. A landing orbit would be established with an apogee of 100 kilometers and a perigee of 20 kilometers—requiring corrections on the fourth and the fourteenth orbits. Then one cosmonaut would EVA from the orbiter (LOK) to the lander (LK). LK would decelerate so it could separate and land. Block E would provide a soft landing. The cosmonaut could have maneuvered the lander—but not much—to pick a good spot near the backup.

After the landing one rover would be directed by the Earth control center (the cosmonaut in orbit can't do this because he is in the radio shadow of the cosmonaut on the Moon) to move over to the lander. The cosmonaut would have spent six hours on the Moon. During this time a computer check from Earth control would have been made of the relative position of the lander and the circumlunar spacecraft. Block E would launch the lunar cabin on top of LK back to rendezvous. The lunar cosmonaut would EVA to the orbiter LOK, then both cosmonauts would crawl into the Earth reentry vehicle for return to Earth.

On the thirty-eighth orbit the lunar cabin gets dumped, and on the thirty-ninth orbit Block I ignites (on the dark side of the Moon) and returns to Earth with BO, the cosmonauts' compartment and SA, the Earth reentry vehicle. It takes eighty-two hours to get back to Earth. Block I makes two corrections, at twenty-four hours, and at forty-four hours after leaving circumlunar orbit. Two hours before reentry, the spacecraft orients itself into a favorable reentry mode, the cosmonauts crawl into the reentry vehicle, separate from the cosmonauts' cabin and Block I, and after penetrating the atmosphere soft land in the Indian Ocean, or on a test site in the USSR.

The N-1/L-3M: Two Launches, Lunar Orbit Rendezvous, Three Cosmonauts on the Moon

The pragmatic Mishin recognized that even if N-1/L-3 were successful in placing one cosmonaut on the Moon for six hours, the achievement would not compare favorably to that of *Apollos 11, 12, 14, 15, 16,* and *17.* Those six missions had allowed twelve astronauts to spend days, not hours, exploring the lunar surface, and to return some 400 kilograms of lunar material for examination by Earth scientists. Hence Mishin opted for a bolder variation of N-1/L-3, termed N-1/L-3M. It was, he wrote, " . . . a more advanced version of a lunar mission than that of the United States and which could have been carried out (with the 'amassed stockpile' of N-1 and L-3 hardware that was later destroyed) with only a small increase in spending in 1975–1976."

In early 1972 all of the chief designers involved in the development had signed off on a detailed plan for carrying out the mission. Also, he says, "We finally managed to get technical tasking from the USSR Academy of Sciences. . . . " It must be noted that no such specifications had ever been received from the Academy for the first version of the mission.[65]

The objective would be to land three cosmonauts, sometime during 1978–1980, "in any region of the lunar surface, where they would stay for up to 14 days (with a subsequent increase up to 30 days), with a direct return from the Moon's surface at any time."[66] In fact, writes I. B. Afanasiev, "The designers calculated that by 1978–80, with financing that would not exceed the routine financing of the standard N-1/L-3 program, the USSR would be capable of steadily developing the necessary infrastructure for creating a lunar base and for conducting lunar missions of average duration (up to three months)."[67]

Johnson details the mission profile:

> . . . one N-1 would place an unmanned 104 metric ton assembly named GB-1 into low Earth orbit, followed by a second N-1 with the manned GB-2 complex weighing 103 metric tons. Each assembly would leave Earth orbit and travel independently to the Moon and there enter lunar orbit.
>
> The GB-2 assembly would deliver a large, multi-man lunar lander and return module which was to be joined in lunar orbit with a powerful descent stage carried as part of GB-1. From this point on the landing profile was similar to that anticipated for the N-1/L-3. After performing the de-orbit maneuver, the descent stage would be jettisoned, and the lunar lander would perform the final deceleration for a soft touchdown. On the Moon, the mass of the lunar lander would be nearly 24 metric tons.
>
> Early stays on the Moon might be restricted to 5–14 days, but longer expeditions were planned on later flights. The duration of the mission would be tied to the number of cosmonauts on board, e.g., 5 days for three cosmonauts and 14 days for two cosmonauts. At the end of the exploration program, the lunar lander's 19.5 metric ton ascent stage would blast-off the Moon for a direct return to Earth.[68]

Afanasiev elaborates on the landing and exploration sequences:

> A number of versions of the lunar spacecraft had been examined. For example, one version put the cosmonauts at launch from Earth in the reentry vehicle, which was secured to the upper part of the craft. When performing in-flight operations and just before touching down on the Moon, they would move through a crawlway-chute to the living/working quarters of the craft, which was mounted beneath the reentry vehicle. In a different version, above the propulsion system of the spacecraft

was a cocoonlike living compartment, inside the upper part of which the reentry vehicle was secured. A large part of the service equipment of the spacecraft was located in a sealed, cylindrical instrument section inside the living compartment. During various operations in flight and on the lunar surface, the cosmonauts would leave the reentry vehicle and work in the interior of the living compartment, which provided not only free access to the control instruments, but also a good view, which would make choosing the landing site easier. Upon return to Earth, the living compartment would separate and the reentry vehicle would exit it just before reentry.[69]

UR-700/LK-700: One Direct Launch, Two Cosmonauts on the Moon

This scenario, by Chelomei in OKB-52, never got past the conceptual stage. It was based on a proposed new launch vehicle, the UR-700, which could lift 145 tons into orbit. This is the way it would have worked, as described by Afanasiev:

> The profile for a "direct" mission with the UR-700/LK-700 was virtually the same as the "direct" version of the Nova/Apollo, which the Americans had rejected in early 1961. It called for placement of a spacecraft and an upper stage into near-Earth orbit and a subsequent launch from orbit to the Moon. Braking and entry of the spacecraft into circumlunar orbit, as well as the descent from orbit and decrease of the main velocity, would be done with a special retrorocket unit. Several kilometers above the lunar surface, the retrorocket unit would be jettisoned, and a soft touchdown of the spacecraft on landing legs would be achieved by throttling the LFRE of the ascent stage, as would be done in the lunar lander of the N-1/L-3 program. The LFREs of all the units, as well as the engines of the UR-700 launchers, would have to burn nitrogen tetroxide/unsymmetrical dimethyl hydrazine.
>
> The two cosmonauts of the LK-700 spacecraft would be in a return module similar to the module developed for the UR-500K/LK-1 flyby program.
>
> After the mission tasks associated with the visit of the crew to the lunar surface had been performed, the landing attachments would be separated and LFRE of the ascent stage would fire, operating at full thrust. After lifting off from the Moon, the LK-700 could either first enter a circumlunar orbit and then leave from it for Earth (one version of the project) or it could immediately enter a flight trajectory for Earth (second version). After a flight trajectory correction had been made with the LFRE of the ascent stage, the return module would have to separate just before arrival at Earth, with subsequent reentry into the atmosphere, controlled descent, and a parachute-assisted landing.[70]

IRONIC EPILOGUE

18

> Looking back, I regret that the quarter of a century of my service as Ambassador to Washington fell mainly during the complex period of Soviet-American rivalry, even hostility. If only it had been possible then to build a sensible foundation of trust between our two nations, how much could have been done and could still be done by both sides to bring our nations closer together.[1]

THOSE ARE THE words of Anatoli Dobrynin, whose tenure as Soviet ambassador to Washington, from 1962 to 1986, blanketed some of the most intensive years of Soviet-American space competition.

Now, it seems, much is being done to bring both sides together in that same arena. The first elements of the International Space Station (ISS), a behemoth project costing perhaps $34 billion, are scheduled to be placed in low Earth orbit beginning in 1997. The lead Russian organization on the project, which is being led by the United States, although it involves many other nations as well, is none other than the S. P. Korolev Rocket-Space Corporation Energia (RSC Energia).

How did this unlikely scenario come about? It did not start well at all. Following the Moon race, the two nations—while aiming thousands of nuclear bomb-carrying rockets at each other—kept reflexively upping their space technology capabilities to stay apace. The USSR's conviction that NASA was cloaking military space ventures in civilian clothing led to a misguided copycatting of the U.S. Space Shuttle. The Shuttle, begun in the early 1970s, had been meant to reduce the cost of carrying payloads into orbit by its reusability and its frequent launches. There was talk of perhaps fifty flights a year, turnaround time of a few weeks, getting the cost of putting a pound of payload into orbit below $100. The actual cost has been of the order of $15,000 per pound, depending on the bookkeeping technique, and not considering the enormous expenditure for research and development.

The frequency of flights has been about seven a year since the disastrous Challenger explosion of January 1986. The level of reusability hoped for was never achieved, nor was the turnaround time between missions and, ironically, the military missions never materialized at anywhere near the level feared by the Russians and hoped for by NASA.

Although some of these missions were far-fetched, they were given credibility by the Soviet military and so it was inevitable that a Russian shuttle replace the N-l as the USSR's next big project in space. Gyorgi Vetrov put it this way:

N-1 had been proposed as a universal launcher, but the military tasks they envisioned—like SDI, and a global surveillance system—were too advanced for the N-1 technology. The main logic used for the N-1 cancellation, regardless of the Moon program failure, was the development of the U.S. Shuttle. The military decided to imitate the U.S. shuttle. It was stupid logic but it's a fact.[2]

Mishin points once again to his bête noire, Glushko, as the villain in the shuttle decision. Glushko, he wrote, " . . . managed to convince D. F. Ustinov and other leaders of the wisdom" of " . . . developing a reusable space transport system (like the Space Shuttle) with an oxygen-kerosene LPRE with a thrust of 700–800 tons."[3] However, this seems an unfair accusation since it was the military, rather than Glushko, which demanded the shuttle initiative while Glushko had been pushing for a renewed attempt at the lunar landing.

Another slant on the Soviet military's motivation for building their own shuttle was described to me by Efraim Akim:

> When the U.S. Shuttle was announced we started investigating the logic of that approach. Very early our calculations showed that the cost figures being used by NASA were unrealistic. It would be better to use a series of expendable launch vehicles. Then, when we learned of the decision to build a shuttle launch facility at Vandenberg AFB for military purposes we noted that the trajectories from Vandenberg allowed an overflight of the main centers of the USSR on the first orbit. So our hypothesis was that the development of the shuttle was mainly for military purposes. Because of our suspicion and distrust we decided to replicate the shuttle without a full understanding of its mission.
>
> When we analyzed the trajectories from Vandenberg we saw it was possible for any military payload to reenter from orbit in three and a half minutes to the main missile centers of the USSR, a much shorter time than an SLBM [submarine-launched ballistic missile] could make possible (ten minutes from off the coast). You might feel that this is ridiculous but you must understand how our leadership, provided with that kind of information, would react. Scientists have a different psychology than military. The military, very sensitive to the variety of possible means of delivering the first strike, suspecting that a first strike capability might be the Vandenberg Shuttle's objective, and knowing that a first strike would be decisive in a war, responded predictably. That's why confrontation is so dangerous. Fortunately, it's all academic now.
>
> *Slava bogu* [thank God].[4]

Out went the manned Moon expedition, in came the development of the Soviet shuttle, named *Buran* (blizzard), and its new, giant launch vehicle, *Energia*. The latter became a central project at the Korolev bureau under its new Chief Designer, Glushko.

"The main program for *Energia* involved about 1,200 organizations and many tens of thousands people," I was told by Boris Gubanov, a major figure in the development. Sergei Afanasiev was the minister in charge. The factories reported to him, and he personally acted as administrator. He had an even stronger role than Glushko." Gubanov continued:

> The prevailing opinion was that the development of the *Energia* launcher would be very difficult, and some said it would never fly. The skeptics were headed by Mishin. After the first flight, Mishin said, "One flight means nothing, wait until the next one. The *Energia* designers should not be decorated, but imprisoned."
>
> That remark appeared in the press. We were so angry his antagonism actually helped us. Mishin was upset because his program (the Moon expedition) had been cancelled and now we had the established program.[5]

The "established program" to build *Buran* was even further stimulated by the fear of a U.S. program, approved in 1983 by President Ronald Reagan, to develop a Strategic Defense Initiative (SDI). The Shuttle, already considered a possible first-strike hydrogen bomb–deliverer as per Akim's account, was then thought of as a key component of SDI. Gubanov reviewed the Soviet logic:

> When the SDI program emerged it was described as a high-energy launch vehicle, capable of reacting to communications satellites and strike weapons, that would be in orbit for a long time. The space shuttle should have the capability of rendezvousing with what would be a 20-ton SDI vehicle and returning it safely to Earth. Nobody had yet developed such a system. That was what *Buran* and the U.S. Shuttle were supposed to do.
>
> Now that we have peace we have to ask why we have these systems.[6]

The same question was asked by the Russian government after the downfall of Communism, and the answer was that the *Buran* and its *Energia* booster were not needed at the price it would have cost to keep them alive, especially since the Cold War was over. *Energia* flew just twice, in 1987 and 1988, the second flight carrying *Buran*, unmanned. It was *Buran*'s only flight. Two flight-ready *Burans* and three *Energia* launchers now languish in gigantic buildings in Baikonur, only a few miles from the facilities that test-launched them. An entire cosmonaut contingent, formed in 1978, was being trained to fly *Buran* at the time the program was mothballed in 1993. It was never formally cancelled, but funding dried out and eventually it simply stopped. Gubanov told me wistfully that, "*Energia*'s future is bleak if it is not used for the Space Station." In fact, the future of both *Energia* and *Buran* is, indeed, bleak, as they will get no jobs in the ISS program.[7]

The future of the U. S. Shuttle, although it costs perhaps $4 billion a year to perform its seven missions, is assured, since it will carry most of the payloads to ISS. It continues to do important work: conducting experiments in microgravity; determining the long-term effects of human weightlessness; putting up large satellites, and repairing them in orbit, like Hubble Space Telescope; and—in preparation for ISS—learning how to assemble structures in space and carrying astronauts, as well as a few cosmonauts, to dock with the Russian space station *Mir*.

RSC Energia is the firm that designed and developed *Mir,* which has been in orbit since 1986, housing crews of Russian and foreign astronauts the whole time. It is the largest space firm in Russia and is getting most of the $472 million that NASA is paying to the Russian Space Agency for the Shuttle-*Mir* missions, and for some of the ISS elements, including the docking system. The current General Designer of RSC Energia is Yuri Semenov, successor to Glushko. Oldtimers still call the town where the company's massive plant is located Podlipki—the railroad stop—although it was officially renamed Kaliningrad, for a former Party leader. Then, in 1996, Boris Yeltsin gave the city a name much more appropriate to its history: Korolev. In the middle of the town stands an imposing, larger-than-life statue of a formidable, stern-looking Sergei Pavlovich. The community's entire livelihood—support services, school system, municipal transport, banks—revolves around the rocket-space facility. A supersecret place for decades, the enterprise had some 35,000 employees throughout the former Soviet Union at its peak of operation in the 1980s. Korolev's office in the central headquarters building, now occupied by Semenov, is visited regularly by American, European, and Japanese space engineers and government officials collaborating on ISS, which is expected to take five or more years to construct.

Although other former Soviet companies will be involved in ISS as well, RSC Energia is by far the leader. Its responsibilities include the development of specialized modules; transport, cargo, and perhaps rescue spacecraft; and docking systems. Between ISS and its commercial initiatives, the company has projects with organizations all over the world. They include such American firms as Boeing Company, Lockheed-Martin Corporation, Loral Space Communications, McDonnell-Douglas Corporation, Rockwell International Corporation, as well as European and Japanese firms. Talks with the Chinese have even taken place.

Who would have thought, in Korolev's time, that such substantive space technology cooperation would ever be possible? Even the principal cooperative project that took place after Korolev's death—the 1975 *Apollo-Soyuz* "handshake in space"—involved minimal technological revelations. Exchanges in softer subjects—space medicine, weather observations, planetary investigations—were carried on quietly throughout the Cold War, and there were many well-meaning toasts to cooperation over the decades at international space gatherings, such as the congresses of the International Astronautical Federation (IAF) and the Committee on Space Research (COSPAR). But the informal remarks of Hugh Dryden, the late Deputy Administrator of NASA, to his Russian counterpart, the late Anatoli Blagonravov, at an IAF congress in Washington in 1961, best characterize the realpolitik of the situation in those times: "If we had cooperation neither of us would have a space program."

The fact is that the space and missile initiatives were prime instruments of the vicious competition between the two countries at virtually all military and political levels. Funds for space exploration and new space weapons flowed abundantly on both sides. Space scientists complained, not without justification, that the manned programs were consuming the greater part of the budgets. But the fact is that space science and applications funding, as well as commercial space investments, were greatly stimulated by the general increase in NASA, DoD, and Soviet missile and manned space projects.

The wisdom of spending about $24 billion on Apollo in the 1960s, and perhaps a like amount, at least relative to the Soviet gross national product and science and technology manpower and resources, on Korolev's N-1/L-3, surely deserves the questioning it got by the public, the pundits, and the politicians in the 1970s and 1980s. But the fact is that those programs look like solid investments in retrospect, having done what John Kennedy said they would do when he asked the Congress, in 1961, to back their funding. With his decision, endorsed by the Congress, to send a U.S. astronaut to the Moon and carry him safely back to Earth "before this decade is out," the young, pragmatic President made a positive point out of the fact that the effort would be difficult. It would be worth doing because it was difficult, he said, since it would lever beneficial results that far exceeded the lunar landing in value to the country.

Indeed, the Apollo program force-fed the development of technologies that have had enormous benefit on the economy. Apollo (and N-1/L-3 in the USSR) pushed hard on the maturation of computers; microminiaturized electronics; new materials of high strength, heat resistance, and low weight; and rocket engine technology. Spinoffs from space developments, although sometimes overblown by space adherents, have been very significant, including important medical instrumentation and knowledge of human physiology. Some specialists have a bullish outlook, as well, for products manufactured in the unique microgravity of space, including pharmaceuticals, crystals, and new metals and alloys, although results so far have been ambiguous.

Of equal or greater moment, not only for the United States and Russia, but for the rest of the world, were the satellites created out of the government-industry space infrastructure, which are now operational for business and personal communications and for weather, weapons, and earth resources observation. The claim has been made that NASA's total expenditure over the years for research and development on communications satellites is a tiny percentage of just one year's tax revenues from the business generated by those satellites. How were they put into orbit? Mostly by rocket-carriers of manned and unmanned spacecraft conceived for military purposes or for the exploration of the unknown.

As reported earlier in this book, Lyndon Johnson once said that the entire expenditure for the space program was more than justified by just one development—reconnaissance satellites. There seems no question that the ability of the USA and USSR to monitor each other's weapons arsenals has made the idea of nuclear war unthinkable.

It is very satisfying to learn that these satellites are now being used for peaceful purposes. The *New York Times* has reported that some of the twelve $1 billion U.S. spy satellites in orbit at any given time will soon monitor some 260 sites on Earth, and eventually 500, to record data on changes in the environment. Candidate sites include coniferous-deciduous forest in Russia, forested mountain slopes in the Czech Republic and Slovakia, glaciers in Switzerland, prairies in Kansas, the Mojave desert in California, the coastal zones around the Galapagos Islands, tropical rain forests in Costa Rica.[8]

Going the other way, commercial Earth-observation satellites whose sensors have greatly improved resolution now have the ability to monitor military movements with a high degree of accuracy. Ted Nanz, the president of Spot Image USA, has written that " . . . with the onset of the Bosnia peacekeeping mission . . . we [have] seen them [Spot Image and Landsat satellites with 10-meter and 30-meter resolution] used during virtually every stage of the military effort and in a variety of applications." They have enabled Bosnia peace negotiators "to reach agreement on boundaries and other geographic demarcations, helping to generate a peace that is much easier to enforce." They have also been used to allow each faction involved in the negotiations to "fly over" a three-dimensional landscape covering the areas under contention; to enable pilots to plot flight paths around hostile areas, previewing their flights on software onboard the airplanes; to manage the clearance of mine fields and to evaluate the condition of local transportation systems.[9]

Other applications and scientific satellites, not only from the USA and USSR but from other nations, have made startling discoveries of the nature and evolution of the Moon, the Sun, every planet except Pluto (and that is on the agenda), the asteroids; the composition of the Earth and its atmosphere, its radiation and magnetic fields, its geodesy, its ozone leakage, its pollution, its soil and coastline erosion, the health or sickness of its crops and its timber, its resources of oil, coal and minerals, its volcanoes, its earthquake faults, its wetlands, its urban sprawl, its oceans, even the early history of its civilizations by detecting from orbit outlines of millennia-old construction.

Astrophysical satellites and the much-troubled Hubble Space Telescope—repaired by shuttle-orbiting astronauts—have told us more about the universe in months than Earth-based telescopes had revealed over decades.

What is unfortunate, however, is the fact that the massive Apollo capability was allowed to deteriorate. Certainly some of the saddest sights in the United States are the huge Saturn V vehicles visible in outdoor museums at NASA's Cape Canaveral, Huntsville, and Houston centers. Some of the surviving hardware was used in 1973 to put up *Skylab*, America's first space station. *Skylab* itself was lofted by a Saturn V and then three crews were orbited by Saturn IBs to occupy the station for 28, 59, and 84 days, all records at the time. Much was learned during these stays about the abilities of astronauts to live and work in space, and to make important Earth and solar observations. Especially heroic was the emergency repair work by the first *Skylab* crew of Pete Conrad, Paul Weitz, and Joe Kerwin. Vibration at launch had torn away the station's meteoroid/thermal shield, which had then ripped off one of the craft's solar panels. Inside temperatures had soared to 190°F. After many EVA hours of grueling work, the crew managed to put a large parasol over the workshop, cooling it down enough for the mission and the subsequent two *Skylab 2* and *3* missions. Sadly, *Skylab* was allowed to decay in orbit and to burn on reentry into the Earth's atmosphere in 1979.[10] One scenario would have seen the Space Shuttle nudge *Skylab* into a higher orbit for prolonged occupation. The Shuttle's development, alas, was delayed by design changes caused by NASA budget cuts and that scenario could not be played out. A second *Skylab* workshop was never launched, and now sits in the National Air and Space Museum where it is regularly toured by thousands of visitors, most of whom do not realize that it is bona fide hardware, not a mock-up.

During the 1970s and 1980s the Soviets, in contrast, continued their space station work. Almost continually since *Salyut 1,* built by the Korolev bureau and launched in 1971, there has been an operating Russian space station in orbit. *Salyut 2, 3,* and *5* were cover names for military space stations whose real name was *Almaz,* built by the Chelomei bureau. Cosmonauts, including some from Communist bloc or Western nations, broke endurance records one after another in *Salyut 4, 6,* and *7* up through early 1991. Then, on February 20, 1986, the first module of the *Mir* space station was orbited. Since then other modules have been added. Soviet and multinational crews, including Americans, have carried out experiments continuously since that time, being supplied periodically by a cargo vehicle called *Progress,* which makes a rendezvous with *Mir* automatically. In practice for the building of the International Space Station, the United States has been docking the Shuttle to *Mir* and rehearsing the kind of intense multinational operations that construction of ISS will call for.

ISS is mobilizing a virtual United Nations of countries working together—not only the United States and Russia, but Japan, Canada, and nine countries of Europe (working through the European Space Agency, they are Germany, France, Italy, Belgium, Switzerland, Spain, Denmark, the Netherlands, and Norway)—in what is called the largest scientific and technological initiative in history.[11]

Numerous Russian firms are active in ISS development besides RSC Energia. The Khrunichev enterprise in Moscow is getting some $215 million from NASA, passed on by Boeing, the space station prime contractor, for construction of the first ISS element—the core module for supplying power, propulsion, and attitude control. It is expected to be put in place by the former Chelomei bureau's[12] *Proton* launch vehicle in November 1997.

Another irony is that the core module, called FGB, being built by Khrunichev, seems to be based on the supply module for Chelomei's *Almaz,* which may have been a takeoff from a McDonnell Aircraft design for the supply module of the Manned Orbital Laboratory, cancelled in 1969 by the U.S. Air Force.[13]

It would please Korolev to know that the docking system and probably the astronaut-cosmonaut rescue vehicle for ISS are being designed in his bureau by some of the same men who worked with him as young engineers. Vladimir Syromiatnikov, for example, was in his mid-twenties when he worked on the design of the first *Soyuz* docking system in Korolev's time. Now in his early sixties, he is integrating a new-generation docking module, which he helped design, with Rockwell International engineers in California. Syromiatnikov, recently honored by becoming the first Russian elected a Fellow of the American Institute of Aeronautics and Astronautics,[14] is one of dozens of Korolev colleagues who are still at RSC Energia more than thirty years after the Chief Designer's death.

Their continued participation in these programs, however, has not been without personal sacrifice, as a consequence of Russia's serious economic problems. Syromiatnikov, for example, who would probably rate a six-figure salary if he worked for an American company, is being paid the equivalent of $500 a month. His plight, unfortunately, is not unusual. Many Russian scientists and engineers, particularly young ones, are leaving the space field for more attractive jobs in other businesses. Production in the rocket and space industry in 1994 dropped precipitously to 30 percent of its 1989 level and is still going down.[15]

All of the space organizations are hurting badly. Typical is the Institute of Medical and Biological Problems in Moscow, which was the principal source of research and application of Russian space medicine knowledge for more than forty years. Its former director, Oleg Gazenko, told American visitors in 1994:

> We are afloat and we do not sink, but we have had a 20–30 percent cut. Our young specialists are leaving the sciences and going to work in banks, where the salaries are three to four times what they are here. . . . We had 2,000 specialists working here and in our other laboratories in Moscow, Kiev, Krasnoyarsk, Sukhumi in the peak years, and we are now down to about 1,500. We are not used to searching for money and we don't know how to do it. We are like children from another world, like zoo animals set free who can't find food for ourselves.[16]

Dr. Gazenko said that the agreements that IMBP negotiated with the United States, France, and Germany to finance *Mir* experiments on long-term effects of space flight have helped ease the situation. They have been of "vital importance" in preparing for ISS and in supporting the services of the *Mir* space station. *Mir* may stay up for another decade or so as various countries have made deals with RSC Energia to update the aging station's capabilities to take on new assignments and new teams of cosmonauts.

But it will be ISS that will infuse the world's space organizations with the most substantive work—the building of the structure itself, the continued experimentation on human physiology and manufacturing in microgravity and low vacuum, and the testing of entrepreneurial ideas for making commercial products in that unique environment. Eventually, it is hoped, ISS will be a jumping-off station for cooperative manned and robotic exploration of the Moon and planets.

Keeping the Russian space capability alive while ISS and the other space initiatives mature is a substantial challenge for the problem-plagued government. But it will certainly be in the country's long-term interest to meet that challenge. The legacy of Sergei Pavlovich Korolev and his successors—rockets, spacecraft, launch facilities, the *Mir* space station, and related hardware—comprises technology assets that have a world market and are, therefore, potential dollar-earners. Yet, the exploration of space seems irrelevant, or at least of very low priority, to citizens who hardly have the sustenance to live. The same people who used to worship cosmonauts and their exploits now, by and large, resent their consumption of national monies.

On one of my most recent research trips to Moscow, I visited the Exhibition of Economic Achievements (VDNKh), a sprawling complex about twenty minutes by metro from Red Square, to check out the status of one its star attractions—the Kosmos pavilion, which I was told was built on Korolev's initiative to allow Soviet citizens to see models of the spacecraft that made history. In earlier visits there had been large crowds, especially children, examining the *Vostok* launch vehicle outside the building, and the *Sputniks,* dog capsules, Venus, Mars, and Moon spacecraft, and other artifacts, plus giant photos of Yuri Gagarin and Korolev inside.

To my shock I saw that virtually the entire building was devoted to a show of American, European, and Japanese automobiles. Only the *Sputnik* and *Apollo-Soyuz* spacecraft hanging from the ceiling, the blowups of Gagarin and Korolev photos, and a few other space curios were still in place. The Chief Designer would be apoplectic, I thought.

Then I wandered over to the nearby Korolev Home-Museum, a ritual stop for space buffs. Just a short walk from the VDNKh metro stop and the soaring titanium rocket memorial to Soviet space conquests, and not far from the Kosmos pavilion, it is convenient to visitors. Alas, as my notes recall, it was deserted.

> Today is Sunday, normally a good visitors day, but there's no one else here. The property, probably about an acre, has a 7 ft high fence around it. A plaque at the entrance declares the place to be the Home-Museum of Academician S. P. Korolev, open from 11 to 4 except Mondays. The house is a modest frame structure, with one out-building that seems to house the caretaker. There are tall trees, birdhouses, and pretty wildflowers on the grounds, which are not manicured but kept looking decent. A gazebo is off in one corner of the property. Nearby

one can see the stereotypical Moscow high-rise apartments which, I'm told, didn't exist when Korolev first came here in 1959 with his second wife, Nina Ivanovna. Two Russian friends and I are the only watchers of an absorbing film on Korolev's life, although it is somewhat worshipful in tone and language.

It is early October, but it's cold as the dickens inside the house. I keep my jacket and gloves on.

We're given a tour of the place after seeing the film. Downstairs is a living room with stuffed chairs and a potted palm; a tiny kitchen; and a dining room with table and eight chairs, where, we're told, Sergei Pavlovich and Nina Ivanovna sometimes entertained cosmonauts and space colleagues. Upstairs are a simple bedroom, a library, and Korolev's home office. The books on his desk are "as they were on his last day," the guide tells us. On the shelf behind his chair are piles of journals, including copies of 1962 and 1963 issues of the American space journals *Astronautics, Aviation Week*, and *Missiles & Rockets*.

Bookcases contain a large collection of cultural literature—Dostoyevsky, Tolstoi—as big as his extensive technical library. On the walls are photos of Tsiolkovsky, Tsander, and there is one showing the now famous "three Ks" of Soviet science and technology from the '50s and '60s era—Mstislav Keldysh, Igor Kurchatov, and Korolev.

There's a large map of the Moon on the wall which, we're told solemnly by the guide, "interested him greatly during his last days."

That is a considerable understatement.

One hopes that a return to the Moon, and the fulfillment of Korolev's other dreams for space exploration, will be realized by upcoming generations of Russians, this time working with the rest of the world.

Appendix
A Chronology of the Moon Race, 1957–1974

	U.S.	USSR
10/4/57		*Sputnik 1,* 83.6 kg
10/9/57	AF urges development of second-generation ICBM as booster for manned lunar mission	
11/3/57		*Sputnik 2,* 508 kg
12/6/57	Vanguard failure	
/58		Korolev issues "Study Program of the Moon"
1/25/58		Lunar probe failure
1/31/58	*Explorer 1,* 4.8 kg, first U.S. satellite, discovers Van Allen belts	
2/3/58		*Sputnik 3* failure
2/5/58	*Vanguard* failure	
3/58	ABMA (von Braun team) proposes clustered engine booster	
3/5/58	*Explorer 2* failure	
3/17/58	*Vanguard 1,* 1.47 kg launched	
3/26/58	*Explorer 3,* 5 kg, gives micrometeoroid data	
4/58	AF refuses Army's "Man High" program	
4/28/58	Vanguard failure	
5/2/58	AF adopts "Man in Space Soonest" plan	
5/15/58		*Sputnik 3,* 1,327 kg
5/27/58	Vanguard failure	
6/25/58		Lunar probe failure
6/26/58	Vanguard failure	
7/26/58	*Explorer 4,* 8 kg, maps radiation	

8/58	Eisenhower transfers funds for AF "Man in Space Soonest" to proposed NASA Von Braun team gets $10 million from ARPA for big booster studies	
8/17/58	Pioneer lunar probe failure	
8/24/58	*Explorer 5* failure	
9/23/58		Lunar probe failure
10/1/58	NASA formed from NACA	
10/11/58	*Pioneer 1* fails to reach Moon	
10/12/58		Lunar probe failure
11/8/58	*Pioneer 2* lunar probe failure	
12/4/58		Lunar probe failure
12/6/58	*Pioneer 3* lunar probe failure	
1/2/59		*Luna 1* flies by Moon at 5,998-km distance
1/19/59	NASA gives North American Aviation contract for F-1 engine for proposed Nova direct-ascent vehicle	
2/5/59	NASA forms working group on lunar exploration	
2/17/59	*Vanguard 2,* in orbit 18 days	
3/3/59	*Pioneer 4* passes within 37,300 mi of Moon	
4/59	NASA forms Research Committee on Manned Space Flight under Harry Goett of NASA Ames	
6/18/59	NASA asks von Braun group to study Saturn launch vehicle for lunar missions	Lunar probe failure

8/12/59	NASA Space Task Group starts study on 3-man capsule for near-lunar return velocities	
9/15/59		*Luna 2* impacts Moon
10/59	Goett committee asks for lunar landing mission	
10/4/59		*Luna 3* photos far side of Moon (photos brought to U.S.)
3/15/60	ABMA von Braun group transferred to NASA (MSFC)	
4/12/60		*Luna 4* misses Moon by 8,500 km
5/15/60		*Sputnik 4*, Vostok prototype, 2,500 kg, deorbit fails
6/23/60		Decree issued authorizing work on N-1 for 60–80 tons to LEO. Stated missions are primarily for military purposes, such as manned maneuvering space stations
7/25/60	Apollo name chosen, and studies underway for manned lunar landing	
8/19/60		*Sputnik 5*, dogs Belka and Strelka recovered on 18th orbit
9/25/60	Pioneer lunar probe fails	
10/17/60	NASA working group headed by George Low plans lunar landings, launch vehicle development	
10/25/60	Convair/Astro, GE, Martin, contract for studies of launch vehicles	
12/1/60		*Sputnik 6*, recovery of canine cabin unsuccessful

12/15/60	Pioneer lunar probe fails	
/61		Chelomei proposes competitive approach to N-1, the UR-500 (Proton), a 3-stage booster using storable propellants (N_2O_4, UDMH in Glushko RD-253 engines), to take 20 tons to LEO, LK-1 to lunar orbit
1/9/61	New Manned Lunar Landing Task Group, led by Low, prepares FY 62 budget pitch for 60s landing using direct ascent or Earth orbit rendezvous	
1/15/61		Korolev asks Moskalenko to speed up military proposals for use of N-1, gets no response
1/19/61	Douglas, Chance Vought get studies for EOR manned landing	
3/9/61		*Sputnik 9* (7, 8 Venus probes) recovers dog Chernushka
3/25/61		*Sputnik 10* recovers dog Zvezdochka after 1 orbit
4/12/61		*Vostok 1*, Gagarin, first man in space, 1 orbit
4/14/61	JFK calls White House meeting with Webb, Dryden to discuss ways to surpass Russians in space	
4/25/61	*Mercury-Atlas 3* fails to orbit	
5/5/61	Apollo Space Task Group issues first draft of requirements	
5/5/61	*Mercury-Redstone 3*, Shepard suborbital flight	

5/25/61	JFK asks Congress to fund lunar landing mission within decade	
6/10/61	Lundin committee (NASA) prefers EOR; Fleming Committee says landing before 1970 feasible	
7/20/61	Golovin committee (NASA/DoD) studies launch vehicle for direct ascent	
7/21/61	MR-4, Grissom suborbital flight	
7/28/61	NASA invites 12 companies to bid on Apollo spacecraft contract	
8/61	Task group prefers EOR but Houbolt pitches LOR to Space Task Group	
8/6/61		*Vostok 2*, Titov, 17 orbits
8/23/61	*Ranger 1* lunar probe fails	
11/15/61	Houbolt pitches LOR to Seamans	
11/18/61	*Ranger 2* lunar probe fails	
11/20/61	Rosen WG (NASA) prefers direct ascent using Nova	
11/28/61	NASA chooses NAA to build Apollo	
1/15/62	STG becomes Manned Spacecraft Center in Houston	
1/26/62	*Ranger 3* misses Moon by 22,862 mi	
2/62		Meeting at Khrushchev's dacha in Pitsunda okays continued development of N-1, authorizes Chelomei's UR-500
2/20/62	*Mercury-Atlas 6*, Glenn first U.S. man in space, 3 orbits	

3/2/62	NASA says LOR feasible using single Saturn C-5 with 3 stages	
3/10/62		Korolev approves L-1 (*Soyuz* predecessor) for circumlunar mission using Vostok to assemble 3 stages launched independently in LEO
4/11/62	JFK gives Apollo DX category	
4/23/62	*Ranger 4* impacts Moon, instruments fail	
5/24/62	MA-7, Carpenter, 7 orbits	
6/7/62	Von Braun recommends LOR	
7/62		Preliminary N-1 design completed for 75 tons to LEO
7/11/62	NASA announces LOR decision for 2 men on the Moon	
7/25/62	NASA issues RFP for lunar lander	
8/11/62		*Vostok 3*, Nikolayev, 64 orbits
8/12/62		*Vostok 4*, Popovich, approaches 5 km from *Vostok 3*, does 48 orbits
9/17/62	NASA adds 9 new astronauts	
9/24/62		Decree directs Korolev to aim for flight testing N-1 by 1965
10/3/62	MA-8, Schirra, 6 orbits	
10/18/62	*Ranger 5* misses Moon by 450 mi	
11/7/62	Grumman gets LEM contract	
2/2/63		Lunar probe fails to orbit
3/3/63		Lunar probe fails to orbit
3/28/63	*Saturn SA-4* launched	

4/2/63		*Luna 4* misses Moon by 8,499 km
5/15/63	MA-9, Cooper, 22 orbits	
6/14/63		*Vostok 5,* Bykovsky, 81 orbits
6/16/63		*Vostok 6,* Tereshkova, first woman in space, approaches *Vostok 5,* 3 mi, 48 orbits
10/18/63	14 more astronauts	
11/22/63	JFK assassinated	
1/29/64	Saturn SA-5, with powered second stage, launched, second stage put in orbit	
1/30/64	*Ranger 6* hits Moon, TV fails	
3/21/64		Lunar probe fails
4/8/64	*Gemini-Titan 1* (Gemini boiler plate) orbits	
4/20/64		Lunar probe fails
5/28/64	Saturn SA-6 (*Saturn 1,* boiler plate Apollo) orbits	
6/4 or 6/18/64		Lunar probe fails
6/30/64	Atlas Centaur test fails	
7/28/64	*Ranger 7* returns 4,308 photos before impact	
8/3/64		Decree issued ordering detailed plan for N-1 test launches
9/18/64	Saturn SA-7 (boiler plate Apollo plus second stage)	
10/12/64		*Voskhod 1,* Komarov, Feoktistov, Yegorov, 16 orbits; Khrushchev ousted
2/17/65	*Ranger 8* impacts Moon (2/20), 7,137 photos	
3/12/65		*Kosmos 60,* probable lunar probe failure
3/18/65		*Voskhod 2,* Leonov (first EVA), Belyaev, 17 orbits

3/21/65	*Ranger 9* impacts Moon (3/24), 5,814 photos	
3/23/65	*Gemini 3,* Grissom, Young, 3 orbits	
5/9/65		*Luna 5* hits Moon but fails soft landing
6/3/65	*Gemini 4,* White (first U.S. EVA), McDivitt, 66 orbits	
6/8/65		*Luna 6* misses Moon by 100,000 mi
7/16/65		Chelomei's *Proton 1* orbits 18.5-ton physics payload
7/18/65		*Zond 3* flies by Moon, takes photos
8/21/65	*Gemini 5,* Cooper, Conrad, 128 orbits	
10/4/65		*Luna 7* crashes on Moon, fails soft landing
10/20/65		Afanasiev directs Korolev's N-1 subcontractors to prepare to use their systems for possible application to Chelomei's UR-700
10/25/65	*Gemini 6* target fails to orbit	
12/3/65		*Luna 8* crashes on Moon, fails soft landing
12/4/65	*Gemini 7,* Borman, Lovell, 220 orbits	
12/15/65	*Gemini 6,* Schirra, Stafford, comes within 1 ft of *Gemini 7*	
12/15/65	Phillips tells Atwood of NASA dissatisfaction with Apollo and Saturn S-II stage	
12/15/65		Smirnov commission gives circumlunar mission to Korolev's OKB-1 with UR-500K/L-1, rather than to Chelomei's UR-500K/LK-1

1/14/66		Korolev dies, Mishin takes over design bureau
1/31/66		*Luna 9* soft-lander returns photos for 3 days
2/26/66	Suborbital test of Saturn 1B with CSM as payload	
3/16/66	*Gemini 8* target	
3/16/66	*Gemini 8,* Armstrong, Scott, dock with target	
3/31/66		*Luna 10* circumlunar, 2 months of data
4/30/66		Luna fails to orbit
5/17/66	*Gemini 9* target A, fails to orbit	
5/30/66	*Surveyor 1* soft-lands on Moon, takes 11,150 photos	
6/3/66	*Gemini 9,* Stafford, Cernan, rendezvous and EVA tests, 47 orbits	
7/5/66	Saturn 1B, with CSM payload, suborbital test	
7/18/66	*Gemini 10* target	
7/18/66	*Gemini 10,* Young, Collins, dock with target, 46 orbits	
8/10/66	*Lunar Orbiter 1* returns photos till 8/29, crashes on Moon 10/29	
8/24/66		*Luna 11* gives data until 10/1, selenocentric orbit
8/25/66	Another Saturn 1B suborbital test	
9/12/66	*Gemini 11* target	
9/12/66	*Gemini 11,* Conrad, Gordon, 47 orbits	
9/20/66	*Surveyor 2* crashes on Moon as vernier fails	
9/25/66		Luna fails
10/19/66	Chaffee, Grissom, White, named to fly first manned *Apollo* 2/12/67	

10/22/66		*Luna 12* photographs Moon
11/66		Expert commission under Keldysh approves draft plan for lunar landing with 95-ton N-1/L-3
11/6/66	*Lunar Orbiter 2* hits Moon, returns 205 frames	
12/21/66		*Luna 13* lands on Moon 12/24/66, takes photos, measures soil density
/67		20 cosmonauts in training for circumlunar mission
1/27/67	Fire in *Apollo 1* capsule, during test, kills Chaffee, Grissom, and White	
2/5/67	*Lunar Orbiter 3* hits Moon 10/9, returns 182 frames	
2/7/67		*Kosmos 140,* Soyuz precursor
3/10/67		*Kosmos 146,* first Proton, with L-1 circumlunar craft, orbits Earth, unmanned, for 8 days
4/8/67		*Kosmos 154*, second Proton with L-1, orbits Earth for 2 days, but Block D fails to ignite
4/17/67	*Surveyor* 3 lands on Moon 4/20, samples soil, takes photos till 5/3	
4/23/67		*Soyuz 1,* 18 orbits, Komarov killed as parachute system fails on reentry, 18 orbits
5/4/67	*Lunar Orbiter 4* crashes on Moon 10/6, records 163 frames	
7/14/67	*Surveyor 4* crashes on Moon 7/17, loses signal 2.5 min before	

8/1/67	*Lunar Orbiter 5* crashes on Moon 1/31/68, end of photo mapping	
9/8/67	*Surveyor* lands on Moon (9/11), returns 19,000 photos, does soil analysis	
9/28/67		Proton explodes on launch in attempt to send 7K/L-1 into highly elliptical orbit in another test for circumlunar flight
10/27/67		*Kosmos 186,* first automatic docking of Soyuz, w/ *Kosmos 188*
11/7/67	*Surveyor 6* lands on Moon (11/10), does first rocket takeoff from lunar surface	
11/9/67	*Apollo 4,* first "all up" test of *Saturn 5* vehicle	
11/22/67		*Zond,* planned lunar flyby, fails to orbit N-1 first full-scale mock-up test
1/7/68	*Surveyor 7* lands on Moon (1/10)	
1/22/68	*Apollo 5* unmanned tests of lunar module in orbit	
2/7/68		*Luna 14* fails to orbit
3/68		Gagarin killed in air crash (was probably to be L-1 or L-3 pilot)
3/2/68		*Zond 4* launches L-1 to high apogee orbit near Moon, attitude control error causes L-1 to self-destruct on return
4/4/68	*Apollo 6,* unmanned test; "pogo" in first stage, malfunctions in second and third stages	
4/7/68		*Luna 14* flyby, studies Earth-Moon gravitation

4/14–15/68		*Kosmos 212–213* docking
4/23/68		Proton fails to launch L-1 (6th L-1 mission)
8/9–17/68	Low calls for *Apollo 8* circumlunar missions if *Apollo* 7 OK (this not made public until 11/12)	
9/15/68		*Zond 5* to 1,950 km from Moon but attitude control failure on return causes spacecraft to make ballistic reentry and land in Indian Ocean
10/11–22/68	*Apollo 7,* Schirra, Cunningham, Eisele, 163 orbits, first manned *Apollo*	
10/25/68		*Soyuz 2,* unmanned, target for *Soyuz 3*
10/26/68		*Soyuz 3,* Beregevoi, 650 ft from *Soyuz 2,* 64 orbits
11/10/68		*Zond 6,* second unmanned flight of UR-500K/L-1, flies to 2,420 km from Moon, gets dramatic view of lunar horizon over half Earth, but on return parachute deploys prematurely and descent is so violent a crew would have been killed
12/21–27/68	*Apollo 8*, Borman, Lovell, Anders, Christmas circumlunar flight with photos of Earth	
1/14/69		*Soyuz 4,* Shatalov, manual docking with *Soyuz 5,* first linkup of two manned spacecraft
1/15/69		*Soyuz 5,* Khrunov, Yeliseev transfer from *Soyuz 5* to *4,* simulating L-3 EVA, and rehearsing

		for first rescue, Volynov lands alone in *Soyuz 4*
1/20/69		*Zond* fails, UR-500 destroyed
2/21/69		First test of N-1/L-3 fails as engines cut off, oxidizer line bursts, engines shut down, booster and L-3 are destroyed but L-1 lands safely nearby
3/3–13/69	*Apollo 9,* McDivitt, Scott, Schweikart, 151 orbits, EVA, LM first manned flight	
5/18–26/69	*Apollo 10,* Stafford, Young, Cernan, second circumlunar flight, LM piloted to 14.9 km of lunar surface, first color TV from space	
5/27/69	Lunar rover approved	
6/4/69		*Luna 15-2* fails to orbit
6/11/69	Missions approved through *Apollo 20*	
7/3/69		Second N-1 test results in explosion which destroys pad for two years after chain of malfunctions
7/13/69		*Luna 15* does 52 lunar orbits, fails to get soil sample on 7/21, Apollo already on Moon
7/16–24/69	*Apollo 11,* Armstrong, Aldrin land on Moon, Collins orbits	
8/7/69		*Zond 7,* third Soviet unmanned circumlunar mission, soft-lands back on Earth in Kustanai
9/23/69		*Kosmos 300,* by Babakin, fails to get translunar injection of "third-generation" craft

10/11/69		*Soyuz 6,* Shonin, Kubasov
10/12/69		*Soyuz 7,* Filipchenko, Volkov, Gorbatko
10/13/69		*Soyuz 8,* Shatalov, Yeliseyev First triple manned launch, welding, multiple maneuvering, 80 orbits
10/28/69	Boeing gets lunar rover contract	
11/14–24/69	*Apollo 12,* Conrad, Bean spend 32 hrs on lunar surface, Gordon orbits 89 hrs, Conrad does 2 EVAs	
11/28/69		Proton explodes in attempt to orbit T1K prototype lunar orbiter
1/4/70	*Apollo 20* cut from NASA budget	
4/11–17/70	*Apollo 13* oxygen tank fails, mission aborted after circling Moon, crew returns to Earth in LM	
6/1–19/70		*Soyuz 9,* Nikolayev, Sevastyanov, 18 days in space
9/2/70	*Apollo 18* & *19* missions cut	
9/12/70		*Luna 16* returns lunar soil samples
9/25/70		*Kosmos 365,* fractional orbital bomb test?
11/10/70		*Luna 17,* unmanned *Lunokhod* roves Moon 11 mos
11/24/70		*Kosmos 379* 3 1/2 day mission (lunar flight time), tests variable throttling for landing
12/2/70		*Kosmos 382,* Proton launch of prototype lunar orbiter T1K, simulates

		orbiter maneuvers, including trans-Earth injection
1/31–2/9/71	*Apollo 14,* Shepard, Mitchell, Roosa in orbit	
2/26/71		*Kosmos 398,* successful T2K (lunar lander w/o legs) test
4/19/71		*Salyut 1,* prototype of manned space station
4/23/71		*Soyuz 10,* Shatalov, Yeliseev, Rukavishnikov, dock with *Salyut,* but fail to board
6/6/71		*Soyuz 11,* Dobrovolsky, Volkov, Patsaev, dock with *Salyut 1,* spend record 23 days in space but are killed 6/29 on descent when valve fails causing decompression of cabin
6/27/71		Third failure to launch N-1, all engines shut down after 51 sec, vehicle destroyed
7/26–8/7/71	*Apollo 15,* Scott, Irwin, lunar rover, 12 days, Worden in CM	
8/12/71		*Kosmos 434,* third and last test of lunar lander in Earth orbit
9/2/71		*Luna 18* crashes on Moon (9/11), after 54 orbits
9/28/71		*Luna 19,* photography
2/14/72		*Luna 20* returns soil sample
4/16–27/72	*Apollo 16*, Young, Duke land at Descartes, Mattingly, in CM, does EVA on return trip	

11/23/72		N-1 explodes in fourth test as six engines shut down after 90 sec, only a few seconds before first-stage separation, pressure overload ruptures lines, fire breaks out
12/7–19/72	*Apollo 17,* Schmitt (geologist), Cernan, Evans in CM	
1/8/73		*Luna 21, Lunokhod 2,* explores Moon for 5 months
4/3/73		*Salyut* 2 space station launched, disintegrates after only 6 weeks
5/14/73	*Skylab* 1 launched, stays in orbit until 1979, is occupied by three crews during 1973	
5/74		Ustinov fires Mishin, replaces him with Glushko. Mishin had two N-1s ready for launch later in 1974, and hoped to have N-1 operational by 1976. Glushko suspends N-1 program (cancels it officially in March, 1976)

NOTES

Prologue: The Anonymous Chief Designer

1. Otrieshka interview with author, Moscow, Dec. 5, 1991
2. In an interview in Moscow, on Sept. 9, 1993, Korolev's former English-language specialist, Vladimir Shevalyov, told me: "Korolev asked me once to put together some information on von Braun. I went to the Lenin Library and found a thick American hard-cover book with a lot of photos and I wrote a short biography of von Braun from it. Korolev was one of the few people allowed to borrow books from the Lenin Library. He was pleased with the bio that I wrote and seemed especially pleased when I told him that there seemed to be some similarities between von Braun's life and his."
3. Rauschenbakh interview with author, Abramtsevo, Russia, Dec. 7, 1991. Academician Rauschenbakh, himself a major contributor to Soviet guidance and control technology, is now head of a committee on the history of space science and technology in the Russian Academy of Sciences.
4. Ivanovsky interview with author, Moscow, Jan. 26, 1993
5. Chertok interview with author, Moscow, Dec. 14, 1991
6. Sakharov, Andrei, *Andrei Sakharov Memoirs* (New York: Alfred A. Knopf, 1990), p. 177
7. Golovanov interview with author, Peredelkino, Oct. 19, 1992
8. Gazenko interview with author, Moscow, Dec. 13, 1991
9. Gitelson interview with author, Princeton, N.J., Jan. 9, 1993
10. Otrieshka interview
11. Gitelson interview
12. Vorobiev interview with author, Kaliningrad, May 20, 1994
13. Giving Korolev one more thing in common with von Braun, who outlined a manned Mars expedition in 1952 in *Das Marsprojekt,* Bechtle Verlag, Esslingen, Germany

Chapter 1: Kibalchich to Korolev

1. Kibalchich, N. I., *A Concept for an Aeronautical Machine,* personal note, Mar. 23, 1881
2. Biryukov, Y. V., "The History of the Designing of Liquid-Propellant Rocket Engines as a New Type of Rocket Engine," *Viniti,* USSR Academy of Sciences, Moscow, 1966, English translation in NASA TT F-540, p. 3
3. Chernyak, A., "Nikolai Kibalchich," *Spaceflight* (July 1962), p. 130
4. McDougall, Walter A., . . . *the Heavens and the Earth* (New York: Basic Books, 1985), p. 19
5. Rynin, Nikolai A., *Interplanetary Flight and Communication (Mezhplanetynie soobsheniya),* 9 vols., English version by Israel Program for Scientific Translations, NASA Technical Translations F-640 to F-648, 1970–71

6. Tokaty-Tokaev, Gregory A., "Foundations of Soviet Cosmonautics," *Spaceflight,* Oct. 1968, p. 336
7. Sokolsky, Viktor, "Work on Theoretical Foundations of Cosmonautics and Evolution of Rocket Technology in the USSR Prior to 1945," in *History of the USSR: New Research. 5,* Social Sciences Today, Moscow, 1986, pp. 64–92
8. Ibid., p. 66
9. Ibid.
10. Wallace, Sir Donald Mackenzie, Alexander II, *Encyclopedia Brittanica,* Chicago, 1962 ed., Vol. 1, p. 565
11. Chernyak, *Nikolai Kibalchich,* p. 130
12. Winter, Frank H., "Nikolai Alexeyevich Rynin (1877–1942), Soviet Astronautical Pioneer: An American Appreciation," *Earth-Oriented Appl. Space Techn.* Vol. 2, No. 1, p. 73
13. Chernyak, *Nikolai Kibalchich,* p. 130
14. *The Papers of Robert H. Goddard,* Esther C. Goddard, Editor, G. Edward Pendray, Associate Editor, Vols. 1–3 (New York: McGraw-Hill, 1970)
15. Rynin, *Interplanetary Flight . . . ,* p. 2
16. Holquist, Michael, "The Philosophical Bases of Soviet Space Exploration," *The Key Reporter,* Winter 1985–86, p. 3
17. Rynin, *Interplanetary Flight . . . ,* p. 3
18. Ibid., p. 5
19. Ibid., p. 8
20. Tokaty-Tokaev, *Foundations . . . ,* p. 336
21. Sokolsky, *Work on . . . ,* p. 68
22. Korolev, Sergei P., "The Practical Significance of Konstantin Tsiolkovsky's Proposals in the Field of Rocketry," paper presented at a meeting commemorating Tsiolkovsky's 100th birthday, Sept. 17, 1957, USSR Academy of Sciences, Moscow, reprinted in English in *History of the USSR: New Research. 5,* Social Sciences Today, Moscow, 1986, p. 52
23. Ibid., p. 55
24. Ibid., p. 56
25. Ibid., p. 59
26. Ibid., p. 338
27. Tokaty-Tokaev, *Foundations . . . ,* p. 338
28. Ibid., p. 346
29. Tsiolkovsky was among the 150 members of the Society for the Study of Interplanetary Travel, formed in Moscow in 1924. Later, in the Western world, would come the German Society for Space Travel (1927), the American Interplanetary Society (1930), and the British Interplanetary Society (1933). The latter two still exist, BIS retaining its original name, and AIS becoming the American Rocket Society (ARS) in 1934, and then the American Institute of Aeronautics and Astronautics (AIAA) in 1963. Even in the formation of space flight societies the Russians were first.
30. Romanov, Alexander, *Spacecraft Designer* (Moscow: Novosti, 1976), pp. 13–14

31. Golovanov, Yaroslav, *Korolev Fakti i Miti* (Moscow: Nauka, 1994), pp 103–13
32. Vetrov interview with author, Moscow, Oct. 16, 1992
33. Rudenko interview with author, Sept. 14, 1993. A space writer for *Air Transport (Vozdushny Transport),* Moscow, Rudenko is writing a book about Glushko.
34. Gubanov interview with author, Moscow, Oct. 15, 1992
35. Avduyevsky interview with author, Moscow, Jan. 21, 1993

Chapter 2: Growing Up in Ukraine: Broken Family, Bolsheviks, Gliders

1. Ishlinsky, A.Y., *Akademik S. P. Korolev, Uchyonii, Inzhenyer, Chelovek* (Moscow: Nauka, 1986), p. 29
2. Vetrov interview with author, Moscow, Oct. 16, 1992
3. Golovanov, *Korolev Fakti i Miti*, p. 12
4. Vetrov, communication with author, June 13, 1995
5. Ishlinsky, *Akademik* . . . , p. 29
6. Ibid.
7. Golovanov, Yaroslav, *Sergei Korolev The Apprenticeship of a Space Pioneer,* (Moscow: Mir, 1975), p. 26
8. Ishlinsky, *Akademik* . . . , p. 29
9. Golovanov, *Apprenticeship* . . . , pp. 30–31
10. Atashenkov, *Akademik S. P. Korolev* (Moscow: Mashinostroyenniye, 1969), p. 10
11. Ibid.
12. Ibid.
13. Golovanov, *Apprenticeship* . . . , p. 33
14. Ishlinsky, *Akademik* . . . , p. 34
15. Golovanov, *Apprenticeship* . . . , pp. 37–38
16. Ibid.
17. Ibid., pp. 40–41
18. Ibid., pp. 41–42
19. Ishlinsky, *Akademik* . . . , p. 38–39
20. Golovanov, *Apprenticeship* . . . , p. 45. Stilianudi was a pupil of the famous artist Repin and a friend of the artists Serov and Vrubel.
21. Ishlinsky, *Akademik* . . . , p. 39
22. Ibid.
23. Ibid.
24. Ibid., pp. 30–31
25. Golovanov, *Apprenticeship* . . . , p. 53
26. Ibid., p. 54
27. Ibid., p. 57
28. Ishlinsky, *Akademik* . . . , p. 31
29. Golovanov, *Apprenticeship* . . . , p. 61
30. Atashenkov, *Akademik S. P. Korolev,* p. 16

31. Taras Shevchenko (1814–1861) was a Ukrainian poet whose liberation ideas caused him to be exiled to a Caspian military base. His statue, erected by Ukrainian-Americans, stands at 22nd and P Streets in Washington, D.C.
32. Ishlinsky, *Akademik* . . . , pp. 35–36
33. Golovanov, *Apprenticeship* . . . , p. 59
34. Ibid., pp. 64–66
35. Ibid., p. 75
36. Ibid., p. 76
37. Ibid., p. 87
38. Ibid., pp. 92–93
39. Ishlinsky, *Akademik* . . . , p. 31
40. Golovanov, *Apprenticeship* . . . , p. 103
41. Ibid., pp 111–14

Chapter 3: To Moscow: Tupolev, Tukhachevsky, First Rockets

1. Golovanov, *Apprenticeship* . . . , pp. 133–34
2. Ibid., p. 134
3. Ibid., p. 135
4. Ibid. This was only a few months after the Wright Brothers built their first wind tunnel in Dayton, in November 1901. Two English engineers, Francis Wenham and John Browning, are credited with having performed the first primitive wind tunnel experiments in 1871. Crouch, Tom D., *The Bishop's Boys* (New York: W. W. Norton & Co., 1989).
5. Zhukovsky founded in 1918 what is still Russia's major center of aeronautical research, the Central Aerohydrodynamic Research Institute (TsAGI). TsAGI is not only named for Zhukovsky, but it is located in the town of Zhukovsky, outside of Moscow.
6. Golovanov, *Apprenticeship* . . . , p. 136
7. In 1930 the aeronautical department of MVTU split off to become Moscow Aviation Institute. MAI has been an important source of the Soviet Union's aeronautical engineers ever since. In 1994 it had more than 14,000 students, down from a peak of 17,000 in the 1980s.
8. Golovanov, *Apprenticeship* . . . , pp. 137–38
9. Ishlinsky, *Akademik* . . . , p. 36
10. This activity, too, was shared by MIT students of the same era. Paul Baker, a 1929 graduate of MIT in aeronautical engineering, recalls that he and 8 or 9 classmates designed and built a training glider and took it to Cape Cod to fly in the summer before senior year. Author's telephone interview with Baker, May 17, 1995.
11. Sokolsky, *Work on* . . . p. 81.
12. Golovanov, *Apprenticeship* . . . , p. 137–39
13. Ibid., p. 148
14. Ibid., p. 154
15. Ibid., p. 162
16. Ibid., pp. 171–72

17. Ibid., p. 175
18. Ibid., p. 176
19. Ibid., p. 168–69
20. Ibid., p. 196
21. Ishlinsky, *Akademik . . .* , p. 37
22. Golovanov, *Apprenticeship . . .* , p. 200
23. Rebrov, Col. M., "Notes on Career of Designer V. P. Glushko," *Krasnaya Zvezda* (Russian), Aug. 26, 1989, p. 4, translated into English in JPRS-USP-90-001, Mar. 15, 1990, p. 54
24. Sokolsky, *Work on . . .* , p. 75
25. Golovanov, *Apprenticeship . . .* , pp 230–31
26. *Kosmonavtika Encyclopedia* (Russian), edited by V. P. Glushko (Moscow: Moskva Publishing House, 1985), p. 434, and Golovanov, *Apprenticeship . . .* , pp. 225–26
27. Von Braun, Wernher, Ordway, Frederick I., III, and Dooling, Dave, *Space Travel, a History* (New York: Harper & Row, 1985), p. 62
28. Glushko, V. P., *Rocket Engines GDL-OKB* (Moscow: Novosti Publishing House, 1975)
29. Ibid., p. 5
30. Ibid., p. 6
31. Ibid.
32. Ishlinsky, *Akademik . . .* , p. 192
33. Rebrov, *Notes . . .* , p. 56
34. Ishlinsky, *Akademik . . .* , p. 192
35. Ibid., p. 7
36. Ibid., p. 9
37. Golovanov, *Apprenticeship . . .* , p. 217
38. Ibid., p. 261
39. Keldysh, M. V., editor, and Vetrov, G. S., compiler, *Creative Legacy of Academician Sergei Pavlovich Korolev* (Russian), (Moscow: Nauka, 1980), p. 54
40. Ibid., p. 56
41. Ishlinsky, *Akademik . . .* , p. 167
42. Goddard, Esther C., and Pendray, G. Edward, *The Papers of Robert H. Goddard,* Vol. II, 1925–1937 (New York: McGraw-Hill, 1970), pp. 588–89
43. Ibid., p. 864
44. Ibid., Vol. III, pp. 1340–45
45. Ibid., Brett letter to Goddard, pp. 1360–61
46. Ibid., Goddard letter to C. N. Hickman, p. 1364
47. Ibid., Goddard letter to Chief, Navy Bureau of Aeronautics, pp. 1420–21
48. Grey, Jerry, *Enterprise* (New York: William Morrow, 1979), p. 25
49. It was not until the 1960s, long after Goddard's death in 1945, that the efforts of his widow, Esther Goddard, his sponsor Harry Guggenheim, and the former president of the American Rocket Society, G. Edward Pendray, resulted in proper recognition of the reclusive inventor. Mrs. Goddard is reported to have received from the U.S. government more than $1 million in patent royalties. NASA named

its space science center in Beltsville, MD, the NASA Goddard Space Flight Center, and numerous organizations established awards in the Goddard name (ARS was the first, creating what is now the Goddard Astronautics Award in 1951).

50. Neufeld, Michael J., *The Rocket and the Reich* (New York: The Free Press, 1995), p. 10
51. Keldysh and Vetrov, *Creative Legacy* . . . , p. 57
52. Rebrov, *Notes* . . . , p. 57
53. Atashenkov, *Akademik S. P. Korolev,* p. 55
54. Golovanov, *Apprenticeship* . . . , p. 258
55. Keldysh and Vetrov, *Creative Legacy* . . . , pp. 59–60
56. Golovanov, *Apprenticeship* . . . , pp. 279–90
57. Ibid., p. 283
58. Keldysh and Vetrov, *Creative Legacy* . . . , p. 64
59. Sokolsky, *Work on* . . . , pp. 79–80
60. Neufeld, *The Rocket* . . . , p. 54
61. Sokolsky, *Work on* . . . , p. 80
62. Koroleva, Natasha, interview with author, Moscow, Dec. 8, 1991
63. Ishlinsky, *Akademik* . . . , p. 140
64. Ibid.
65. Sokolsky, *Work on* . . . , p. 80
66. Ishlinsky, *Akademik* . . . , p. 141
67. Rauschenbakh interview with author, Abramtsevo, Dec. 7, 1991
68. Ishlinsky, *Akademik* . . . , p. 141
69. Sokolsky, *Work on* . . . , p. 81

Chapter 4: The Gulag and Sharaga Years

1. Goddard and Pendray, *The Papers of Robert H. Goddard,* Vol. III, p. 1168
2. Korolev letter to the Prosecutor, May 30, 1955
3. Vetrov, Gyorgi, *Korolev i Ego Delo* (Moscow: Nauka, 1996)
4. Ibid.
5. Pallo interviews with author, Moscow, May 15 and Sept. 14, 1993
6. Balanina letter to Stalin, July 22, 1938. Excerpted from *Voenno-Istoricheskii Zhurnal,* No. 10, 1989
7. Volkogonov, Dmitri, *Stalin Triumph and Tragedy* (Rocklin, CA: Prima Publishing, 1991), p. 307
8. *National Geographic,* Mar. 1990, p. 48
9. Conquest, Robert, *The Great Terror* (New York: Macmillan, 1968), p. 532
10. Leonov interview with author, Moscow, Dec. 11, 1991
11. Alexandrov, Viktor, *The Tukachevsky Affair* (Englewood Cliffs, N.J.: Prentice-Hall, 1964), p. 92
12. Ibid., p. 81
13. *Current Digest of the Soviet Press,* Vol. XIII, No. 52, Jan. 24, 1962, p. 28
14. Schulz, Heinrich E., et al., *Who Was Who in the USSR* (Metuchen, N.J.: The Scarecrow Press, 1972)
15. Yezhov was executed himself eventually.

16. Tokaty-Tokaev, G. A., *Spaceflight,* Oct. 1968, p. 341
17. Leonov interview
18. Ulam, Adam, *Stalin, the Man and His Era* (New York: Viking, 1973), p. 473
19. Gitelson interview with author, Princeton, N.J., Jan. 9, 1993
20. Volkogonov, *Stalin* . . . , remarks by Harold Shukman, in Preface, p. xv
21. Rauschenbakh interview with author, Abramtsevo, Russia, Dec. 7, 1991
22. Burlatsky, Fedor, *Khrushchev and the First Russian Spring* (New York: Macmillan, 1988).
23. Vetrov interview with author, Moscow, Oct. 16, 1992
24. Volkogonov, *Stalin* . . . , p. 556
25. Solzhenitsyn, Aleksandr, *The First Circle* (New York: Harper & Row, 1958)
26. Ibid., p. ix
27. Burlatsky, *Khrushchev* . . . , p. 11
28. Volkogonov, *Stalin* . . . , p. 374
29. Sakharov, Andrei, *Andrei Sakharov Memoirs* (New York: Alfred A. Knopf, 1990), p. 177
30. Kerber interview with author, Moscow, Dec. 10, 1991
31. Ibid.
32. Sakharov, *Memoirs,* p. 178
33. Koroleva, Natasha, interview with author, Moscow, Dec. 8, 1991
34. Vetrov interview
35. Ibid.
36. Ibid.
37. Biryukov, Yuri V., and Komarov, Vikenty M., "Korolev's Kazan University," *Man in Space 1992,* No. 1, Moscow, pp 54–58
38. Ibid
39. Ibid.
40. Ibid.

Chapter 5: The German V-2: Bedrock Technology

1. Chertok interview in *Izvestia,* Nos. 54, 55, 56, 57, 58, Mar. 4–9, 1992
2. Ibid.
3. Vetrov interview with author, Moscow, Sept. 8, 1993
4. Chertok interview in *Izvestia*
5. Neufeld, Michael J., "The Guided Missile and the Third Reich: Peenemunde and the Forging of a Technological Revolution," in *Science, Technology and National Socialism* (New York: Cambridge University Press, 1993)
6. Thiel did most of his work on the engine at Kummersdorf. Later, at Peenemunde, von Braun made a major contribution to the V-2 engine development with his idea of incorporating eighteen injector nozzles rather than a single injector plate. "Towards the end of the war the single plate solution worked well, but it was too late to introduce it because the SS had the production already running and did not want a major change like this. The statement that von Braun tried to claim some of Thiel's achievements as his own is absolutely false." Letter to the author from Ernst Stuhlinger, June 9, 1995.

7. Schulze, Heinrich A., *Technical Data on the Development of the A-4 V-2,* F&D 876-1126, a private report prepared for Wernher von Braun at NASA Marshall Space Flight Center, Huntsville, AL, Feb. 1965, and Stuhlinger, Ernst, letter to the author, June 9, 1995
8. Ordway, Frederick I. III., and Sharpe, Mitchell R., *The Rocket Team* (New York: Thomas Y. Crowell, 1979), pp. 251–52
9. Rothmund, C., *The History of the Viking Engine,* IAA Paper 2.2-93-675, 44th International Astronautical Congress, Graz, Austria, Oct. 1993
10. Chertok interview with author, Moscow, Dec. 4, 1991
11. Magnus and Hoch were from the University of Gottingen.
12. Magnus, Kurt, letter to the author, Feb. 21, 1995
13. Scientific Intelligence Research Aid #74, Central Intelligence Agency, Washington, D.C., *Scientific Research Institute and Experimental Factory 88 for Guided Missile Development, Moskva/Kalingrad*, p. 2, Mar. 4, 1960
14. Huzel, Dieter K., *Peenemunde to Canaveral* (Englewood Cliffs, N.J.: Prentice-Hall, 1961) p. 81
15. Magnus letter. The platform was "developed by Kreisel GmbH, Berlin, and was not used in the V-2."
16. Mishin interview with author, Sept. 6, 1990
17. Chertok interview in *Nezavisimaya Gazeta,* Aug. 19, 1993, as translated in JPRS-USP-93-005, Oct. 5, 1993, p. 30, Washington, D.C.
18. Chertok interview with author, Moscow, Dec. 4, 1991
19. Mishin interview with author, Moscow, Sept. 7, 1990
20. Tyulin, Gyorgi Alexandrovich, *Krasnaya Zvyezda,* Apr. l, 1989, pp 3–4. Tyulin was named by General Gaidukov to head the so-called Soviet Technical Commission, which became known as the "Tyulin facility," on Bismarckstrasse in the Oberschoeneweide area. It included Chertok and a number of other engineers and designers.
21. Ibid.
22. Ordway and Sharpe, *The Rocket Team . . . ,* pp. 294–309
23. Pickering telephone interview with author, Dec. 2, 1993
24. Ordway and Sharpe, p. 306
25. Koroleva interview with author, Moscow, Dec. 9, 1991
26. Magnus, Kurt, *Raketensklaven* (German), pp 29–30 (Stuttgart: Deutschen Verlags-Anstalt, 1993)
27. Tyulin, *Krasnaya Zvyezda*
28. Stuhlinger letter to author, Oct. 21, 1995
29. Thiel telephone interview with author, Oct. 17, 1995
30. Stuhlinger letter
31. Neufeld, Michael J., *The Rocket and the Reich . . . ,* p. 138
32. Ibid., p. 156
33. The designer of the experiment was Vladimir Syromiatnikov, who had worked for Korolev as a young engineer in the 1950s and 1960s.
34. Stuhlinger letter
35. Mrs. Grottrup has also written of her Russian experience in *Rocket Wife* (London: Andre Deutch, 1969).

36. *Nova* television show No. 2004, "Nazis and the Russian Bomb," WGBH, Boston, MA, aired on Feb. 2, 1993
37. Magnus, *Raketensklaven,* pp. 44–45
38. Chertok interview with author, Moscow, Dec. 4, 1991
39. Kryukov interview with author, Moscow, Sept. 8, 1993
40. Tarasenko, Maxim, communication with author, Mar. 2, 1996
41. Tarasenko, Maxim, *Evolution of the Soviet Space Industry,* IAA-95-IAA.2.1.01, 46th International Astronautical Congress, Oslo, Oct. 2–6, 1995, p. 1. This paper shows emphatically that retracing the history of the Soviet missile and space organizations remains a black art. Tarasenko's pioneering work, still hampered by secrecy, has benefited from sponsorship by the U.S. Federation of American Scientists and Princeton University.
42. Mozzhorin, Yuri, "From the First Ballistic Rockets to . . . ," *Aviatsia i Kosmonautika,* No. 7, 1991, pp. 34–35
43. Siddiqi, Asif A., "Soviet Space Programme, Part 1, Organizational Structure 1940s–1950s," *Spaceflight,* Vol. 36, Aug. 1994, p. 283
44. Zak, Anatoli, writes that the plant had been built by "probably several thousand" German workers between 1929 and 1932, " . . . only one of the numerous examples of questionable 'cooperation' between the young Soviet republic and the growing claws of the German monster," *Nezavasimaya Gazeta,* Apr. 13, 1993, p. 6, translated into English in JPRS-USP-93-003, June 28, 1993, p. 35.
45. Chertok interview with author, Moscow, Dec. 4, 1991
46. Ibid.
47. Magnus letter to the author, Feb. 21, 1995
48. Magnus, *Raketensklaven,* pp. 107–8
49. Ivanovsky interview with author, Moscow, Jan. 26, 1993
50. Chertok interview with author, Moscow, Dec. 4, 1991
51. Tyulin, *Krasnaya Zvyezda*
52. Ibid.
53. *Nova* program
54. Ibid.
55. Chertok interview, *Nezavisimaya Gazeta*
56. Tyulin, *Krasnaya Zvyezda*
57. Chertok interview, *Nezavisimaya Gazeta*
58. Ibid.
59. Magnus, *Raketensklaven,* pp. 130–31
60. Ibid.
61. Ibid., pp. 133–34
62. Chertok interview with author, Moscow, Dec. 4, 1991
63. *From First Satellite to Energia-Buran and Mir* (Kaliningrad, RSC Energia, 1994), p. 2
64. Chertok interview with author, Moscow, Dec. 4, 1991
65. Magnus letter. "As far as I know, this concept has already been discussed in Peenemunde."
66. Chertok interview in *Nezavisimaya Gazeta*
67. Magnus, *Raketensklaven,* pp. 223–24

68. *From First Satellite . . .*, p. 7
69. Not to be confused with the R-12 IRBM later developed by Yangel at Dnepropetrovsk
70. Not to be confused with the R-14 developed at Dnepropetrovsk
71. Magnus letter
72. Magnus, *Raketensklaven,* pp. 220-22
73. Magnus letter. "I do not remember that any of the Gorodomliya Germans have been working on such a project as R-113."
74. Magnus, *Raketensklaven,* pp 313–14
75. Magnus letter

Chapter 6: The World's First ICBM: Aimed at the USA

1. The warhead itself did not survive reentry. The first successful warhead reentry took place the following month.
2. *Home Service,* Moscow, July 22, 1958
3. *Pravda,* Nov. 29, 1957
4. *Tass,* Sept. 8, 1958
5. Martin, Richard, letter to the author, August 18, 1994
6. Varfolomeyev, Timothy, "Soviet Rocketry that Conquered Space," *Spaceflight,* Aug. 1995, pp. 260–63
7. In Department 3, headed by Korolev, of the branch of NII-88 devoted to ballistic missiles
8. Vetrov, *Korolev i Ego Delo*
9. Ibid.
10. Ibid.
11. Ishlinsky, *Akademik . . .*, p. 76
12. Golovanov, Yaroslav, *Korolev Fakti i Miti,* pp. 383–84, 409–10
13. Koroleva, Natasha, interview with author, Moscow, Dec. 8, 1991
14. Astashenkov, *Akademik S. P. Korolev,* p. 105
15. Ibid., p. 106
16. Ibid., p. 108
17. Ivanovsky interview with author, Moscow, Jan. 26, 1993. "King" in Russian is "Korol.' "
18. Kryukov interview with author, Moscow, Sept. 9, 1993
19. Vetrov, *Korolev i Ego Delo.*
20. Ibid.
21. Chertok, *Izvestia,* Mar. 4–9, 1992
22. It is interesting that the Bossart team, which began to study ICBM concepts in San Diego in 1946, identified the separation of the warhead nose cone from the vehicle as an important design solution. Also called out: tanks integral with outer skin; swivelling of rocket nozzles for control; radio-inertial guidance with ground tracking update; and the use of small vernier engines for precise guidance cutoff. Letter from Richard Martin, Aug. 18, 1994.
23. Zaloga, Steven J., *Target America* (Novato, CA: Presidio Press, 1993), p. 127
24. Vetrov, *Korolev i Ego Delo*

25. Ibid.
26. Ibid.
27. Ibid.
28. Ibid.
29. Kruykov interview
30. Utkin interview with author, Moscow, Jan. 29, 1993
31. Lardier, Christian, *L'Astronautique Sovietique* (Paris: Armand Colin, 1992), pp. 82-83, and Zaloga, p. 130
32. Ibid., Lardier
33. *From First Satellite . . .*, p. 7
34. Avduyevsky interview with author, Moscow, Jan. 21, 1993
35. Myers, Dale, "The Navaho Cruise Missile—A Burst of Technology," *Acta Astronautica,* Vol. 26, No. 8, 1992, pp. 741–48
36. Zaloga, pp. 133–34
37. Ibid.
38. Lenorovitz, Jeffrey M., "Russians Detail 1950s Cruise Missile Effort," *Aviation Week & Space Technology,* Nov. 2, 1992, p. 50
39. Rozhdostvensky interview with author, Moscow, Jan. 24, 1993
40. Myers, p. 746
41. Avduyevsky interview
42. Natenzon, Yakov M., *Space Bulletin,* Vol. 1, No. 2, 1993, p. 27
43. Kryukov interview
44. Not to be confused with the N-1 moon rocket of later years
45. Vetrov, Gyorgi and K. A. Krasnova, *S. P. Korolev, A Scientific Biography* (Moscow: Nauka, 1996)
46. Kryukov interview
47. *From First Satellite . . .*, p. 9
48. Kryukov interview
49. Grechko interview with author, Moscow, May 16, 1993
50. Khrushchev, Nikita, *Khrushchev Remembers The Last Testament*, translated and edited by Strobe Talbott (Boston: Little, Brown, 1974), p. 45
51. Ibid.
52. Avduyevsky interview
53. Kryukov interview
54. Zaloga, p. 141
55. *From First Satellite . . .*, p. 8
56. Zaloga, p. 255
57. Martin letter
58. *Khrushchev Remembers . . .*, pp. 47–48
59. Martin letter. The first successful demonstration of an all-inertial Atlas over a 6,350-nautical-mile range was on Feb. 24, 1961.
60. *From First Satellite . . .*, p. 7. R-11 was the first Soviet tactical missile using a storable propellant (nitric acid and a hydrocarbon fuel). It had an Isaev engine of 8 tons thrust, was first launched in 1953, and was put into service in 1955.

Later there were variations: R-11FM, introduced in 1955 for launch from submarines, and R-11M, which carried a "military nuclear charge," put into operation in 1958.

61. Chertok interview with author, Moscow, Dec. 4, 1991
62. Varfolomeyev, *Soviet Rocketry* . . . , p. 261
63. Varfolomeyev letter to *Spaceflight,* Vol. 38, Jan. 1996, p. 31
64. Varfolomeyev, *Soviet Rocketry* . . . , p. 261
65. Dornberger, Walter B., *V-2* (New York: Viking Press, 1972)
66. Chertok interview
67. Keldysh and Vetrov, *Creative Legacy* . . . , pp. 208–90
68. Ostashov interview with author, Moscow, Sept. 15, 1993
69. Peebles, Curtis, "Tests of the SS-6 Sapwood ICBM," *Spaceflight,* Nov.–Dec. 1980, p. 340
70. Ibid., p. 341
71. Khrushchev, Sergei, interview with author, Providence, R.I., Dec. 9, 1993
72. *From First Satellite* . . . , p. 13
73. Communication from Maxim Tarasenko, Mar. 23, 1996
74. Mishin interview with author, Oct. 13, 1992
75. Ibid.
76. Evidently an oft-quoted phrase but with variations. Chertok says that Glushko's words were "Put an engine on the gates and the gates will fly."
77. Vetrov interview with author, Oct. 13, 1992.
78. Author's interview with Irina Strazheva, wife of Yangel, Barvikha, Oct. 10, 1992. She died in 1995.
79. Ibid.
80. Golovanov, *Korolev Fakti i Mifi* . . . , p. 439
81. Mishin interview with author, Moscow, Oct. 13, 1992
82. Strazheva interview, Barvikha. In Russian "TASS" rhymes with "nass," which means "us."
83. Ibid.
84. Kovtunenko died in July 1995
85. Kovtunenko interview with author, May 17, 1993
86. *From First Satellite* . . . , p. 13
87. Ostashov interview with author, Sept. 16, 1993
88. Kovtunenko interview with author
89. Vetrov communication with author
90. Kovtunenko interview with author
91. Averkov, S., "Top Secret: Explosion at Baikonur Cosmodrome," from *Rabochnaya Tribuna,* Moscow, Dec. 6, 1990, p. 4, translated from Russian in JPRS-USP-91-002, Apr. 16, 1991
92. Associated Press bulletin, Oct. 24, 1990, Moscow
93. Averkov, "Top Secret"
94. Ibid.
95. Kovtunenko interview with author

96. Zaloga, *Target America,* p. 235
97. Ibid.
98. *From First Satellite . . . ,* p. 63

Chapter 7: Sputnik: No Big Deal to Khrushchev—at First

1. Rauschenbakh comment to author at 41st International Astronautical Congress, Montreal, Oct. 11, 1991
2. *Pravda,* Oct. 5, 1957, p. 1. The generic *sputnik,* which means "traveling companion" or "satellite," became a proper name in the rest of the world, as *Sputnik* was followed by 2, 3, and so on.
3. In *Rockets and Cosmic Space,* a monograph published by Tsiolkovsky in 1903
4. *Pravda,* Oct. 6, 1957, pp. 1–2
5. *Pravda,* Oct. 8, 1957, p. 3
6. Haviland telephone interview with author, May 22, 1995
7. Brill, Yvonne, one of the participants in the Rand study, letter to the author, June 15, 1995
8. Project RAND Report SM-11827, May 2, 1946, from Rand 25th Anniversary Volume (Santa Monica, CA: Rand Corp., 1973), p. 3
9. Golovanov, Yaroslav, "The Beginning of the Space Era," *Pravda,* Oct. 4, 1987, p. 3, translated into English in JPRS-USP-88-001, Feb. 26, 1988, p. 48
10. Vetrov, Gyorgi, private communication, June 13, 1995
11. Keldysh and Vetrov, *Creative Legacy . . . ,* p. 343, and Tarasenko, Maxim, *Military Aspects of Soviet Cosmonautics* (Moscow; Nikol 1992) p. 16. Tarasenko reports that the recommendation called for "development of a 2-3 ton satellite, a recoverable satellite, a satellite for a long orbital stay of 1-2 people and an orbital station with regular Earth ferry communication."
12. Mishin interview with author, Sept. 2, 1992
13. Keldysh and Vetrov, *Creative Legacy . . .*
14. Sagdeev, Roald, *The Making of a Soviet Scientist* (New York: John Wiley, 1994), p. 155
15. For a detailed account of the genesis of Project Orbiter see Stuhlinger, Ernst, and Ordway, Frederick I. III, in *Wernher von Braun Crusader for Space* (Malabar FL: Krieger Publishing, 1994), pp. 123–31
16. The full name seems a satire on Soviet bureaucracy. It was the Interdepartmental Commission for the Coordination and Control of Scientific-Theoretical Work in the Field of Organization and Accomplishment of Interplanetary Communications of the Astronomical Council of the USSR Academy of Sciences.
17. Krieger, F. J., *Behind the Sputniks* (Washington, D.C: Public Affairs Press, 1958), p. 330
18. Feoktistov interview with author, Moscow, Dec. 12, 1991
19. Avduyevsky, V. S., *M. V. Keldysh Selected Works Rocket Technology and Cosmonautics* (Russian), (Moscow: Nauka, 1988), p. 235
20. Ibid., pp 235–40
21. Vetrov, *Korolev i Ego Delo*

22. Grechko interview with author, Moscow, May 16, 1993
23. Gerchik, K. "Breakthrough into the Cosmos" (Russian), Central Council of Veterans of Baikonur Cosmodrome, Moscow, 1992, pp. 77–78. The State Commission was chaired by Vassily Ryabikov. It included, besides Korolev himself, Marshal Mitrofan Nedelin, Gyorgi Pashkov, Valentin Glushko, Nikolai Pilyugin, Viktor Kuznetsov, Mikhail Ryazanski, Vladimir Barmin, Alexander Mrykin, S. Shishkin, I. Bulychev, Alexei Nesterenko, and A. Maksimov.
24. Ivanovsky interview with author, Moscow, Jan. 26, 1993
25. Strekalov interview with author, Moscow, Jan. 28, 1993
26. The Mir-Shuttle rendezvous was Strekalov's fourth orbital flight, although his fifth space launch. He and Vladimir Titov were successfully ejected by the launch escape system when their Soyuz T-10 spacecraft was engulfed by flames in 1983.
27. Zheleznov, Nikolai, "Hello—(Bip-Bip) Scientist, Designer and Cosmonaut Speaking," *Soviet Life,* Oct. 1982, p. 35
28. Floriansky, Mikhail, *Oct. 4—For the First Time in the World*, Moscow News Supplement, No. 40 (3288), 1987
29. Ishlinsky, *Akademik . . .* , p. 62
30. Rebrov, Col. Mikhail, *Sputnik No. 1,* Moscow News Supplement, No. 40 (3288), 1987, p. 3
31. U.S. IGY officials had known of the Soviet plans since 1955 when *Pravda* had alerted its readers, and the delegates to the 1955 International Astronautical Congress in Copenhagen were informed as well. In June 1957, an article appeared in the Soviet journal *Radio,* by one V. Vakhnin, alerting radio specialists on how to tune in on an orbiting satellite's signal.
32. Major General John B. Medaris, von Braun's boss at the Army Ballistic Missile Agency
33. *Astronautics,* Nov. 1957, p. 6
34. Ibid., p. 17
35. Summerfield, Martin, "Problems of launching an earth satellite," *Astronautics,* Nov. 1957, pp. 18–21, 86–88
36. *Le Figaro,* Paris, Oct. 7, 1957, pp 4–5
37. Kennan, George F., *George F. Kennan Memoirs 1950–1963* (New York: Pantheon Books, 1972), p. 140
38. Grechko interview with author, and Mishin interview, *Pravda,* Oct 20, 1989
39. Van Allen, James A., *Origins of Magnetospheric Physics* (Washington, D.C.: Smithsonian Institution Press, 1983), p. 49. Stuhlinger and Van Allen had begun discussions on the use of a satellite to investigate cosmic rays above the atmosphere more than three years earlier, in 1954, when Van Allen was at Princeton University. Also, Pickering communication with author, Mar. 18, 1996.
40. Ibid., p. 93
41. Ivanovsky interview with author, Moscow, Jan. 26, 1993
42. Wilson, Glen P., *Prologue,* Quarterly of the National Archives, Winter 1993, pp 364–70

43. These objectives are virtually the same as those developed at a conference at the University of Michigan on January 27, 1956, the results of which were published in *Scientific Uses of Earth Satellites* (University of Michigan Press, 1956)
44. Sagdeev, *The Making* . . . , pp. 156–57
45. Van Allen, *Origins* . . . , p. 82
46. Ibid., p. 66
47. Ibid.
48. Ibid., p. 72. Van Allen reports that it was physicist Robert Jastrow who first used the term at a meeting of the International Atomic Energy Agency in Europe.
49. *Pioneers 1, 2, 3, and 4* were attempted lunar probes launched in 1958 and early 1959.
50. Van Allen, *Origins* . . . , pp. 129–30. Letter from Van Allen to Robert Toth, New York *Herald Tribune,* Mar. 13, 1959.
51. Principal source is *TRW Space Log 1957–1991,* Redondo Beach, CA, pp 50–51

Chapter 8: Unmanned Firsts: Hitting the Moon and Venus

1. Vetrov interview with author, Moscow, Oct. 21, 1992
2. Keldysh and Vetrov, *Creative Legacy* . . . , pp. 400–4
3. Ibid.
4. Maximov interview with author, Moscow, Dec. 13, 1991
5. Ibid.
6. Ibid.
7. Natasha Koroleva interview
8. Maximov interview
9. Pallo interview with author, Moscow, Sept. 14, 1993
10. Dryden, Hugh L., NASA statement prepared for President Eisenhower, Sept. 14, 1959
11. Ibid.
12. Harvey, Bryan, *Race Into Space* (London: Ellis Horwood Ltd., 1988), p. 37
13. Rauschenbakh interview with author, Abramtsevo, Dec. 7, 1991
14. DeBra telephone interview with author, Mar. 12, 1994
15. Ishlinsky, *Akademik* . . . , pp. 351–55
16. Tarasenko, Maxim, communication with author, July 14, 1995. Ai-Petri, now abandoned, was an extremely spartan station using a wheeled trailer-like vehicle with an antenna. It was probably a modified air defense radar station. The radar was located on the side of a mountain that projected into the sea.
17. Ishlinsky, *Akademik* . . . , pp. 351–55
18. Ibid.
19. Ibid.
20. Ibid.
21. Akim interview with author, Moscow, Sept. 10, 1993
22. Ivanovsky interview with author, Moscow, Dec. 9, 1991

23. Memorable was the look of disbelief on the face of Sedov when I handed him the five-inch-thick Sunday *New York Times* for perusal. "This is for only one day?" he gasped. Later, when the Eastern Airlines Constellation approached National Airport it banked sharply, giving passengers a vivid view of the area. Sedov beckoned Blagonravov to the window. "Look! Pentagon!" he exclaimed excitedly.
24. Glennan, T. Keith, *The Birth of NASA* (Washington, D.C.: NASA History Office, 1993), p. 31
25. Sergei Khrushchev, telephone interview with author, July 11, 1995
26. Semyon Kosberg, an aircraft engine designer just three years older than Korolev, was asked by Korolev to pitch in on rocket engine development for the lunar spacecraft. He would later design numerous rocket engines, not only for Korolev but for Korolev's rival, Vladimir Chelomei. He died of injuries from an automobile accident on Jan. 3, 1965.
27. *Novosti Kosmonavtiki* No. 26/18–31 Dec., 1993, p. 46
28. *From First Satellite* . . . , p. 9
29. In fact, the launch vehicle, too, is named Molniya, and it is still being used.
30. Maximov, G. Y., *From the History of Constructing and Testing of the First Soviet Automatic Interplanetary Stations,* IAA-91-690, 42nd Congress of the International Astronautical Federation, Montreal, Oct. 5–11, 1991, p. 1
31. Ishlinsky, *Akademik* . . . , p. 227
32. Maximov, *From the History* . . . , p. 1
33. Ibid., p. 2
34. Ishlinsky, *Akademik* . . . , p. 95
35. Natasha Koroleva interview
36. Golovanov interview with author, Peredelkino, Oct. 19, 1992
37. Chertok interview with author, Moscow, Dec. 4, 1991
38. Bryushinin, V. M., *Breakthrough Into the Cosmos* (Russian), (Moscow: Veles Ltd., 1994), p. 105
39. *Jane's Spaceflight Directory 1986,* p. 175
40. Broad, William J., "Spy Satellites Early Role As 'Floodlight' Coming Clear," *The New York Times,* Sept. 12, 1995
41. Ivanovsky interview
42. *Zond 3* went on to repeat its transmission of photos from 2.2 million km and again from 31.5 million km, in what was presumably a test of the transmission of photos from subsequent Venus probes, a transmission that never actually occurred. Source: *Jane's Spaceflight Directory 1986,* p. 236
43. Rozhdostvensky interview with author, Moscow, Jan. 24, 1993
44. Ibid.
45. Vetrov interview with author, Moscow, Sept. 8, 1993
46. Rozhdostvensky interview
47. Ibid.
48. Harvey, *Race Into Space,* p. 103
49. Rauschenbakh interview with author, Abramtsevo, Dec. 7, 1991
50. Harvey, *Race Into Space,* p. 103

Chapter 9: Gagarin First, Shepard an Anti-climactic Second

1. Leonov interview with author, Moscow, Dec. 11, 1991
2. Ibid.
3. A "perfect" suborbital launch of eight minutes, reaching 180 km altitude, but carrying no chimpanzee
4. Shepard, Alan, and Slayton, Deke, *Moon Shot* (Atlanta: Turner Publishing Inc., 1994), p. 91. Shepard deplored the use of chimpanzees in the first place. "Scientifically it was a huge waste of money, time and manpower," he wrote (p. 89), "but the medical team didn't want to subject a man to weightlessness and high acceleration without first using a primate."
5. Shevalyov interview with author, Moscow, Sept. 9, 1993
6. *The New York Times,* Apr. 12, 1961
7. Ibid., Apr. 13, 1961. It is somewhat ironic that the editorial went on to predict, rhapsodically, that the flight "was a stepping stone toward orbiting a space-station as a jumping-off point for trips to the moon and beyond," ironic because in recent years the *Times* has consistently editorialized against the International Space Station being built for launch starting in 1997 by the Russians and Americans together with the Europeans, Japanese, and Canadians.
8. *Pravda,* Apr. 13, 1961
9. Finney, John W., *The New York Times,* Apr. 15, 1961
10. Ibid.
11. Gazenko interview with author, Moscow, Dec. 13, 1991
12. Ibid.
13. *Ogonyok,* Moscow, Aug. 18–25, 1990, No. 34, pp. 4–5
14. Vladimir Yazdovsky was the first head of the USSR's research in space medicine.
15. Ezell, Edward C., and Ezell, Linda N., *The Partnership,* NASA History Series, 1978, p. 64
16. Ishlinsky, *Akademik . . . ,* p. 203
17. Severin interview with author, Moscow, Sept. 16, 1993
18. Ezell and Ezell, *The Partnership*, p. 65
19. Murray, Charles, and Cox, Catherine Bly, *Apollo The Race to the Moon* (New York: Simon & Schuster, 1989), p. 190, and telephone conversation with Christopher Kraft, Sept. 10, 1996
20. Rauschenbakh interview with author, Abramtsevo, Dec. 7, 1991
21. Gallai interview with author, Moscow, Sept. 9, 1993
22. Ibid.
23. Ibid.
24. Link, Mae Mills, NASA SP-4003, 1965, p. 53
25. Feoktistov interview with author, Moscow, Dec. 12, 1991
26. Ivanovsky interview with author, Moscow, Jan. 26, 1993
27. Gallai interview
28. Tikhonravov, M. K., Rauschenbakh, B. V., Skuridin, G. A., and Vaisberg, O. L., "Ten Years of Soviet Space Research," *Kosmicheskie Issledovaniya,* Vol. 5, No. 5., pp. 643–79, Sept.–Oct., 1967

29. *From First Satellite . . .*, p. 11.
30. *TRW Space Log* lists the May 15 flight as *Sputnik 4*, while the official Soviet name is *Korabl-Sputnik*. The July 28 flight got no designation. The Aug. 19 flight is called *Sputnik 5* in *TRW Space Log*, and *Korabl-Sputnik 2* officially.
31. *Novosti Kosmonavtiki* No. 5, 1994
32. The Pchelka and Mushka flight on Dec. 1 is listed as *Sputnik 6* in *TRW Space Log*, but *Korabl-Sputnik 3* in official Russian records. Chernushka's flight was *Sputnik 9* in *TRW*, *Korabl-Sputnik 4* officially. Zvezdochka, called *Sputnik 10* in *TRW*, was *Korabl-Sputnik 5*.
33. Severin interview
34. Ivanov, V. L., editor, Strelnikov, A. T., compiler, *Principal Milestones of the Space Era* (Russian), Strategic Rocket Forces, Dzerzhinski Military Academy, Moscow, 1993, pp. 77–78
35. Belyanov, V., et al., "Tomorrow is Space Program Day: The Classified Documents on Gagarin's Spaceflight," *Rabochaya Tribuna*, Apr. 11, 1991, pp. 1, 4 English translation in JPRS-USP-91-004, Sept. 20, 1991, pp. 71–77
36. Ibid., p. 73
37. Tarasenko, Maxim, communication with author, Dec. 1, 1995
38. Harvey, *Race into Space*, p. 53
39. Belyanov, "Tomorrow" . . . , pp. 71–77
40. Ibid., p. 73
41. Ibid.
42. Ibid.
43. Clark, Philip, E-mail communication to FPSpace, Mar. 6, 1996
44. Tarasenko, Maxim, E-mail communication, Mar. 6, 1996
45. Gallai interview with author, Moscow, Sept. 9, 1993
46. Shevalyov interview with author, Moscow, Sept. 9, 1993
47. Ishlinsky, *Akademik . . .*, p. 339
48. Low interview at NASA headquarters, Washington, D.C. John F. Kennedy Oral History Project, May 1, 1964
49. Ibid.
50. Kennedy press conference, Apr. 12, 1961
51. Ibid.
52. Titov interview with author, Moscow, Dec. 11, 1991
53. Sources are *TRW Space Log* and *Jane's Spaceflight Directory 1986*
54. Mishin interview with G. Salakhutdinov in *Ogonyok*, No. 34, Aug. 18–25, 1990, pp. 4–5. English translation in JPRS-USP-91-002, Apr. 16, 1991, p. 67
55. Ibid., p. 68
56. Ibid.

Chapter 10: Voskhod: A "Circus Act"

1. Tikhonravov et al., *Ten Years of Soviet Space Research*
2. Mishin interview with Salakhutdinov.
3. Vetrov communication with author, Sept. 4, 1995, and Sergei Khrushchev telephone conversation with author, Sept. 5, 1996

4. Feoktistov interview with author, Moscow, Dec. 15, 1991
5. Ibid.
6. Mishin interview with Salakhutdinov
7. Ezell and Ezell, *The Partnership,* p. 80
8. Severin interview with author, Kaliningrad, Sept. 16, 1993
9. Ezell and Ezell, *The Partnership*
10. Severin interview
11. Mishin interview with Salakhutdinov
12. Tikhonravov et al., *Ten Years of Soviet Research*
13. The first two *Geminis* were unmanned tests.
14. Vetrov communication with author
15. Kamanin, Nikolai, "Pages from a Diary," *Sovetskaya Rossiya,* Mar. 19, 1990, p. 6
16. Ishlinsky, *Akademik,* p. 421, and Haeseler, Dietrich, "Leonov's Way to Space," *Spaceflight,* Aug. 1994, pp. 280–81. Haeseler reports that one of the airlocks was sold at auction at Sotheby's, New York, for $80,000 in 1993.
17. Ed White would make America's first EVA in June 1965, three months after Leonov's. White was killed in the *Apollo 1* fire in 1967.
18. Severin interview
19. Kamanin, "Pages from a Diary"
20. Ishlinsky, *Akademik . . . ,* pp. 486–87
21. Leonov interview with author, Moscow, Dec. 11, 1991. Also, Kamanin diary. Also, Ishlinsky, pp. 254–55.
22. Ishlinsky, *Akademik . . . ,* pp. 254–55
23. Leonov interview
24. Ishlinsky, *Akademik . . . ,* p. 336, and private communication from Maxim Tarasenko, Sept. l, 1995
25. Shepard, Alan, and Deke Slayton, *Moon Shot* (Atlanta: Turner Publishing, 1994), p. 182
26. Ibid., p. 177
27. Hacker, Barton C., and Grimwood, James M., *On the Shoulders of Titans: A History of Project Gemini,* NASA SP-403, Washington, D.C., 1977, p. 483
28. Popovich interview in *Izvestia,* Dec. 29, 1965
29. Schirra statement at Dec. 30, 1965, NASA press conference

Chapter 11: Spy Sats and Com Sats

1. National Intelligence Estimate, *The Soviet Space Program,* Dec. 5, 1962, p. 3
2. The Soviets' practice of giving the same name to different vehicles is continually vexing. *Zenit* is not only a reconnaissance satellite but a rocket booster. *Vostok, Soyuz,* and *Molniya,* too, are not only spacecraft but rocket boosters.
3. Then, in May 1995, *Corona*'s history was divulged in detail at a briefing conducted by the CIA, George Washington University, and the Smithsonian Institution. See Day, Dwayne, "Corona: America's First Spy Satellite Program," *Quest* (Grand Rapids, MI: Cspace Press, Summer 1995), pp. 4–21; McDowell,

Jonathan, "US Reconnaissance Satellite Programs," pp. 22–33, and McDonald, Robert A., "Corona: Success for Space Reconnaissance . . . ," *PE & RS (Photogrammetric Engineering and Remote Sensing)*, June 1995, pp. 689–720.

4. Mosley, Leonard, *Dulles: a Biography of Eleanor, Allen and John Foster Dulles and Their Family Network* (New York: Dial Press, 1978), p. 432
5. Wheelon telephone interview with author, Nov. 2, 1995
6. Bundy, McGeorge, *Danger and Survival, Choices About the Bomb in the First Fifty Years* (New York: Random House, 1988), p. 338
7. Tarasenko, Maxim, communication with author, Nov. 2, 1995
8. Burrows, Willam E., *Deep Black, Space Espionage and National Security* (New York: Random House, 1986), p. vii
9. Wheelon, Albert D., "Corona: A Triumph of American Technology," keynote address at a conference, *Piercing the Curtain, Corona and the Revolution in Intelligence*, sponsored by George Washington University and the CIA, May 23, 1995, Washington, D.C.
10. *Discoverer 17* (Nov. 12, 1960), *18* (Dec. 7), *25* (June 16, 1961), *26* (Jul. 7), *29* (Aug. 30), *30* (Sept. 12), *32* (Oct. 13), *35* (Nov. 15), *36* (Dec. 12), and *38* (Feb. 27, 1962). Source: *TRW Space Log.*
11. Johnson, Nicholas, letter to the author, Dec. 27, 1995
12. Secret memo to the Deputy Secretary of Defense from General Nathan F. Twining, Chairman of the Joint Chiefs of Staff, Apr. 25, 1959
13. Frumkin, Yuri, *Development of First Soviet Photoreconnaissance Satellite "Zenit,"* Priroda (Russian), Apr. 1993, pp. 72–78, English translation in JPRS-USP-93-005, Oct. 5, 1993, pp. 15–20
14. Tarasenko communication with author, Dec. 12, 1995
15. Frumkin, *Development* . . . , p. 19
16. Ibid.
17. Ibid., p. 16
18. Johnson, letter to the author
19. Frumkin, *Development* . . . , p. 16
20. Ibid.
21. Johnson letter
22. Frumkin, *Development* . . . , pp. 17–18
23. Ibid.
24. *Vostok* was the name of both the manned spacecraft and the rocket launcher.
25. Ishlinsky, *Akademik* . . . , pp. 423–24
26. *Izvestia,* May 27, 1986, p. 3, as communicated by Nicholas Johnson, Dec. 27, 1995
27. Sevastyanov interview with author, Moscow, Dec. 11, 1991
28. Johnson letter
29. Keldysh and Vetrov, *Creative Legacy* . . . , p. 437
30. Hilton, William F., and S. R. Dauncey, *Communications Satellite Orbits,* Paper 71, Eleventh International Astronautical Congress, Stockholm, 1960, and *Communication Satellite Systems Suitable for Commonwealth Communications,* Symposium on Communications Satellites, British Interplanetary Society,

May 12, 1961; *Communications Satellites* (New York: Academic Press, 1962), pp. 95–112
31. Keldysh and Vetrov, *Creative Legacy* . . .
32. Tarasenko communication, Dec. 12, 1995, and *From First Satellite* . . . , p. 12
33. Shevalyov interview with author, Sept. 9, 1993
34. Ishlinsky, *Akademik* . . . , p. 68
35. Otrieshka interview
36. Golovanov interview with author, Oct. 19, 1992
37. Vetrov interview with author, May 11, 1993
38. Unfortunately I was unable to interview Nina Ivanovna, who turned me down each time I phoned her. She talked with a very weak voice, supporting her statement that she was too ill to go through an interview.

Chapter 12: The Organization: Korolev Up and Down

1. Ivanovsky interview with author, Moscow, Dec. 9, 1991
2. Postyshev, Vladimir, chief specialist of the Russian Federation's Commission on Transportation, Communications, Information Science, and Space, *Rossiskaya Gazeta,* June 23, 1992, p. 4, English translation in JPRS-USP-92-005, Aug. 21, 1992, p. 53
3. Ibid.
4. Tarasenko, Maxim, *Military Aspects of Soviet Cosmonautics* (Russian), (Moscow: Nikol, 1992), p. 16
5. Tarasenko, *Evolution* . . . , p. 1
6. Postyshev, p. 52
7. Feoktistov, "The Space Program Without Fanfares or Ambitions," *Novoye v Zhizni,* Apr. 1991, pp. 1–63, English translation in JPRS-USP-91-005, Oct. 1, 1991, p. 4
8. Orieshkin interview with author, Moscow, Jan. 27, 1993
9. Barmin, Vladimir, *Izvestia,* Sept. 29, 1987, p. 3, translated into English in JPRS-USP-88-001, Feb. 26, 1988, p. 39
10. Ibid.
11. Ibid.
12. Sagdeev, *The Making* . . . , p. 187
13. Ibid., p. 186. Sagdeev remarks that, with the existence of a commission with such a name, one wonders how the USSR's propagandists could ever, with straight faces, pillory the U.S. "military-industrial complex."
14. Shakhmatov, V. A., *Breakthrough Into Space* (Russian), (Moscow: 1994), pp. 110–11, and author's telephone conversation with Sergei Khrushchev, Sept. 5, 1996
15. Ishlinsky, *Akademik* . . . , p. 338
16. Ibid., p. 224
17. Ibid.
18. Ibid., p. 225

19. *From First Satellite* . . . , p. 7
20. Fuhrman, R. A., *Fleet Ballistic Missile System Polaris to Trident,* AIAA 78-355, AIAA Annual Meeting, Washington, D.C., Feb. 1978
21. Ishlinsky, *Akademik* . . . , pp. 280–81
22. Golovanov interview with author, Peredelkino, Dec. 13, 1991
23. Tarasenko, Maxim, *From Confrontation to Competition and Cooperation,* IAF Paper IAA-94-IAA.3.2.637, 45th International Astronautical Congress, Jerusalem, Oct. 9–14, 1994
24. Ishlinsky, *Akademik* . . . , p. 372
25. Ibid., p. 257
26. Ibid., p. 255
27. Gitelson interview with author, Princeton, N.J., Jan. 9, 1993
28. Otrieshka interview with author, Moscow, Dec. 5, 1991
29. Ishlinsky, *Akademik* . . . , p. 217
30. Vetrov, *Korolev i Ego Delo*
31. Mishin interview with author, Oct. 13, 1992
32. Ishlinsky, *Akademik* . . . , p. 227
33. Ibid., p. 96
34. Stuhlinger, Ernst, and Ordway, Frederick I, III, *Wernher von Braun, A Biographical Memoir* (Malabar, FL: Krieger Publishing Co., 1994), p. 204
35. Orieshkin interview
36. Vetrov interview with author, Oct. 16, 1992
37. Ishlinsky, *Akademik* . . . , pp. 55–56
38. Otrieshka interview
39. Orieshkin interview with author, Jan. 27, 1993
40. Vetrov interview with author, Moscow, Oct. 21, 1992
41. Ishlinsky, *Akademik* . . . , p. 360

Chapter 13: The Technology: Simple but Reliable

1. Ivanovsky interview with author, Moscow, Jan. 26, 1993
2. Grechko interview with author, Moscow, May 16, 1993
3. Akim interview with author, Moscow, Sept. 10, 1993
4. Grechko interview
5. Gubanov interview with author, Moscow, Oct. 15, 1992
6. Vorobiev interview with author, Moscow, May 20, 1994
7. Ishlinsky, *Akademik* . . . , pp. 449–50
8. Ibid., p. 449
9. Ibid., p. 450
10. Ibid., pp. 245–46
11. Schwinghamer telephone interview with author, Sept. 19, 1995
12. Ishlinsky, *Akademik* . . . , p. 246
13. Ibid., p. 247
14. Ibid., p. 248
15. Ibid., pp. 250–51

16. Ibid., p. 242
17. Ibid., p. 197
18. Ibid., pp. 231–35

Chapter 14: The Party, the Paranoia

1. Chertok interview with author, Moscow, Dec. 4, 1991
2. Ibid.
3. Ishlinsky, *Akademik . . .*, p. 389
4. Vetrov interview with author, Moscow, Oct. 16, 1992
5. Avduyevsky interview with author, Jan. 21, 1993
6. Ivanovsky interview with author, Jan. 26, 1993
7. Ibid., and Sept. 5, 1996, telephone interview with Sergei Khrushchev, who told me that Korolev himself very likely took the initiative to place Ivanovsky in this position. In general, he said, it was important for Korolev to see that his people got key Party jobs. All Party Secretaries in the design bureaus actually came under the Chief Designer.
8. Ishlinsky, *Akademik . . .*, p. 316
9. Orieshkin interview with author, Moscow, Jan. 27, 1993
10. Ishlinsky, *Akademik . . .*, p. 69
11. Koroleva, Natasha, interview with author, Moscow, Dec. 8, 1991
12. Mishin interview with author, Moscow, Sept. 6, 1990
13. Shevalyov interview with author, Moscow, Sept. 9, 1993
14. Ibid.
15. Ishlinsky, *Akademik . . .*, p. 56
16. Smith, Hedrick, *The New Russians* (New York: Random House, 1990), p. 106
17. Ishlinsky, *Akademik . . .*, p. 338
18. Ibid., p. 396
19. Ibid., p. 388
20. Ibid., p. 405
21. Vetrov interview with author, Moscow, Oct. 16, 1992
22. Sheremetyevsky interview in *Nezavisimaya Gazeta,* Aug. 19, 1993, as translated in JPRS-USP-93-005, Oct. 5, 1993, p. 30
23. Karrask, *International Affairs,* Moscow, June 1991, pp. 128–38, as reported in JPRS-USP-91-007, Nov. 22, 1991, p. 55
24. Ishlinsky, *Akademik . . .*, p. 214
25. Golovachev, "History of Kaliningrad Space Design Bureau," *Trud* (Russian), Nov. 22, 1989, p. 4 English translation in JPRS-USP-90-002, May 15, 1990, p. 56
26. Gazenko remarks to author and other AIAA visitors, Moscow, May 16, 1994
27. Grechko interview with author, Moscow, May 16, 1993
28. Khrushchev, Sergei, *Khrushchev on Khrushchev* (New York: Little, Brown, & Co., 1990), p. 4
29. Khariton, Yuli, and Yuri Smirnov, *Bulletin of the Atomic Scientists,* May 1993, pp. 28–29

30. Ibid.
31. Ishlinsky, Akademik . . . , p. 235
32. Frumson interview with author, Oct. 11, 1992
33. Sevastyanov interview with author, Moscow, Dec. 11, 1991
34. *Russian Space History,* Catalog for Sale 6516, Item 33, Sotheby's, New York, 1993
35. Khrushchev, Sergei, *Khrushchev on Khrushchev,* p. 104
36. Akim interview with author, Moscow, Sept. 10, 1993, and author's telephone interview with Sergei Khrushchev, Sept. 5, 1996
37. Shevalyov interview with author

Chapter 15: Racing Apollo: The Odds Were Enormous

1. Rebrov, M. "The Way it Was: the Difficult Fate of the N-1 Project," *Krasnaya Zvyezda* (Russian), Jan. 13, 1990, No. 11, p. 4
2. Keldysh and Vetrov, *Creative Legacy* . . . , p. 5
3. Ibid., pp. 405–8
4. Van Nimmen, Jane, and Bruno, Leonard C., with Rosholt, Robert L., *NASA Historical Data Book,* Vol. I (Washington, D.C.: NASA, 1988), p.6
5. Ibid, p. 13. In just nine years, by 1967, the NASA budget would increase by a factor of 50, to $5 billion. The NASA employment would eventually be 36,000. Total employment, including engineers, scientists, and their support staffs in industry, would reach 411,000 during the peak of the *Apollo* program buildup in 1965.
6. As early as April 1957, however, there were paper studies of big boosters by the von Braun team at the U.S. Army Ballistic Missile Agency, Redstone Arsenal, AL. One study, which expended 50,000 man-hours, dealt with a large clustered engine booster that would generate 1.5 million pounds of thrust. Ertel, Ivan D., and Morse, Mary Louise, *The Apollo Spacecraft: A Chronology,* NASA SP-4009, Vol. 1, p. 8
7. Hansen, James R., *Engineer in Charge, a History of the Langley Aeronautical Laboratory 1917–1958* (Washington, D.C.: NASA Scientific and Technical Information Office, 1987), p. 385
8. Glennan, T. Keith, *The Birth of NASA* (Washington, D.C.: NASA History Office, 1993), p.13
9. Ibid., p. 245
10. Logsdon, John M., *The Decision to Go to the Moon* (Cambridge, MA: MIT Press, 1970), p. 106
11. Ibid., p. 107
12. Ibid., p. 109
13. Memorandum from John F. Kennedy to Lyndon B. Johnson, Apr. 20, 1961, John F. Kennedy Library, Boston, MA
14. Logsdon, *The Decision* . . . , p. 111
15. Von Braun, Wernher, memo to Vice President Johnson, Apr. 29, 1961, NASA Historical Archives

16. Logsdon, *The Decision . . .*, p. 128
17. Von Braun, Wernher, "Crossing the Last Frontier," Mar. 22, 1952, p. 24; "The Journey," Oct. 18, 1952 "Can We Get to Mars?," Apr. 30, 1954, p. 22 *Collier's*
18. Von Braun, *Collier's,* Mar. 22, 1952, p. 39
19. Vetrov, G., "The Difficult Fate of Rocket N-1," *Nauka i Zhizn*, No. 4, 1994, pp. 79–80, and Kryukov, S., "The Brilliance and the Eclipse of the Lunar Program," pp. 81–85
20. Johnson, Lyndon B., speech before the Senate Democratic Caucus, Jan. 7, 1958, as quoted by Kearns, Doris, in *Lyndon Johnson & the American Dream* (New York: Harper & Row, 1976), p. 145
21. *NASA Historical Data Book,* Vol. II, p. 180
22. Vetrov, *The Difficult Fate . . .*, p. 79
23. Filin, Vyacheslav, "Development of Lunar Spacecraft for the Manned Moon Landing Program," *Aviatsya i Kosmnavtika,* No. 12, Dec. 1991, pp. 44–45, translated into English in JPRS-USP-92-005, Aug. 21, 1992, pp. 16–17
24. Vetrov, *The Difficult Fate . . .*, p. 79
25. Maximov interview with author, Moscow, Dec. 13, 1991
26. Kubasov interview with author, Moscow, Jan. 28, 1993. In 1975 Kubasov accompanied fellow cosmonaut Alexei Leonov on the *Soyuz 19* mission in 1975 to rendezvous with U.S. astronauts Tom Stafford, Deke Slayton, and Vance Brand in *Apollo 18* for a "handshake in space."
27. Keldysh and Vetrov, *Creative Legacy . . .*, pp. 489–500
28. Avduyevsky interview with author, Moscow, Jan. 29, 1993
29. Minenko interview with author, Washington, D.C., July 27, 1993
30. Ibid. Antonio Ferri, Dean Chapman, Edward van Driest, and Lester Lees were all outstanding U.S. aerodynamicists.
31. Feoktistov interview with author, Moscow, Dec. 12, 1991
32. Mishin interview with G. Salakhutdinov in *Ogonyok,* No. 34, Aug. 18–25, 1990, pp. 4–5, English translation in JPRS-USP-91-002, Apr. 16, 1991, p. 69
33. Faget interview with Ertel, Ivan, and Grimwood, James, Dec. 15, 1969, as reported in *The Partnership,* Ezell, Edward C., and Ezell, Linda N., NASA History Series, p. 373
34. Clark, Phillip, *The Soviet Manned Space Program,* Orion, 1988, pp. 23–24
35. Kryukov, *The Brilliance . . .*, pp. 81
36. Ibid., p. 82. It is not unusual, of course, for NASA estimates to be far below the eventual actual cost either. However, in the case of *Apollo* the original estimate given to President Kennedy was reportedly $24 billion and NASA actually spent about that amount. Another report, however, is that the original estimate made by James Webb's staff was $12 billion. The politically wise Webb doubled it knowing that he could get the larger figure approved by a Congress that was in strong support of the program.
37. Vetrov interview with author, Moscow, May 17, 1993. SDI stands for Strategic Defense Initiative, a program endorsed in the U.S. by President Reagan to destroy incoming Soviet missiles and orbiting satellites. It caused great concern to the Soviets in the 1980s.

38. Kryukov, *The Brilliance . . .*, p. 81. The Party line that N-1 was not a military program still persists with some. Cosmonaut Valery Kubasov told me that, "N-1 was not primarily for the military. In the manned space program almost nothing was done for the military. Almost all manned missions were for peaceful purposes."
39. Sergei Khrushchev has written interesting memoirs which shed light on this period in two volumes titled *Nikita Khrushchev: Krisizi i Raketi* (Moscow: Novosti, 1994). In 1968, four years after the ebullient Nikita Khrushchev was deposed, Leonid Brezhnev directed Chelomei to remove Sergei from the premises as one of his measures to get the ex-Premier to stop writing his memoirs, an effort in which he was unsuccessful.
40. Vetrov interview with author, Moscow, Oct. 13, 1992
41. Khrushchev, Sergei, interview with author, Dec. 9, 1993
42. Considered higher than the American Ph.D.
43. Khruschev, Sergei, communication with author, Feb. 2, 1996.
44. Tarasenko communication with author.
45. Vetrov interview with author, May 17, 1993. UR-500, said Vetrov, was never offered as an ICBM because the military decided to switch to use of a smaller warhead, requiring less launch capability.
46. Mishin, V. P., "Why Didn't We Fly to the Moon?," *Novoye v Zhizni,* Nauke, Tekhnike, No. 12, 1990, translated from Russian in JPRS-USP-91-006, Nov. 12, 1991, p. 10
47. Johnson, Nicholas L., *The Soviet Reach for the Moon* (Washington, D.C.: Cosmos Books, 1994), p. 7
48. Even though as late as Nov. 20, 1961—six months after the Kennedy announcement—a NASA group working on launch vehicles and propulsion still recommended direct ascent using *Nova* as the "most promising" mode. *NASA Historical Data Book,* Vol. II, p. 182.
49. Afanasiev, I. B., Unknown Spacecraft (From the History of the Soviet Space Program), *Novoye v Zhizni, Nauke, Teknike* (in Russian), No. 12, Dec. 1991, pp 1–64, translated into English in JPRS-USP-92-003, May 27, 1992, p. 3
50. Kryukov, Sergei, *The Brilliance . . .*, p. 84
51. Ishlinsky, *Akademik . . .*, p. 231
52. *From First Satellite . . .*, pp. 13–14
53. Ishlinsky, *Akademik. . .*, pp. 100–114
54. Ibid., p. 121
55. Ibid., p. 125
56. Ibid., p. 124
57. Faget telephone interview with author, Sept. 7, 1995
58. Johnson, *The Soviet Reach . . .*, p. 7
59. Khrushchev, Sergei, communication with author, Feb. 2, 1996, and Tarasenko, Maxim, communication with author, Feb. 28, 1996
60. Tarasenko communication
61. Khrushchev, Sergei, interview with author, Dec. 9, 1993, Providence, R.I., and communication, Feb. 2, 1996
62. Kryukov, *The Brilliance . . .*, p. 84

63. Mishin, "*Why Didn't We . . .* " , p. 8
64. Neither does the U.S., having abandoned the F-1 program after *Apollo.*
65. Gubanov interview with author, Moscow, Oct. 15, 1992
66. Grechko interview with author, Moscow, May 16, 1993
67. Not much is known about the R-56, even today.
68. Vick, Charles, from an exchange with engineers from NPO Mashinostroyen-niya, Paris, 1991
69. The document giving N-1/L-3 a go-ahead was still secret at this writing. It is designated Postanovleniye No. 655-268 of the Central Committee, Communist Party, and the Council of Ministers.
70. Communication from Tarasenko, Sept. 19, 1996
71. Vetrov and Krasnova, *Creative Legacy . . .*, Document 214 dated May 25, 1965
72. Kryukov, *The Brilliance . . .*, p. 84
73. Orieshkin interview with author, Moscow, Jan. 27, 1993
74. Ibid. Sergei Khrushchev believes that Orieshkin has this meeting confused with the 1962 Pitsunda meeting described earlier.
75. Kraft telephone interview with author, Sept. 10, 1996
76. Although it did not get Kennedy's "DX," or highest priority from a procurement standpoint, until Apr. 11, 1962
77. Kennedy, John F., speech prepared for delivery to the Dallas Citizens Council, Nov. 22, 1963, JFK Library. Note the emphasis on the space program's practical programs and the fact that there was no specific mention of Apollo in the very state where a major part of the program was being prosecuted. Already, it seems, Kennedy was becoming sensitive to the criticism of the extraordinary emphasis being given to the Moon landing by some Americans.
78. Kennedy, address to the 18th General Assembly of the UN, Sept. 20, 1963, JFK Library
79. National Intelligence Estimate Number 11-5-63
80. Bundy, McGeorge, memo to President Kennedy, Sept. 18, 1963, JFK Library
81. Khrushchev, Nikita, *Khrushchev Remembers . . .*, p. 54
82. Gilruth, Robert, interview, Apr. 1, 1964, Oral History Project, JFK Library
83. Chertok interview with author, Moscow, Dec. 4, 1991
84. Gazenko interview with author, Moscow, Dec. 15, 1991
85. Kovtunenko interview with author, Moscow, May 17, 1993. Kovtunenko died in 1995.
86. Vetrov, *Creative Legacy . . .*, Document 1.221
87. Ibid.
88. Vetrov, interview with author, May 13, 1993
89. Mishin, "*Why Didn't We . . .*," p. 10
90. Henceforth UR-500 would become known as the *Proton* launch vehicle. In 1993, after some 200 successful launches, it began to be marketed in the West as a commercial rocket by a new collaboration between Lockheed, Khrunichev, and Energia.
91. Afanasiev, *Unknown Spacecraft . . .*, p. 9

92. Mishin interview with author, Moscow, Oct. 13, 1992
93. Johnson, *The Soviet Reach . . .*, p. 9
94. Voinov, Lev, interview with author, Moscow, Oct. 19, 1992
95. Ibid.
96. Sakharov, Andre, *Andre Sakharov Memoirs,* p. 178. Sakharov got his facts wrong here. The only Apollo program-related activity during this period, which must have been late in Dec. 1965, involved two Titan II launches, not a Saturn launch, to effect a rendezvous of *Gemini* 7 and *8*. Total weight of the two spacecraft was about 7 metric tons, not 19.

Chapter 16: The End of Anonymity: Burial in the Kremlin Wall

1. Golovanov, Yaroslav, "Death of Korolev," *Sovershenno Sekretno,* No. 9, 1992, pp. 20–22
2. Ibid.
3. Ibid. Validol is a tranquilizer.
4. Ishlinsky, *Akademik . . .*, p. 410
5. Golovanov, *Death of . . .*
6. Atashenkov, *Akademik . . .*, p. 199
7. Golovanov, *Death of . . .*
8. Ishlinsky, *Akademik . . .*, p. 420
9. Golovanov, *Death of . . .*
10. Ishlinsky, *Akademik . . .*, p. 68
11. Golovanov, *Death of . . .*
12. Ibid.
13. Tarasov, A., in an interview with Vassily Mishin, "Missions Asleep and Awake," *Pravda,* Oct. 20, 1989, p. 4
14. General Shevchenko would be killed in an automobile accident, along with two other generals, about a year later.
15. Natasha Koroleva interview with author, Dec. 8, 1991
16. Ibid.
17. Oberg, James E., *Red Star in Orbit* (New York: Random House, 1981), p. 87
18. Tarasenko, Maxim, communication with author, Moscow, Mar. 5, 1995
19. Golovanov, *Death of . . .*
20. Ibid.
21. Ishlinsky, *Akademik . . .*, pp. 411–12
22. Golovanov, *Death of . . .*
23. Kamanin, L. N., "Notes on Relationship with Korolev," *Ogonyok,* No. 7, 9–16, Feb. 1991, pp 28–31, translated from Russian in JPRS-USP-91-004, Sept. 20, 1991, p 84
24. There were occasional Western reports which identified Korolev as the person whom the Moscow press called the "Chief Designer" but they could never be confirmed officially. One such report was written by the *New York Times* Moscow correspondent Theodore Shabad and printed in the Nov. 12, 1963 issue.

25. *Vestnik Akademii nauk SSSR,* No. 1, Moscow, Jan. 1966, pp 13–19
26. Ibid.
27. Ibid.
28. Ibid.
29. Ibid.
30. Ibid.
31. Ibid.
32. Ibid.
33. Ibid.
34. Ibid.
35. Ibid.
36. Ibid.
37. Ibid.
38. The birth date according to the old calendar, Dec. 30, 1906. It corresponds to January 12, 1907, in the Gregorian calendar, now used in Russia.

Chapter 17: After Korolev: Demise of the Circumlunar and Lunar Landing Missions

1. Mishin, "Why Didn't We Fly. . . " *Novoye v Zhizni,* Nauke, Tekhnike, No. 12, 1990, translated from Russian in JPRS-USP-91-006, Nov. 12, 1991, p. 12
2. Shepard and Slayton, *Moon Shot,* p. 186, and author's telephone conversation with Kraft, Sept. 11, 1996
3. Ibid., p. 187
4. Ibid., pp. 189–90
5. Mishin, "Why Didn't We Fly . . . ," p. 16
6. *The Soviet Space Program,* National Intelligence Estimate Number 11-1-67, Mar. 2, 1967, p. 18
7. Afanasiev, I. B. *Unknown Spacecraft . . . ,* p. 10
8. Cooper, Henry S. F., *The New Yorker,* Mar. 7, 1988, p. 59
9. Afanasiev, *Unknown Spacecraft . . . ,* p. 10
10. Ibid, p. 11
11. Mishin, "Why Didn't We Fly . . . , " p. 16
12. Vick, Charles, communication with author, Mar. 28, 1996, based on interview with Mishin, Oct. 7–11, 1991, Montreal
13. Murray, Charles, and Cox, Catherine Bly, *Apollo The Race to the Moon* (New York: Simon & Schuster, 1989), pp. 309–314
14. Mishin, "Why Didn't We Fly . . . ," pp. 13
15. Afanasiev, *Unknown Spacecraft . . . ,* p. 11
16. Ibid.
17. Johnson, Nicholas, letter to author, Mar. 18, 1995
18. Afanasiev, *Unknown Spacecraft . . . ,* p. 11
19. Kraft, Christopher, telephone conversation with author, Dec. 1994
20. Chaikin, Andrew, *A Man on the Moon* (New York: Viking, 1994), p. 57
21. Afanasiev, *Unknown Spacecraft . . . ,* p. 11
22. Mishin, Vassily, *Roads to Space* (New York: McGraw-Hill, 1995), p. 90

23. Afanasiev, *Unknown Spacecraft . . .* , p. 15
24. Ibid.
25. Mishin, "Why Didn't We Fly . . . ," p. 8
26. Afanasiev, *Unknown Spacecraft . . .* , p. 15
27. Harvey, *Race into Space . . .* , p. 136
28. Afanasiev, *Unknown Spacecraft . . .* , p. 11
29. Ibid. Other sources of data and information on the UR-500K/L-1 launches include: Johnson, Nicholas L., *The Soviet Reach for the Moon* (Washington, D.C.: Cosmos Books, 1994), pp. 15–16; and *Space Log 1957–1991*, published by TRW Space & Technology Group, One Space Park, Redondo Beach, CA 90278
30. Johnson, *The Soviet Reach . . .* , p. 28
31. Ibid.
32. Mishin, *Roads to Space*, p. 91
33. Johnson, *The Soviet Reach . . .* , p. 28
34. Afanasiev, *Unknown Spacecraft . . .* , p. 16
35. Mishin, "Why Didn't We Fly . . . ," p. 16
36. Mishin interview with author, Moscow, Dec. 3, 1991. Tarasenko points out, however, that many of the "500 research institutes" were only marginally involved.
37. *Soviet Space Programs, 1962–65,* Senate Committee on Aeronautical and Space Sciences, Dec. 30, 1966, pp. 419–20
38. Mishin, "Why Didn't We Fly . . . ," p. 16
39. Afanasiev, *Unknown Spacecraft . . .* , p. 16
40. Mishin interview with author, Moscow, Oct. 13, 1992
41. Kryukov, Sergei, "The Rise and Fall of the Lunar Program," *Nauka i Zhizn*, No. 4, 1994, p. 85
42. Mishin, "Why Didn't We Fly . . . ," p. 5
43. Akim interview with author, Moscow, Sept. 19, 1993
44. Golovanov interview with author, Peredelkino, Oct. 19, 1992
45. Gubanov interview with author, Moscow, Oct. 15, 1992
46. Afanasiev, I. B., "N-1: Top Secret," *Krylia Rodiny,* Nos. 9–11, 1993
47. NK-15 was a modification of the Kuznetsov NK-9 engine developed for the R-9 ballistic missile. NK-15 powered the first stage of the N-1 during the four test failures. NK-15V was the high-altitude version of NK-15, used for N-1's upper stages. NK-33 was a modification of NK-15 manufactured to improved reliability standards. NK-33 was to power the fifth N-1 test, which never occurred.
48. There are some seventy NK-33s left over from the N-1 program. They were acceptance-tested for launcher application under an arrangement between a U.S. rocket engine maker, Aerojet, Sacramento, CA, and the Kuznetsov organization, now called "NK Engines," Samara Scientific/Technical Complex.
49. Bekey conversation with author, Frascati, Italy, May 30, 1996
50. Mishin interview with A. Tarasov, *Pravda*, Oct. 20, 1989, p. 4, and Kryukov, *The Brilliance,* p. 83
51. Kryukov, *The Rise and Fall . . .* , p. 83

52. Demyanko interview with author, Moscow, Sept. 10, 1993
53. Mishin interview with author.
54. Glushko, V. P., *Rocket Engines GDL-OKB,* Novosti Publishing House, Moscow, 1975
55. *Kosmonavtika,* V. P. Glushko, Chief Editor, Moscow, 1985
56. Ishlinsky, *Akademik* . . . , p. 194
57. Ibid., p. 195
58. Gallai interview with author, Moscow, Sept. 9, 1993
59. *Izvestia,* Aug. 18, 1989
60. Sagdeev, *The Making* . . . , pp. 178–79
61. The Moon's orbital inclination varies from 17 degrees to 28 degrees, in an 18-year cycle, from the Earth's equatorial plane. Jastrow, Robert, and Thompson, Malcolm H., *Astronomy: Fundamentals and Frontiers*, second Ed. (New York: John Wiley & Sons, 1974), p. 408.
62. Afanasiev, *Unknown Spacecraft* . . . , p. 17
63. Mishin, "Why Didn't We Fly . . . ," p. 17
64. Ibid.
65. Ibid.
66. Ibid., p. 19
67. Afanasiev, *Unknown Spacecraft* . . . , p. 17
68. Johnson, *The Soviet Reach* . . . , p. 28
69. Afanasiev, *Unknown Spacecraft* . . . , p. 17
70. Ibid., pp. 16–17

Chapter 18: Ironic Epilogue

1. Dobrynin, Anatoli, *In Confidence* (New York: Random House, 1995), p. 9
2. Vetrov interview with author, Moscow, Sept. 8, 1993
3. Mishin, "Why Didn't We Fly . . . ," p. 17
4. Akim interview with author, Sept. 10, 1993
5. Gubanov interview with author, Oct. 15, 1992
6. Ibid.
7. The *Buran* and *Energia* vehicles are being maintained, however, in flight-certified condition by continued maintenance. There is still hope that, if economic and political conditions make it possible, the vehicles might be used in such applications as the emplacement of large constellations of communications satellites, or even an international SDI program.
8. Broad, William J., *New York Times*, Nov. 27, 1995, p. 1
9. Nanz, Ted, *Space News,* Feb. 19–25, 1996, pp. 19–20
10. *Jane's Spaceflight Directory 1986*, p. 82
11. Four other ESA members are not participating: UK, Austria, Finland, and Ireland. Sweden has not yet decided either way.
12. Now called NPO Mashinostroyenniye
13. Charles, John, E-mail communication, Mar. 1, 1996, describing an article by David Portree submitted to *Quest* magazine

14. One other Russian, Leonid Sedov, was made Honorary Fellow of AIAA more than 30 years ago.
15. Tarasenko, Maxim, *Current Status of Russian Space Program,* IAF-95-IAA 3.3.01, 46th International Astronautical Congress, Oslo, Oct. 2–6, 1995
16. Gazenko, Oleg, remarks to a group of visitors from the AIAA, Moscow, May 16, 1994

Bibliography

Books

Alexandrov, Victor, *The Tukhachevsky Affair* (Englewood Cliffs, NJ: Prentice-Hall, 1964)

Astashenkov, P. T., *Akademik S. P. Korolev* (Academician S. P. Korolev) (Moscow: Mashinostroyenniye, 1969)

Avduyevsky, V. S., *M.V. Keldysh, Izbranniye Trudi Raketnaya Teknika i Kosmonavtika* (M. V. Keldysh Selected Works Rocket Technology and Cosmonautics) (Moscow: Nauka, 1988)

Berlin, Peter, and John Rhea, *Roads to Space* (New York: McGraw-Hill, Aviation Week Group, 1995)

Beschloss, Michael R., *The Crisis Years: Kennedy and Khrushchev, 1960–63* (New York: Harper Collins, 1991)

Bolonkin, Alexander, *The Development of Soviet Rocket Engines* (Falls Church, VA: Delphic Associates, 1991)

Burlatsky, Fedor, *Khrushchev and the First Russian Spring* (New York: Macmillan, 1988)

Burroughs, William E., *Deep Black* (New York: Random House, 1986)

Chertok, Boris E., *Raketi i Lyudi* (Rockets and People) (Moscow: Mashinostroyenniye, 1994)

Clark, Phillip, *The Soviet Manned Space Program* (New York: Orion, 1988)

Conquest, Robert, *Kolyma The Arctic Death Camps* (New York: Viking, 1979)

Daniloff, Nicholas, *The Kremlin and the Cosmos* (New York: Alfred A. Knopf, 1972)

Ezell, Edward Clinton, and Linda Neuman Ezell, *The Partnership: A History of the Apollo Soyuz Test Project* (Washington, DC: NASA SP-4209, 1978)

Ezell, Linda Neuman, *NASA Historical Data Book Vols II, III* (Washington, DC: NASA SP-4012, 1988)

Gallai, Mark, *S Chelovek na Bortu* (With Man on Board) (Moscow: Moskva, 1985)

Getty, J. Arch, *Origins of the Great Purges* (Cambridge: Cambridge University Press, 1985)

Glennan, T. Keith, *The Birth of NASA* (Washington, DC: NASA History Office, 1993)

Goddard, Esther C., and G. Edward Pendray, *The Papers of Robert H. Goddard*, 3 vols (New York: McGraw-Hill, 1970)

Graham, Loren R., *Science in Russia and the Soviet Union: A Short History* (New York: Cambridge University Press, 1993)

Goldstine, Herman H., *The Computer, from Pascal to von Neumann* (Princeton, NJ: Princeton University Press, 1993)

Golovanov, Yaroslav, *Sergei Korolev, The Apprenticeship of a Space Pioneer* (Moscow: Mir Publishers, 1975)

——— *Korolev, Fakti i Miti* (Facts and Myths) (Moscow: Nauka, 1994)

Hansen, James R., *Engineer in Charge, A History of the Langley Aeronautical Laboratory, 1917–1958* (Washington, DC: The NASA History Series, NASA SP-4305, 1987)

Harvey, Brian, *Race into Space, the Soviet Space Programme* (Chichester, UK: Ellis Horwood Ltd., 1988)

Hyland, William, and Richard W. Shryock, *The Fall of Khrushchev* (New York: Funk & Wagnalls, 1968)

Ishlinsky, A. Y., *Akademik S. P. Korolev, Uchenii, Inzhener, Chelovek* (Academician S. P. Korolev, Scientist, Engineer, Man) (Moscow: Nauka, 1986)

Johnson, Nicholas L., *The Soviet Reach for the Moon* (Washington, DC: Cosmos Books, 1994)

Keldysh, M. V., and Gyorgi Vetrov, *Tvorcheskoye Naslediye Akademika Sergeya Pavlovicha Koroleva* (Creative Legacy of Academician Sergei Pavlovich Korolev) (Moscow: Nauka, 1980)

Khrushchev, Sergei, *Khrushchev on Khrushchev* (Boston: Little, Brown, 1990)

——— *Nikita Khrushchev: Krizisi i Raketi* (Nikita Khrushchev: Crises and Rockets) 2 vols, (Moscow: Novosti, 1994)

Krieger, F. J., *Behind the Sputniks* (Washington, DC: Public Affairs Press, 1958)

Lardier, Christian, *L'Astronautique Sovietique* (Soviet Aeronautics) (Paris: Armand Colin, 1992)

Larina, Anna, *This I Cannot Forget* (New York: W. W. Norton, 1993)

Logsdon, John M., *The Decision to Go to the Moon* (Cambridge: MIT Press, 1970)

Magnus, Kurt, *Raketensklaven* (Rocket Slave) (Stuttgart: Deutschen Verlags-Anstalt, 1993)

McDougall, Walter A., *. . . the Heavens and the Earth, A Political History of the Space Age* (New York: Basic Books, 1985)

Murray, Charles, and Catherine Bly Cox, *Apollo The Race to the Moon* (New York: Simon & Schuster, 1989)

Neufeld, Jacob, *The Development of Ballistic Missiles in the United States Air Force 1945–1960* (Washington, DC: Office of Air Force History, 1989)

Neufeld, Michael J., *The Rocket and the Reich* (New York: The Free Press, 1995)

Nitze, Paul H., *From Hiroshima to Glasnost* (New York: Grove Weidenfeld, 1993)

Oberg, James E., *Red Star in Orbit* (New York: Random House, 1981)

——— *Uncovering Soviet Disasters* (New York: Random House, 1988)

Ordway, Frederick I. III, and Mitchell R. Sharpe, *The Rocket Team* (New York: Crowell, 1979)

Rauschenbakh, Boris V., *Hermann Oberth, the Father of Space Flight* (Clarence, NY: West-Art, 1994

Remnick, David, *Lenin's Tomb The Last Days of the Soviet Empire* (New York: Random House, 1993)

Riabchikov, Evgeny, *Russians in Space* (Garden City, NY: Doubleday, 1971)

Romanov, A., *Spacecraft Designer* (Moscow: Novosti Press, 1976)

Sagdeev, Roald, *The Making of a Soviet Scientist* (New York: John Wiley & Sons, 1994)

Shalamov, Varlam, *Kolyma Tales* (New York: W. W. Norton, 1980)
Sheldon, Charles S. II, *Review of the Soviet Space Program* (New York: McGraw-Hill, 1968)
Shelton, William, *Soviet Space Exploration The First Decade* (New York: Washington Square Press, 1968)
Shepard, Alan, and Deke Slayton, *Moon Shot* (Atlanta: Turner Publishing Inc., 1994)
Shipler, David K., *Russia, Broken Idols, Solemn Dreams* (New York: Penguin, 1984)
Shklovsky, Iosif, *Five Billion Vodka Bottles to the Moon* (New York: W. W. Norton, 1991)
Smith, Hedrick, *The New Russians* (New York: Random House, 1990)
Stine, G. Harry, *ICBM: The Making of the Weapon That Changed the World* (New York: Orion, 1991)
Stoiko, Michael, *Soviet Rocketry: Past, Present and Future* (New York: Holt, Rinehart and Winston, 1970)
Storms, Harrison, *Angle of Attack Harrison Storms and the Race to the Moon* (New York: W. W. Norton, 1992)
Stuhlinger, Ernst, and Frederick I. Ordway III, *Wernher von Braun, A Biographical Memoir* (Malabar, FL: Krieger Publishing Co., 1994)
Tarasenko, M. B., *Voenniye Aspekti Sovietskoi Kosmonavtiki* (Military Aspects of Soviet Cosmonautics) (Moscow: Nikol, 1992)
Thompson, Tina D., *Space Log* (Annual) (Redondo Beach, CA: TRW Space & Technology Group)
Van Nimmen, Jane, Leonard C. Bruno, and Robert L. Rosholt, *NASA Historical Data Book, Vol I* (Washington, DC: NASA SP-4012, 1988)
Vetrov, Gyorgi, *Korolev i Ego Delo, Svet i Teni v Historii Kosmonavtiki* (Korolev and His Business. Light and Shadows in the History of Cosmonautics) (Moscow: Nauka, to be published)
Vetrov, Gyorgi, and K. A. Krasnova, *S. P. Korolev, Nauchnaya Biografiya* (S. P. Korolev, A Scientific Biography) (Moscow: Nauka, 1996)
Vick, C. P., *The Soviet Civil/Military Space Missile & Aircraft Industry* (Fredericksburg, VA: to be published)
Von Braun, Wernher, Frederick I. Ordway III, and Dave Dooling, *Space Travel A History* (New York: Harper & Row, 1985)
Zaloga, Steven J., *Target America The Soviet Union and the Strategic Arms Race, 1945–64* (Novato, CA: Presidio Press, 1993)

Interviews

Isaak Abramov, Kaliningrad, 9/16/93
Vladimir Agapov, Moscow, 5/12/93
Efraim Akim, Moscow, 9/10/93
Oleg Alifanov, Moscow, 5/18/94
Vsyevolod Avduyevsky, Moscow, 1/21/93

Vyacheslav Balebanov, 5/23/94
Gorimir Chernyi, Moscow, 10/16/92
Boris Chertok, Moscow, 12/4, 12/18/91, 3/31/92, 9/17/93
Yuri Demyanko, Moscow, 9/10/93
Herbert Efremov, 5/16/94
Konstantin Feoktistov, Moscow, 12/12/91
Vladislav Filin, Moscow, 9/18/90
Viktor Frumson, Moscow, 10/11/92
Mark Gallai, Moscow, 9/9/93
Oleg Gazenko, Moscow, 12/13/91, 5/16/94
Josef Gitelson, Princeton, NJ, 1/9/93
Yaroslav Golovanov, Moscow, 12/13/91, Peredelkino, 10/19/92
Gyorgi Grechko, Moscow, 5/16/93
Boris Gubanov, Moscow, 10/15/92
Alexander Gurshstein, Moscow, 10/18/92
Yevgeny Ilyin, Moscow, 5/16/94
Alexander Ivanchenkov, Moscow, 12/9/91
Oleg Ivanovsky, Moscow, 1/26/93
Vassily Karpi, Moscow, 9/14/93
Vladimir Karrask, 5/23/94
Lev Kerber, Moscow, 12/10/91
Sergei Khrushchev, Providence, RI, 12/9/93
Kyril Kondratyev, Moscow, 12/11/91
Natasha Koroleva, Moscow, 12/8/91, 10/16/92, 1/30/93
Vladimir Kovtunenko, Moscow, 5/17/93
Sergei Kryukov, Moscow, 9/8/93
Valery Kubasov, Moscow, 1/28/93
Alexei Leonov, Moscow, 12/11, 12/12/91
Vladimir Lobachev, Kaliningrad, 12/5/91
Boris Lokhmachev, Baikonur, 5/25–26/94
Gleb Lozino-Lozinsky, Moscow, 10/19/92
Mikhail Marov, Cannes, 5/22/92, Moscow, 10/21/92
Gleb Maximov, Moscow, 12/13/91
Viktor Minenko, Moscow, 7/27/93
Vassily Mishin, Moscow, 9/90, 12/3/91, 10/13/92
Yuri Mozzhorin, Moscow, 9/14/90
Yuri Orieshkin, Moscow, 1/27/93
Arkady Ostashov, Kaliningrad, 9/16/93
Antonina Otrieshka, Kaliningrad, 12/5/91
Arvid Pallo, Moscow, 5/15/93, 9/14/93
Vladimir Poliatchenko, Moscow, 10/14/92
Boris Rauschenbakh, Torremolinos, Spain, 10/89, Abramtsevo, Russia, 12/7/91, Graz, Austria, 10/14/93
Mikhail Rozhdostvensky, Moscow, 1/24/93
Mikhail Rudenko, Moscow, 9/14/93

Vitaly Sevastyanov, Moscow, 12/11/91
Gai Severin, Kaliningrad, 9/16/93, 5/17/94
Vladimir Shevalyov, Moscow, 9/9/93
Viktor Sigaev, Khimki, 5/19/94
Irina Strazheva, Barvikha, 10/10/92, Moscow, 10/13/92
Gennadi Strekalov, Moscow, 1/28/93
Vitaly Svertschek, Kaliningrad, 9/16/93
Vladimir Syromiatnikov, Moscow, 12/91, 1/93, Princeton, NJ, 5/6/95
Gherman Titov, Moscow, 12/11/91
Vladimir Utkin, Moscow, 1/29/93
Gyorgi Vetrov, Moscow, 10/13, 10/16, 10/21/92, 5/11, 5/13/93
Lev Voinov, Moscow, 10/19/92
Pavel Vorobiev, Kaliningrad, 5/20/94
Yuri Zonov, Moscow, 12/12/91

INDEX

A-4. *See* V-2 technology
A-9/A-10 missile, 74-75
Abbott, Charles G., 42
Abezguz, N. Y., 21, 22
Academy of Sciences, Soviet, 14, 45, 144, 206, 217, 238, 247, 275, 282-85; and access to journals, 244; Commission on Interplanetary Communications of, 125; computer capabilities of, 219; Korolev as member of, 98, 280; and N-1/L-3M program, 311; responsibilities of, 206; and *Sputnik*, 125-26; and unmanned space exploration, 144, 147
Aerojet-General Corporation, 190
Aeronautical and Space Sciences Committee (U.S. Senate), 252
Afanasiev, A. V., 162
Afanasiev, I. B.: and L-1 program, 293, 294; and N-1 program, 289, 294, 296, 298, 300, 303, 304, 308; and N-1/L-3M program, 311-12; and scenarios for manned Moon flights, 308; and UR-700/LK-700 program, 312; and *Zond* program, 292, 297
Afanasiev, Sergei Alexandrovich: and Korolev's funeral, 283; Korolev's relationship with, 276; Mishin's relationship with, 278; and N-1 program, 266, 272-73; and reasons for Soviet failures in manned space exploration, 303; and space shuttle, 315; and technology issues, 221
Agena vehicle, 188, 286-87
Ai-Petri control center, 144-45
Air Force, U.S., 147, 154, 192, 260, 263, 320
Akim, Efraim, 146, 221-22, 243, 302, 303, 314, 315
Albina (dog), 161
Aldrin, Edwin "Buzz," 246, 287-88, 295
Aleksandrov, A. G., 21
Aleksandrova, L. A., 23
Almaz space station, 320
Alsop, Joseph, 131
American Institute of Aeronautics and Astronautics (AIAA), 130, 244, 321
American Rocket Society (ARS), 129, 130, 147, 244
Anders, William, 293
Andreev, V. A., 41
Andropov, Y. V., 284
animals in space exploration, 159, 161, 162, 167, 168, 176, 177, 183, 245
ANT-9 plane (*Soviet Wings*), 33-34
anti-ballistic missiles, 87, 259
anti-satellite weapons, 262-63
anti-Semitism, 115, 230
Apartseva, Bella, 230
Apollo 1 mission, 163, 289, 301
Apollo 4 mission, 290
Apollo 5 mission, 291
Apollo 6 mission, 291-92
Apollo 7 mission, 293
Apollo 8 mission, 35, 293
Apollo 11 mission, 12, 246, 287, 295-96, 300, 306, 308, 310
Apollo 12 mission, 296, 310
Apollo 13 mission, 287, 296
Apollo 14 mission, 297, 310
Apollo 15 mission, 299, 310
Apollo 16 mission, 299, 310
Apollo 17 mission, 300, 310
Apollo program, 3, 14, 102, 131, 155, 187, 188, 191, 245, 288, 322; and *Apollo-Soyuz* mission, 51, 96, 317; benefits of, 317-18; and cooperative U.S.-Soviet programs, 270-71; deterioriation of capabilities of, 319; and Eisenhower administration, 249; funding for, 3, 245, 268, 271, 291, 298, 299, 301, 317; and N-1 program, 253, 264, 301-2; and organizational structure, 302; problems with, 293, 301; and rendezvous in space, 302; Shepard-Slayton book about, 287-88; Soviet views about, 302; and *Soyuz* program, 255, 256; and UR-700/LK-700 program, 312; U.S. support for, 269-70, 271; and U.S.-Soviet competition, 253, 274, 288, 293, 295-96; and V-2 technology, 66. *See also Saturn V/Apollo* mission; *specific* Apollo *mission*
Appazov, Refat Fazylovich, 215, 220, 221, 222, 223, 224
Ariane launch vehicle, 66
Armstrong, Neil, 246, 286-87, 295
Army Air Corps, U.S., 42-43, 122-23
Army, U.S., 66, 133, 252. *See also* Jupiter missile
artificial gravity environment, 7, 13, 186
Astashenkov, P. T., 94-95
Astronautics magazine, 129, 130
AT&T, 190, 198
Atlas missile: and communication satellites, 198; cruise missiles versus, 103; deployment of, 112; development of, 92, 102, 109; and *Gemini 9* mission, 287; H-bomb payload of, 91, 113; and manned space exploration, 160; and *Mercury* program, 160, 177, 178, 249, 260; R-7 missile compared with, 108, 109, 112-13; storable propellant in, 113; and U.S.-Soviet competition, 110
atomic bomb, 91, 100, 234, 240
Avduyevsky, Vsyevolod, 101-2, 103, 232, 254
Averkov, S., 119
Avro 504K biplane, 33

Babakin, Gyorgi Nikolaevich, 157
Baikonur launch complex, 6, 81, 217, 219, 222, 224, 234, 264, 307; and ballistic missile development, 151-52, 168, 191; and Chelomei spacecraft, 262-63; and *Luna* program, 143; and manned space exploration, 167, 170, 178; and *Molniya* communications satellites, 290; and N-1 program, 257, 267, 298, 301; and Nedelin disaster, 167-68; and R-7 launcher, 109-10, 112, 191; R-16 tests at, 119; and scenarios for manned Soviet Moon flights, 307; and space shuttle, 315; and *Sputnik* program, 129, 135, 226; and unmanned space exploration, 151-52
Balanin, Grigory Mikhailovich (stepfather), 19, 20, 24-25, 28, 31, 34, 46

Ballistic Missile Early Warning System, U.S., 152
ballistic missiles: CIA reports about, 191; funding for, 201, 246; German development of, 46; guidance and control systems for, 143; Korolev's contributions to development of, 15; as Korolev's responsibility, 251; organizational structure for development of, 78, 202; RNII work on, 46; Soviet development of, 151-52, 168, 189, 191; and space shuttle program, 314; and *Sputnik*, 131; Stalin's knowledge about, 234; U.S. development of, 46; U.S. estimations of Soviet, 191; and U.S.-Soviet competition, 191; and V-2 technology, 66. *See also* anti-ballistic missiles; *specific missile*
Barmin, Vladimir, 2, 52; and manned space exploration, 167-68; and N-1 program, 266; and N-1/L-3M program, 300; and organizational structure, 203, 204-5; and technology issues, 221; and V-2 technology, 71
Bauman, Nikolai, 29
Bauman Institute. *See* Moscow Higher Technical School (MVTU)
Bay of Pigs, 250
Beacon 1 program, 137
Bean, Alan, 296
Bekey, Ivan, 304
Belka (dog), 167, 168
Bell Aircraft, 190
Belyaev, Pavel, 182, 184-85, 189, 283
Bendix Aviation, 190
Beregevoi, Gyorgi, 292
Beria, Lavrenti, 21, 49, 57, 58, 202, 214, 240
Beria, Sergei, 106
Berry, Charles, 187
Bestuzhev-Ryumin, A., 29
BICH-11 glider, 39, 48
bioreactors: development of, 5
Blagonravov, Anatoli, 123, 147, 229, 317
Bleicherode, Germany. *See* Institut Rabe
Block B engines, 298
Block D engines, 274, 289, 292, 297, 309
Block E engines, 264, 309, 310
Block I engines, 310
Bodungen, Klein, 69
Boeing Company, 190, 243, 316, 320
Bogomolov, Alexei, 135
Bolkhovitinov, Viktor, 64-65, 69
Bollay, William, 102
Bolsheviks, 20, 29-30
Bomarc missile, 243
Bondarenko, Valentin, 242
Bondaryuk, Mikhail, 103
Borman, Frank, 188, 189, 293
Bosnia, 318
Bossart, Karel, 2, 92, 111
Branch 1. *See* Gorodomliya (island)
Brett, George H., 43
Brezhnev, Leonid I., 114, 120, 201, 208, 295, 300; and Gagarin (Yuri) flight, 175; and Korolev's anonymity, 278-79; and Korolev's death/funeral, 278, 284; Korolev's relationship with, 202, 206, 233, 268-69
Bryushinin, V. M., 151-52
Budnik, Vassily Sergeevich, 71, 96, 117
Bundy, McGeorge, 191, 270
Buran program, 314-16
Burlatsky, Fedor, 57
Burya cruise missile, 103-4
Bushuyev, Konstantin, 2, 96, 99, 104, 107, 112, 272
Butirskaya prison, 58, 59, 60-61
Bykovsky, Valery, 159, 165, 178, 261, 283, 289-90

C-119 aircraft recovery, 192
California Institute of Technology. *See* Jet Propulsion Laboratory
cameras, 192, 193, 194-95, 196
Cape Canaveral, 207, 267, 289, 307, 319
Cape Kennedy, 264
Carpenter, Scott, 177
Center for Deep Space Radio Communications (Yevpatoriya, Crimea), 149-50
Center of Flight Control (TsUP), 156, 215, 217
Central Aerohydrodynamic Institute (TsAGI), 30-31, 33-34, 37, 48, 88
Central Committee (Communist Party), 123, 127, 167-68, 170, 206, 217, 231, 282
Central Design Bureau of Experimental Machine Building (TsKBEM), 307
Central Design Bureau No. 29 (KB-29), 58-59, 60
Central Intelligence Agency (CIA): ballistic missile reports of, 191; and Bay of Pigs, 250; and ICBM development, 113, 152; and Moon flights, 293; and reconnaissance satellites, 192; and satellite development, 190; and Soviet economic situation, 270; Soviet protections against, 222, 223; and U.S.-Soviet competition, 293; and V-2 technology, 68, 85, 86, 88-89
Central Scientific Research Institute for Machine Building (TSNIIMash), 100, 107
Central Specialized Design Bureau (TsSKB), 196
Cernan, Gene, 287
Chaffee, Roger, 163, 289
Chaika (dog), 167
Challenger explosion, 313
Charomsky, A. D., 62
Chelomei, Vladimir, 2, 14, 15, 101, 106; and cruise missiles, 258; Korolev's relationship with, 258, 259-60, 262-63, 265, 266-67, 268; management style of, 258; and manned systems, 273, 308; and Moon flights, 257-58, 259-60, 262-63, 266-67, 268, 274; and organizational structure, 203; and *Polyot* program, 262-63; reputation of, 14-15, 258, 262-63; and Taran project, 259. *See also* UR-500K launcher; UR-700 program
Chelomei design bureau. *See* OKB-52
Cheranovsky, Boris, 39
Chernushka (dog), 168
Chertok, Boris, 2, 77, 230, 232; and ICBM development, 99, 109-10; on Kapustin Yar proving grounds, 81; as NII-88 deputy chief engineer, 78; perception of Korolev by, 3, 70, 79; and technology issues, 221; and unmanned space exploration, 151-52; and U.S.-Soviet competition, 271; and V-2 technology, 64, 65, 67, 68, 69, 70-71, 73, 74, 75, 77, 80-82, 84-85, 86, 89; and Yangel, 115
chimpanzees. *See* monkeys/chimpanzees
Chkalov, Valery, 58, 59
Chrysler Corporation, 190
Churchill, Winston, 64
Clark, Phillip, 174

Collins, Mike, 287, 295
Commission on the Cultivation of the Scientific Heritage of the Pioneers of the Development of Cosmic Space, 47
Commission on Interplanetary Communications (Soviet Academy of Sciences), 125
Committee No. 2, 205, 206, 216
Committee on Science and Astronautics (U.S. House of Representatives), 176
Committee on Space Research (COSPAR), 317
communication satellites: and benefits of space exploration, 318; first broadcasting from, 138; funding for, 197; Korolev as developer of, 2; and SDI, 315; Soviet development of, 149, 189, 196-98, 200; U.S., 198, 269-70, 315, 318; and U.S.-Soviet competition, 190. See also *Molniya* communications satellites
communications: and manned space exploration, 163-64, 171, 172; Soviet timetable for development of, 248; and *Sputnik*, 125-26, 128; and unmanned space exploration, 144-45, 149-50, 152, 153, 154, 156
Communist Party: Congresses of, 22, 54, 142, 236; and funding, 232; and ICBM development, 98-99; interference in Korolev's projects by, 98-99; Korolev as member of, 230-32, 233-34, 236; and Korolev's funeral, 284; Korolev's professional relationship with, 231-32, 234-36; and "mail boxes," 237-38; and manned space exploration, 160, 167-68, 169-70; membership in, 30, 230-32; and N-1/L-3 program, 299; and organizational structure, 206, 216, 217, 231-32; paranoia of, 229, 237-45, 307; powers of, 231, 237-38, 248, 254-55; Presidium of, 126, 127, 233; and reasons for Soviet Moon flight failure, 307; and satellite development, 123, 125, 126, 127; Sixteenth All-Union Conference of, 33; and Soviet military, 231; and *Soyuz* program, 254-55, 262; and *Sputnik*, 125, 126, 127; and technology issues, 227, 254-55; and unmanned space exploration, 142; and U.S.-Soviet competition, 245. *See also* Central Committee; *specific person*
computers, 219-21, 222-24, 317
Conquest, Robert, 51
Conrad, Pete, 188, 189, 287, 296, 319
Consolidated Vultee Corporation, 115
Cooper, Gordon, 178, 188, 189
Cooper, Henry S. F., 290
Corona program, 190, 191, 192
Cosmonaut Training Center (TsPK), 217
cosmonauts: apprehensiveness of, 168; burial of, 285; and Communist Party paranoia, 242; deaths of, 242, 283, 285, 289-90, 298, 302; at international meetings, 239; and Korolev's funeral, 283; selection of, 164, 165, 271-72; and space shuttle, 315; training of, 164-65. *See also specific person*
Council of Chief Designers, 1, 79, 118, 203-4, 205, 251-52, 259
Council of Labor and Defense, 45
Council of Ministers: decree of 1960 of, 253; and ICBM development, 104; and Korolev's funeral, 282; Korolev's relationship with, 232; and N-1/L-3 program, 299; and organizational structure, 202, 205, 206, 207, 208, 216, 217, 232; Praesidium of, 233; and space exploration, 253; and *Sputnik*, 125; and U.S.-Soviet competition, 253. *See also* Military-Industrial Commission
Crawley, Edward, 306
cruise missiles, 15, 74, 86, 92, 101, 102-4, 258
Cuba: and Bay of Pigs, 250; and Cuban missile crisis, 116, 151, 152
Cunningham, Walter, 293
Curtiss-Wright, 190
Cuxhaven, Germany: V-2 demonstrations at, 72

D-1 ballistic rocket, 63
D-2 winged guided missile, 63
Damka (dog), 168
DeBra, Daniel, 143
Debus, Kurt, 2
Deep Space Network, U.S., 150
Delone, N. B., 26
Dementiev, Pyotr Vasilievich, 63, 70, 259
Demichev, P. N., 284
Demyanko, Yuri, 305
Department 3. *See* NII-88
Department of Defense Industries, 217
Department of Defense, U.S., 317
Dezik (dog), 161
digital signal processing, 155
dirigibles, 11
Discoverer missions, 143-44, 190, 191, 192, 193, 196
Dnepropetrovsk, Ukraine. *See* NPO Yuzhnoye
Dobrovolsky, Gyorgi, 242, 285, 298, 302
Dobrynin, Anatoli, 313
dogs: and space exploration, 125-26, 129, 132, 161, 167, 168, 183
Dolgov, Pyotr, 242
Dornberger, Walter, 110
Douglas Aircraft, 115, 122-23, 190
Draper, C. Stark, 2, 109, 143
Dryazgov, M. P., 47-48, 63
Dryden, Hugh, 2, 134, 142, 143, 249, 317
Duke, Charles, 299
Dulles, Allen, 190
Dynasoar project, 263

Eastman Kodak, 190, 196
Echo communication satellite, 198
Eisele, Donn, 293
Eisenhower, Dwight D., 124-25, 134, 142, 147, 191, 198, 249
ejection systems, 168, 174, 175, 180, 181-82, 242
electronic systems, 224-25, 264, 308, 317. *See also* Ministry of Electronic Industry (MEP)
Elektron satellites, 212
Energia launch vehicle, 314-15, 316
Enos (chimpanzee), 177
Erlikh, Igor, 261-62
escape systems, 166, 168, 174, 175, 180, 181-82
Esnault-Pelterie, Robert, 10
EVA. *See* extra-vehicular activity (EVA)
Exhibition of Economic Achievements (VDNKh), 322
Explorer program, 72, 133, 135, 136, 137, 249
extra-vehicular activity (EVA): and *Gemini* missions, 269, 287; and manned space exploration, 182-83, 184-85, 186-87, 189; and N-1/L-3 program, 269, 310; and *Skylab* program, 319; and *Soyuz 4* mission, 294; and *Voskhod 2* mission, 277

F-1 engines: development of, 160-61, 252, 260, 263; and Moon flights, 250; and reasons for Soviet failures in space exploration, 303m, 161; and *Saturn* missions, 187, 265, 267, 291; Soviet views about, 264
Factory 22: (OPO-4, Fourth Experimental Section), 34-35
Faget, Maxime, 2, 256, 262
Fairchild Camera, 190
Federation Aeronautique International, 175
Fedorov, Alexander, 35
Feoktistov, Konstantin, 2, 242, 243; and Korolev's funeral, 283; and manned space exploration, 165-66, 167, 181-82; and organizational structure, 203; and *Soyuz* development, 254, 255; and *Sputnik*, 125, 128
FGB core module, 320
Filin, Vyacheslav, 253
Filipchenko, Anatoli, 296
film: of Korolev life, 323; of Mercury launches, 245; for Soviet military, 242; about Soviet space exploration, 241-42
Finney, John, 160
Flerov, Petya, 31
Floriansky, Mikhail, 128
Fort Monmouth (NJ) Observatory, 142
Friendship 7 mission, 177
Frolov, Boris, 41, 69
Frumkin, Yuri, 192, 193, 194-96
Frumson, Viktor, 241-42
Frunze, Mikhail, 22
funding: for ballistic missiles, 201, 246; for communication satellites, 197; and Communist Party, 232; and cooperative projects, 317; estimations for, 257; for GDL, 44; Korolev's access to, 4, 5, 6, 202, 276; and organizational structure, 210, 276; and reasons for Soviet Moon flight failure, 307; for reconnaissance satellites, 318; and Russian economy, 321; for Soviet military projects, 246, 253, 270; for Soviet Moon flights, 246, 272; Soviet priorities for, 246; for Soviet space exploration, 149, 210, 253, 270, 298-99, 322; for space medicine, 321; for space shuttle, 313, 314; in U.S., 205-6, 215, 245, 248, 318; and U.S.-Soviet competition, 301. *See also specific agency or program*
Fyodorov, Nikolai, 9, 10

G-1 missile. *See* R-10 missile
Gagarin, Yuri: awards and honors for, 120, 322; death/burial of, 283, 285; health of, 170; and Korolev's arrest/imprisonment, 51, 52; and Korolev's eulogy, 284-85; and Korolev's funeral, 283; personal account of flight by, 170-75; selection of, 159, 170, 186; training/preparation of, 163-64, 165, 170, 171. *See also* Gagarin (Yuri) flight
Gagarin (Yuri) flight, 149, 158, 232; and Communist Party paranoia, 242-43; and divergence in U.S. and Soviet approaches, 163; Gagarin's account of, 170-75; and Korolev's management style, 215; Korolev's reaction to, 175-76; and reentry, 170-71, 173-75; Shepard's flight compared with, 159-60; Soviet reaction to, 175-76, 284; thirtieth anniversary of, 171; U.S. reaction to, 160-61, 176; and U.S.-Soviet competition, 177, 249; and *Vostok* spacecraft, 162, 170
Gaidukov, Lev Mikhailovich, 71, 76
Gallai, Mark, 128, 163-65, 166-67, 175, 306
Gas Dynamics Laboratory (GDL), 36, 37-39, 44-45. *See also* Reaction Propulsion Institute
Gazenko, Oleg, 4-5, 161, 202, 239, 271-72, 321
GB-1 and GB-2 assemblies, 311
GDL. *See* Gas Dynamics Laboratory; Reaction Propulsion Institute; *specific scientist*
Gemini 3 mission, 182, 189
Gemini 4 mission, 186-88, 189, 269
Gemini 5 mission, 188, 189
Gemini 6 mission, 188, 189
Gemini 7 mission, 188, 189
Gemini 8 mission, 286-87
Gemini 9 mission, 287
Gemini 10 mission, 287
Gemini 11 mission, 287
Gemini 12 mission, 287
Gemini program: and *Apollo* project, 188; life-support system in, 163; return module in, 273; and rotor concept, 262; and *Soyuz* program, 180, 254, 255, 256; and U.S.-Soviet competition, 178, 180, 181, 183, 243. *See also specific mission*
General Electric Company, 190, 256
George Washington University, 190
German Democratic Republic (GDR), 178
Germans: accomplishments using V-2 technology of, 85-86; and Communist Party paranoia, 240; and ICBM development, 110; Korolev's views about, 81; "most valuable contribution" to Soviets of, 88-89; returned to Germany, 89-90; sabotage of V-2 technology by, 83; in Soviet Union, 56, 75 89, 240; and Soviet work on V-2 technology in Germany, 66-75; and Soviet work on V-2 technology in USSR, 75-89; and Stalin's purges, 56
Germany: ballistic missile development in, 46; glider competitions in, 27-28; Korolev in post-war, 60, 61, 63, 70-73; Soviet scientists in post-war, 64-65, 67-75. *See also* V-2 technology
Gilruth, Robert, 2, 271
GIRD. *See* Group for Studying Reaction Propulsion; Reaction Propulsion Institute
GIRD-09 engine, 40-41
GIRD-X, 41
Gitelson, Josef, 5, 6, 56, 210-11
glasnost, 55, 238, 306
Glazunov, Mikhail, 210
Glenn, John, 163, 177, 245, 249, 260
Glennan, T. Keith, 147, 148, 249
gliders, 25, 27-28, 32-33, 34-35, 48. *See also* Koktebel; *specific glider*
Glushko, Valentin, 2, 54, 106, 206, 249, 258; and anonymity, 305; arrest/imprisonment of, 49, 61, 71; biographical information about, 38; character of, 114-15; contributions of, 14-15; death of, 306; at GDL, 36, 38-39; as head of OKB-1, 301, 305; and hydrogen engines, 305; and ICBM development, 97, 100-101, 102, 108, 114-15, 117; Korolev as deputy to, 62; Korolev eulogy by, 305-6; and Korolev's arrest/imprisonment, 49; Korolev's relationship with, 14, 62, 102, 114-15, 282; liquid-fuel rocket work of, 38-39; and manned space exploration, 167-68; Mishin's relationship with, 305, 308, 314; and

N-1 program, 272, 273, 301, 304, 305; and N-1/L-3M program, 300; and OKB-456 rocket engine, 78; and organizational structure, 203; and R-7 launcher, 291; and reasons for Soviet failures in space exploration, 303, 304, 305; and rocket-powered gliders, 48; and scenarios for manned Soviet Moon flights, 308; and space shuttle, 314, 315; and Stalin purges, 48; on Tikhomirov, 37; and V-2 technology, 69, 71, 72, 79; and *Vulkan* program, 305; writings of, 305. *See also* OKB-456; *specific engine*
Glushko design bureau. *See* NPO Energomash; OKB-456
Goddard, Robert H., 10, 35, 41-43, 44, 45, 49, 62, 139
Golovachev, V., 238
Golovanov, Yaroslav, 4, 34, 45, 115, 290; on aviation, 22-23; and Chelomei-Korolev relationship, 258; and Korolev's health/death, 276, 279, 280; on Korolev's personal life, 16, 17-18, 20, 25, 26, 39, 151, 200; on MVTU, 29-30, 31; and organizational structure, 209-10; and possible Tsiolkovsky-Korolev meeting, 14; and reasons for Soviet failures in space exploration, 303; and satellite development, 123; on Tsander, 37; and Utochkin's flight, 17-18
Gonor, Lev, 78, 93, 99, 115
Gorbachev, Mikhail, 238, 306
Gorbatko, Viktor, 159, 296
Gordon, Richard, 287, 296
Gorodomliya Island, 79-90, 97
Great Britain, 64, 66, 72, 130-31, 134-35, 144. *See also* Jodrell Bank
Grechko, Gyorgi, 105, 126, 132, 219-21, 222-24, 240, 255, 264-65
Grechka, Nina, 255
Grek, Valentina, 281
Grigorovich, Dimitri P., 23, 34
Grinfeld, Lidia Mavrikievna, 18-19
Grishin, V. V., 284
Grissom, Gus, 163, 176, 177, 182, 189, 289
Grizodubova, Valentina, 57
Gromov, Mikhail, 33-34, 39, 57, 58, 59
Grottrup, Helmut: and ICBM development, 97; return to Germany of, 90; in Soviet Union, 80, 83, 84-85, 86, 87; and V-2 technology, 67, 68, 69, 75, 80, 83, 84-85, 86, 87, 97
Grottrup, Irmgaard, 76, 77, 81
Group for Studying Reaction Propulsion (GIRD), 36, 40-41, 44-45, 71. *See also* Reaction Propulsion Institute; Tsander, Friedrikh Arturovich
Gubanov, Boris, 14, 118, 224, 264, 303, 315, 316
Guggenheim Foundation, 42, 43, 49
guidance and control systems: for ballistic missiles, 143; and *Gemini 8* mission, 286-87; and ICBM development, 108-9, 120; and Korolev's management style, 213; and *Kosmos 146* mission, 289; and manned space exploration, 162, 168, 187, 247; and N-1 program, 266; and organizational structure, 202; RNII development of, 47-48; and Tsiolkovsky's work, 13; and unmanned space exploration, 143, 149, 152; and U.S. organizational structure, 209; and *Zond* program, 292, 297. *See also specific engineer*
gulag camps, 50-51, 52, 57, 59
gyroscopes, 13, 47, 143-44, 195, 203, 298

Haise, Fred, 296
Ham (chimpanzee), 159, 176
Harvey, Brian, 157, 296
Haviland, Robert P., 122
Hearst, William Randolph, Jr., 92
Hilton, William F., 198
Hoover, George, 124
Houbolt, John, 9, 187
House of Representatives, U.S.: Committee on Science and Astronautics, 176
Hughes Aircraft, 190
hydrogen bomb, 4, 91, 113, 123, 127, 315
hydrogen engines, 305, 308

ICBMs. *See* intercontinental ballistic missiles (ICBMs)
Ilyushin, S. V., 34-35
Ilyushin-2 attack aircraft, 58
Institut Rabe (Bleicherode, Germany), 67-68, 69, 70, 71, 76, 78, 80. *See also* Nordhausen, Germany
Institute of Aeronautical Sciences, 130
Institute of Applied Mathematics, 221. *See also* Keldysh, Mstislav
Institute of Medical and Biological Problems (IMBP), 217, 271-72, 321
Institute of Theoretical Astronomy, 220
Institute of Thermal Processes. *See* Scientific Research Institute for Thermal Processes
instrumentation and test equipment, 88-89
intercontinental ballistic missiles (ICBMs): CIA reports about Soviet, 152; and cruise missiles, 92, 101, 102-4; funding for, 103; and guidance and control systems, 108-9, 120; Khrushchev announces Soviet possession of, 92; Korolev as developer of, 2; NII-88 development of, 93-109, 117; and organizational structure, 205; and reconnaissance satellites, 113, 191; and satellite development, 123-24; skepticism about, 98; Soviet deployment of, 191; Soviet space exploration conflicts with development of, 151-52; as Soviet top priority, 90; and *Sputnik*, 126, 127; and storable propellants, 113-14, 118-19; U.S. development of, 91, 92, 102, 103, 108, 109, 110, 112-13, 143; and U.S.-Soviet competition, 110, 190, 234; and V-2 technology, 97, 108; winged, 104; and Yangel-Korolev relationship, 115-16. *See also specific missile, especially* R-7 missile
Intercosmos, 238-39
intermediate-range ballistic missiles, 115. *See also specific missile*
International Astronautical Federation (IAF), 47, 124-25, 161, 317
International Geophysical Year, 121, 122, 124, 136
International Space Station (ISS), 169, 313, 316, 320, 321-22
Isaev, Alexei Mikhailovich, 65, 67, 86, 103, 236, 263, 308
Ishlinsky, Alexander Y., 52, 118, 283
Itek, 190
Ivanovsky, Oleg Genrikhovich: and Communist Party, 233; and computer capabilities, 219; and funding, 201; Korolev's relationship with, 167; and manned space exploration, 164, 166, 167, 171; perceptions of Korolev by, 95;

personal/professional background of, 232-33; and *Sputnik*, 3, 127, 133-34; and unmanned space exploration, 146-47, 155; and V-2 technology, 80
Izvestia, 197, 242, 306

J-2 engines, 263-64, 291-92
jet propulsion, 8, 12, 36, 41-43
Jet Propulsion Laboratory (California Institute of Technology), 46, 72, 133, 154-55
Jodrell Bank (United Kingdom), 141, 142, 144, 296
Johnson, Lyndon, 134, 152, 191, 249, 250, 252, 318
Johnson, Nicholas L., 311
Jupiter C launch vehicle, 133
Jupiter missile, 2, 66, 209

K-5 glider, 25
Kaliningrad, 5, 116, 301, 316
Kaluga: Tsiolkovsky Museum at, 11-12, 241
Kamanin, Nikolai, 170, 183, 184-85, 251
Kammler, Hans, 69
Kapitonov, I. V., 284
Kapitsa, Pyotr, 283
Kaplan, Joseph, 122
Kapustin Yar: A-4 test launches from, 81-83; cruise missile testing at, 101; and ICBM development, 94-95, 98, 101, 109; and organizational structure, 216, 217; and R-1 program, 84, 94-95; R-2 launches at, 98; and RT-1 missile development, 213
Karin, Vladimir, 210
Karman, Theodore von, 72
Karrask, Vladimir, 237-38
Katyusha rockets, 38, 106
Kavyzin, Mikhail, 225-26
Kazan sharashka, 60, 61, 62, 63, 97, 98
Keldysh, Mstislav: and communication satellites, 197; and Communist Party, 232; as head of Soviet Academy of Sciences, 206; and ICBM development, 106, 107, 118; and Korolev's funeral, 283, 284; Korolev's relationship with, 107, 261, 284; and manned space exploration, 167-68; and N-1 program, 260-61, 306-7; and organizational structure, 203, 206; reputation of, 323; role in space exploration program of, 221, 261; and *Sputnik*, 125-26; and technology issues, 221, 222; and unmanned space exploration, 139, 144, 146, 149, 151. *See also* Institute of Applied Mathematics; NII-1 (Scientific Research Institute-1)
Keldysh Commission, 288
Keldysh Research Center. *See* NII-1 (Scientific Research Institute-1)
Kennan, George, 131-32
Kennedy, John F.: and *Apollo* program, 271; assassination of, 234; and cooperation between U.S. and Soviet union, 270; and Cuban missile crisis, 116, 151; and manned space travel, 176, 179, 187, 249-50, 269-70, 302, 317; and *Sputnik*, 131; and spy satellites, 191; and U.S.-Soviet competition, 269-70
Kerber, Lev, 58-60
Kerrebrock, Jack, 306
Kerwin, Joe, 319
KGB, 49, 239, 243
Khariton, Yuli, 240
Khomyakov, Mikhail, 127
Khrunichev enterprise, 320
Khrunov, Yevgeni, 289-90, 293-94
Khrushchev, Nikita: and anonymity of scientists, 91-92, 235, 278; and Chelomei-Korolev relationship, 260; and Communist Party paranoia, 240; and cooperation between U.S. and Soviet Union, 270; and Cuban missile crisis, 116, 151, 152; deposing of, 182; and funding for space exploration, 246; and ICBM development, 92, 105-6, 109, 118, 119, 120; Korolev's relationship with, 105-6, 132, 151, 180-81, 202, 206, 207, 233, 268; and manned space exploration, 175-76, 178, 179, 180, 182, 260; and Moon flights, 260, 268; and N-1 program, 246; and organizational structure, 202, 206, 207; and reconnaissance satellites, 191; and satellite development, 124; and *Sputnik*, 122, 132-33; and Stalin's purges, 54-55, 56-57; and Tukhachevsky's rehabilitation, 54-55; U.N. trip of, 235; and unmanned space exploration, 142, 148, 152; and U.S.-Soviet competition, 181, 235, 243, 269, 270
Khrushchev, Sergei: on Chelomei, 258; and Chelomei space projects, 263; and Chelomei-Korolev relationship, 258-59, 260; and Communist Party paranoia, 240; and manned space exploration, 181; and Moon flights, 258; and N-1 program, 304-5; and organizational structure, 208; and R-7 funding, 113; and reasons for Soviet space failures, 304-5; and unmanned space exploration, 148; and U.S.-Soviet competition, 243; and *Voskhod* missions, 243
Kibalchich, Nikolai, 8, 9-10
Kiev Polytechnical Institute, 25-28
Killian, James, 134
Kirensky, Leonid, 5
Kirillov, Anatoli Semyonovich, 151-52
Kleimenov, Ivan Terentievich, 45, 48, 49, 50, 71
Koktebel: glider competitions at, 26-28, 32-33, 34-35
Koktebel (glider), 33, 34
Kolesnikov, A. A., 62
Kolyma prison, 1, 50-51, 52-53, 57, 59, 71
Komarov, Vladimir, 165, 182, 189, 242, 243, 283, 285, 289-90, 302
Komsomol (Young Communist League), 30, 283
Kondratyuk, Yuri, 9
KORD system, 294, 295, 298
Korneev, Leonid, 39-40
Korolev museum, 114, 322-23
Korolev, Pavel Yakovlevich (father), 16-17
Korolev, Sergei Pavlovich: anonymity of, 1-7, 15, 91-92, 146, 175-76, 235, 238, 240-41, 278-79, 305; arrest/imprisonment of, 1, 49, 50-54, 57-58, 59, 60-62, 71, 81, 97, 98, 115, 230, 234, 279, 280; awards and honors for, 280, 282, 306, 316, 322; birth of, 16; burial of, 278; character of, 19, 20-21, 24, 94, 120, 213-14; childhood/youth of, 16-28; as Communist Party member, 230-32, 233-34, 236, 247; contributions of, 2-3, 5-7, 14-15, 247-48, 281-82, 284-85, 305-6, 323; as creator of space organization, 209-10; death of, 1, 51, 52, 157, 275, 278, 279-85; early flights/pilot's license of, 23, 27, 36; education of, 19, 20, 21-22, 25-28,

29-34; eulogies for, 247, 284-85; film about life of, 323; goals of, 110, 111; health of, 6, 20, 35, 39, 51, 183, 198-99, 275, 276-79; influences on, 8, 12-14, 35, 139, 198, 284, 323; memorabilia of, 278-29; mishaps of, 41; personal life of, 25, 26, 39, 46, 94, 151, 199-200; as Red Army colonel, 63; rehabilitation of, 91; responsibilities of, 251, 261; restrictions on, 8; and U.S.-Soviet competition, 3, 188, 200, 245, 261, 265, 271, 275; writings of, 3, 44, 247-48

Korolev, Sergei Pavlovich-management style of: as autocratic, 48, 212; and awards and honors, 146-47; Chelomei compared with, 258; and Communist Party, 231, 236, 254-55; and dedication and motivation of people, 3, 71, 78-79, 225; and ICBM development, 93-94, 95, 98-99, 100, 104; and manned space exploration, 166-67; at meetings, 5, 221; and organizational structure, 203, 205, 207-8; and Soviet leadership, 207-8; and *Soyuz* program, 254-55; and *Sputnik*, 3, 127, 128; and staff relations, 48, 208-9, 210-15, 222—24, 225-29; and technology issues, 48, 221, 224, 225-29; and unmanned space exploration, 140, 145, 146, 156; von Braun's style contrasted with, 2, 212-13

Koroleva, Maria Nikolaevna Moskalenko (mother, later Balanina). *See* Moskalenko, Maria Nikolaevna

Koroleva, Natalya "Natasha" (daughter), 67, 72, 140-41; birth of, 46; and divorce of parents, 94, 151; education of, 39, 151; and Korolev's arrest/imprisonment, 49, 51, 60-61; and Korolev's contributions, 278-79; and Korolev's death, 278-79, 280; Korolev's relationship with, 94, 150-51

Koroleva, Nina Ivanovna Kotenkova (second wife), 16, 52, 94, 112, 151, 200, 276, 277, 323

Koroleva, Xenia Vincentini "Lyalya" (first wife): courtship/marriage of, 25, 26, 39; death of, 94, 278; divorce of, 94, 151; education of, 25, 26, 39, 61; and family life, 28, 46; and Korolev's arrest/imprisonment, 49, 60, 61-62; Korolev's relationship with, 61-62, 94

Korytov, Gyorgi, 227-28

Kosberg, Semyon, 149, 282

Koshits, Dmitri, 35, 39

Kosmos 4, 7, and 20 missions, 190, 195-96

Kosmos 27 mission, 154

Kosmos 47 mission, 182

Kosmos 60 mission, 156

Kosmos 96 mission, 154

Kosmos 146, 154, 186, and 188 missions, 289, 290, 291

Kosmos 212, 213, and 238 missions, 292

Kosmos 379, 382, 398, 434 missions, 297, 298, 299

Kosmos 637 mission, 198

Kosmos program, 229, 322. *See also specific mission*

Kostrov, Alexander, 238

Kostyukevich, Pavel, 176, 208, 213-14, 235, 236

Kotelnikov, Vladimir, 283

Kovtunenko, Vyacheslav, 116-17, 118-19, 120, 272

Kozlov, Dimitri, 196

Kraft, Christopher, 2, 269, 287, 293

Krasavka (dog), 168

Krasnapolyana sharaga, 60

Krichevsky, Sergei, 32, 33

Krupp Industries, 67

Kryukov, Sergei, 2, 301; and funding estimates, 257; and ICBM development, 99, 101, 104-5, 107, 108; and manned Moon flights, 252; and N-1 program, 256, 261, 267-68; on NII-88 activities, 95-96; professional background of, 95-96; and R-3 rocket, 99; and technology issues, 77, 228-29; and U.S.-Soviet competition, 257

Kubasov, Valery, 253-54, 296

Kudashev, Alexander, 26

Kuibyshev Design Bureau. *See* Central Specialized Design Bureau (TsSKB)

Kulakov, F. D., 284

Kuloilag camp, 59

Kurchatov, Igor, 4, 100, 220, 225, 240, 323

Kuznetsov, Nikolai: and N-1 program, 254, 265, 268, 271, 303, 305; and R-9 engines, 114, 206; and reasons for Soviet failures in manned space exploration, 305; and scenarios for manned Soviet Moon flights, 308

Kuznetsov, Viktor Ivanovich: and ICBM development, 118; and Korolev's death, 281; and manned space exploration, 167-68; and organizational structure, 203; and technology issues, 221; and V-2 technology, 69

Kuznetsov plant (Samara), 304

L-1 program, 260, 274, 286, 289, 290, 291, 292, 293, 294-95, 296, 297

L-3 program, 268, 307. *See also* N-1/L-3 program; N-1/L-3M program

Laika (dog), 129, 132, 161

Lamkin, Vladimir, 277

Landsat satellite, 191, 318

Langemak, Gyorgi Erikovich, 49, 71

Lavochkin, Semyon A., 34, 103, 112, 157

Lavochkin design bureau, 116, 153, 157, 166, 288, 297, 300

Lavochkin fighter planes, 34, 62

Lavrov, Svyatoslav Sergeevich, 220, 221, 222, 223, 224, 231, 232

Legostaev, Viktor, 2

Lehesten, Germany: and V-2 technology, 69, 71

Lenin, Vladimir, 37, 54, 116, 233

Lenin Prize, 146-47, 282

Leningrad Mechanical Institute, 89

Leonov, Alexei: and Korolev's arrest/imprisonment, 51-54; and Korolev's funeral, 283; and Moon flights, 246; and Stalin's purges, 55; and *Voskhod* 2 EVA mission, 159, 182-85, 186, 189, 269, 277

life-support systems, 7, 56, 162-63, 181, 248. *See also* space suits

Likhachev, Alexei, 45

Likhushchin, Valentin, 107

Lindbergh, Charles A., 42, 129

Lisichka (dog), 167

LK-1 program, 259, 273-74, 312

LK-700 program, 312

Lockheed Aircraft Corporation, 190, 192

Lockheed-Martin Corporation, 316

Lokotilov, Alexei, 225-27

Loral Space communications, 316

Lovell, A.C.B. "Bernard," 122, 141, 144
Lovell, James, 188, 189, 287, 293, 296
Low, George, 2, 176
Lozino-Lozinsky, Gleb, 274-75
Luna 1 mission, 141-42, 148, 222
Luna 2 mission, 142, 148
Luna 3 mission, 131, 143-47, 148
Luna 4 mission, 156, 242-43
Luna 5 mission, 156
Luna 6, 7, and 8 missions, 156
Luna 9 mission, 50, 157-58, 288
Luna 10, 11, and 12 missions, 288
Luna 15 mission, 296
Luna 16 and 17 missions, 297
Luna 21 mission, 300
Luna program, 140-41, 157, 253, 277, 300. *See also specific mission*
lunar probes: Soviet attempts at, 15, 142, 148, 155, 156; U.S. attempts at, 137, 148. *See also specific program or mission*
lunar soft landings, 149, 156, 158, 310
Lunokhod 1 mission, 297
Lunokhod 2 mission, 300
Lupanov, B. V., 21
Lyubyanka prison, 59
Lyulka, A. M., 308
Lyushin, Savva, 32-33

McDivitt, Jim, 186-88, 189
McDonnell Aircraft, 190, 320
McDonnell-Douglas Corporation, 316
McDougall, Walter A., 8
McElroy, Neil, 113
Magadan prison, 52-53, 71
Magnus, Kurt: return to Germany of, 89, 90; in Soviet Union, 76-77, 79, 80, 81, 82, 83, 85, 86, 87; and V-2 technology, 67, 68, 73, 76-77, 79, 80, 81, 82, 83, 85, 86, 87
Makeev, Viktor, 104, 209
Malenkov, Gyorgi, 216, 234, 258
Malinovsky, Ridion, 167-68, 271
Malyshev, Vyacheslav, 104
Manenok, Lev, 211
Manned Orbital Laboratory, 320
Mariner missions, 131, 152, 153, 154, 194
Marov, Mikhail, 221
Mars: Korolev's dreams/responsibilities for flights to, 7, 211-12, 247, 248; and N-1 program, 253-54; robotic spacecraft for flights to, 247, 248, 251; Soviet attempts to reach, 148-49, 150, 151-52, 153-54, 166, 235, 248, 272, 291, 295; technology issues for flights to, 219, 221, 228; Tsander's dreams about flight to, 37;U.S. attempts to reach, 153-54, 251; and U.S.-Soviet competition, 138, 148, 150, 235
Mars missions, 152, 153, 154
Martin, Richard, 92
Martin Company, 190
Massachusetts Institute of Technology (MIT), 30-31, 109, 143, 306
Maximov, Gleb, 139-40, 141, 149, 150, 211, 253, 254
Mayat, V. S., 280
Mazurov, K. T., 284
Mechta moon ship. See *Luna* program
Medaris, John B., 129
media: Soviet, 206, 306, 307. See also *Pravda*; *Izvestia*
Medvedev, Roy, 51
Mercedes-Benz Corporation, 220
Mercury program: and *Atlas* rocket, 160, 177, 178, 249, 260; development of, 162, 163, 178, 187, 249, 254; film of, 245; and *Redstone* project, 159, 176, 177; and Shepard flight, 250; Soviet knowledge about, 255; and *Soyuz* development, 255; and U.S.-Soviet competition, 177, 178. *See also specific astronaut*
Merrill, Grayson, 72
Meyer, Andre, Jr., 187-88
mice: in V-2 rockets, 162
Mikoyan, Anastas, 22, 176, 284
Mikoyan, Artem, 112, 115
Military Industrial Commission, 139, 202, 203, 206, 217, 299, 301
military, Soviet: and Communist Party, 231; films for, 242; funding for, 246, 253, 270; GIRD as part of, 44-45; and Moon flights, 252, 257, 271; and N-1 program, 263, 314; Odessa seaplane detachment of, 22, 23; and organizational structure, 207; and R-2 missile, 231; and R-12 missile, 117; and Soviet priorities, 260; and space exploration, 252, 253, 260; and space shuttle, 305, 314; and *Sputnik*, 126; and technology issues, 223-24; and *Vulkan* program, 305. *See also* Air Force, Soviet (VVS); Reaction Propulsion Institute (RNII); *specific weapon*
military, U.S., 124, 206, 313, 314, 318. *See also specific branch*
Millionshchikov, M. D., 283
Minenko, Viktor, 254-55
Ministry of Aircraft Industries, 242
Ministry of Armaments, 70-71, 78, 85, 202, 205, 216. *See also* Ministry of Defense Industry (MOP); Ustinov, Dimitri
Ministry of Armed Forces, 216
Ministry of Aviation Industry (MAP), 70, 78, 202, 203, 217
Ministry of Communications (Minsvyazi), 217
Ministry of Defense, 207, 217, 252, 253, 282
Ministry of Defense Industry (MOP), 96, 139, 205, 217. *See also* Ministry of Armaments
Ministry of Electronic Industry (MEP), 202, 217. *See also*
Ministry of Means of Communication Industry
Ministry of General Machine Building (MOM), 217, 232, 266, 273, 276, 299
Ministry of Health (Minzdrav), 217
Ministry of Machine Building and Instrument Engineering, 202
Ministry of Means of Communication Industry, 202
Ministry of Medium Machine Building (MSM), 217
Ministry of Radio Industry (MRP), 217
Ministry of Sciences, 217
Ministry of Shipbuilding, 203
Ministry of State Security, 202
Ministry of Weapons, 70
Mir space station, 75, 111, 128, 316, 320, 321
Mishin, Vassily, 2, 77, 234, 244; as administrator, 301; Afanasiev reprimands, 278; and *Apollo* program, 297, 299, 302; and circumlunar flights, 273, 274; firing of, 300; and funding,

298-99; Glushko's relationship with, 305, 308, 314; and ICBM development, 117, 118; and Korolev-Glushko relationship, 114; and Korolev-Ustinov relationship, 202; as Korolev's deputy, 96; and Korolev's health, 278; and Korolev's legacy, 286, 288-89; and Korolev's management style, 211; and manned space exploration, 96, 161, 178-79, 180, 181, 182; and Moon flights, 96, 286, 288-89, 290-91, 294, 308; and N-1 program, 273, 291, 294, 295, 300, 301, 303, 304, 308, 315; and N-1/L-3 system, 288, 299; and N-1/L-3M program, 310-11; and organizational structure, 299; and reasons for Soviet failures in space exploration, 303, 304; responsibilities of, 286, 288-89; and satellite development, 123; and *Soyuz* program, 255, 292, 293-94; and space shuttle, 314, 315; and space stations, 293-94, 298; and *Sputnik*, 132; and technology issues, 65, 69-70, 71, 221; and U.S.-Soviet competition, 286, 291

missiles: and Communist Party, 237-38; and organizational structure, 202, 203, 205, 206; and Soviet priorities, 260; submarine launching of, 208-9; U.S. funding for, 205-6; winged, 203. *See also* ballistic missiles; intercontinental ballistic missiles (ICBMs); *specific missile*

Mitchell, Edgar, 297

Mittelwerk plant (Nordhausen, Germany), 68-69

Molniya communications satellites, 149, 196, 197-98, 200,

Molniya design bureau (Moscow), 274-75

Molotov, Vyacheslov, 49

monkeys/chimpanzees, 159, 162, 176, 245

Montania, Germany, 71

Moon flights: chronology of U.S. and Soviet, 324-39; circumlunar, 248, 257-58, 259-60, 262-63, 273, 274, 277, 286, 288, 290, 293-94, 296, 297, 302-10; and colony on Moon, 218; and Communist Party paranoia, 241, 242-43; and docking in space, 261; Korolev's great desire for, 3, 73, 139, 140-41, 251, 254; and Korolev's management style, 215; Korolev's proposal for, 139, 247; and Korolev's views about surface of Moon, 157-58; list of U.S. and Soviet, 147-48; Mishin as head of manned, 96; and nuclear bombs, 141; reasons for Soviet failures in, 302-7; scenarios for manned Soviet, 307-12; and Soviet military, 252, 257, 271; Soviet priority for, 246, 251; Soviet support for, 268-69; Soviet timetable for, 246, 247, 272; technology for, 219, 221, 222; U.S. priority for, 251; and U.S.-Soviet competition, 138, 188-89, 190, 215, 250-51, 257, 269, 288-89, 291, 293, 296, 301, 324-39. *See also* space exploration; *specific person, program, or mission*

Moscow Aviation Institute (MAI), 31-32, 89, 306

Moscow Engineering and Physical Institute (MIPI), 89

Moscow Higher Technical School (MVTU), 28, 29-34, 89, 110-11

Moscow Physical Technical Institute (MPTI), 89

Moscow State Technical School (MGTU). *See* Moscow Higher Technical School (MVTU)

Moscow State University (MGU), 89

Moskalenko, Anna (aunt), 16

Moskalenko, K. S., 253, 271

Moskalenko, Maria Nikolaevna (mother), 23, 28, 33, 35, 46; education of, 17; and Korolev as Moscow student, 31; and Korolev's arrest/imprisonment, 50, 61; and Korolev's childhood/youth, 22; marriages of, 16-17, 19; as single mother, 17

Moskalenko, Vassily (uncle), 24, 31

Moskalenko, Yuri (uncle), 17, 19, 27

Mozzhorin, Yuri, 78

Mrykin, Alexander Grigorevich, 139

Mueller, George, 2, 302

Mushka (dog), 168

MVTU. *See* Moscow Higher Technical University

Myasishchev, Vladimir, 103, 112

Myers, Dale, 102

N-1 program: acknowledgment of existence of, 306; and *Apollo* program, 253, 264, 273, 301-2; approval for, 260-61, 266; cancellation of, 253, 300-301, 303, 306-7, 314, 315; and Communist Party paranoia, 241; criticisms of, 272-73; declassification of, 306; design/testing of, 252, 256-57, 260-61, 266, 290-91; and docking in space, 308; engines for, 246, 253, 264, 265-66, 267, 272-73, 294-95, 300, 302-5; funding for, 246, 253, 256, 257, 267, 268, 291; and Korolev's contributions, 104, 306; and L-1 spacecraft, 294-95; launchings of, 294-95, 298, 300; legacy of, 301; and Mars mission, 253-54; priority of, 180-81, 263; problems/weaknesses of, 264, 265, 267-68, 271, 294-95, 300, 302-5, 308; purposes of, 246, 253, 257, 263; and reentry, 289; and rescue system, 298, 300; rivals to, 246, 257-58, 260, 265-67, 268, 273; *Saturn V* compared with, 246, 266-67; and scenarios for manned Soviet Moon flights, 307; secrecy about, 246; and Soviet military, 263, 314; and *Soyuz* program, 289; and space shuttle, 314; timetable for, 261, 267, 268, 291; and U.S.-Soviet competition, 288; and Yangel-Korolev relationship, 272. *See also* N-1/L-3 program; N-1/L-3M programN-1/L-3 program: acknowledgment of existence of, 306; authorization for, 273, 277, 288; benefits of, 317; development of, 263, 265, 269, 271, 274; and EVA, 269, 310; funding for, 298-99, 301, 317; legacy of, 301; and lunar landers, 308-10; Mishin as head of, 286; and organizational structure, 299; problems/weaknesses of, 271, 272, 297, 301; and reentry, 309, 310; and rendezvous in space, 263, 308-10; and scenarios for manned Soviet Moon flights, 308-10; Soviet support for, 265, 268-69; and *Soyuz* program, 297; timetable for, 288; and U.S.-Soviet competition, 265, 271, 288

N-1/L-3M program, 300, 308, 310-12

N-2 programs, 104, 256, 266-67

N-3 programs, 104, 256

N-15 engine, 304

Nakhabino proving ground, 40-41

Nanz, Ted, 318

Natenzon, Yakov, 103

National Advisory Committee for Aeronautics (NACA), 248

National Aeronautics and Space Administration (NASA): and benefits of space exploration, 317, 318; and cooperative projects, 320; creation of, 134, 154, 248; and EVA, 287; and F-1 engine,

260; funding for, 206, 215, 252, 268, 314, 319; Johnson Space Center of, 215; Langley Research Center of, 187; and manned space exploration, 162, 177, 179, 187, 249, 260; Marshall Space Flight Center, 226, 227, 267; and military projects, 313; and Moon flights, 142, 252, 260; museums of, 319; projects of, 251, 252; and rotor landings, 262; and Shuttle-*Mir* missions, 316; and space and missile development in U.S., 206; and space shuttle, 314; and *Sputnik*, 134; and technology issues, 225; and training of U.S. engineers, 89; and unmanned space exploration, 154-55. *See also specific program*
National Air and Space Museum, 177, 320
National Science Foundation, 124
National Security Council, 129
National Space Council, 249
Navaho cruise missiles, 92, 103
Naval Research Laboratory: and satellite development, 124
Navy, U.S., 43, 92, 122, 209, 234
Nedelin, Mitrofan, 100, 119, 167-68, 242, 271
Nelyubov, Grigory, 165, 170
Nepmen, 26, 27
Nesterenko, Alexei, 207
Neufeld, Michael, 75
New York Coliseum: Soviet exhibition at, 239
nihilists, 9
NII-1 (Scientific Research Institute-1), 77, 106-7
NII-4 (Scientific Research Institute-4), 216, 221
NII-10 (Scientific Research Institute-10), 203
NII-88 (Scientific Research Institute-88): anti-Semitism at, 115; and Communist Party, 235; computer capabilities at, 221; creation/expansion of, 77-79, 95-96; funding for, 96-97; Germans working at, 79; ICBM development at, 93-109, 117; Korolev as chief designer of, 78-79; and organizational structure, 202, 216; physical condition of, 93-94; political support for, 96-97; and R-1 program, 84; and technology issues, 221; and unmanned space exploration, 146; and V-2 technology, 64, 65, 72, 77-90; Yangel as director of, 115; young Soviet engineers at, 89. *See also* OKB-1
NII-885 (Scientific Research Institute-885), 78
Nikishin, Leonard, 242
Nikolaev, Andrian, 165, 170, 177, 178, 277, 283, 296-97
NK-15V engine, 304
NK-33 engines, 263, 265, 304
NKVD, 49, 55, 58, 59, 61, 62
Nordhausen, Germany, 67, 68-69, 70, 71, 72, 78, 79, 83
North American Aviation Corporation, 102, 190, 252, 260, 265, 267, 301
North Atlantic Treaty Organization (NATO), 112, 114, 132
Nova program, 260, 312
Novatsky, Pavel, 122
Novocherkassk prison, 50
Novozhilov, Pavel, 208-9
NPO Energia, 305
NPO Energomash, 265-66
NPO PM (Scientific Production Association "Applied Mechanics"), 198
NPO Yuzhnoye (Dnepropetrovsk, Ukraine), 100, 116, 117, 264, 272. *See also* Yangel, Mikhail
nuclear propulsion, 225
nuclear weapons, 132, 220, 141, 248

OAVUK. *See* Society of Aviation and Aerial Navigation of Ukraine and the Crimea
Oberg, James, 279
Oberth, Hermann, 10, 35, 44, 45, 161
Odessa, Ukraine, 20-21, 22, 23, 38
Office of Naval Research, U.S., 124
OKB-1: anti-Semitism at, 230; bureaucracy at, 211; and Communist Party, 230-32, 233, 234-35, 236, 237, 238, 241; and computer capabilities, 219-21; exhibits at, 209; Glushko as head of, 301, 305; and ICBM development, 100, 101, 110, 120; Korolev named head of, 100; Korolev's reporting obligations at, 206; and manned space exploration, 180-81; Mishin as head of, 286; museum at, 241, 301; and organizational structure, 210; projects at, 211-12; and reconnaissance satellites, 194, 196; and technology issues, 219-21, 223, 225; typical day at, 199; and unmanned space exploration, 140; young engineers at, 89, 105. *See also* NPO Energia
OKB-52 (Chelomei design bureau), 240, 263, 273, 298, 320
OKB-456 (Glushko design bureau), 78, 114, 202, 203, 204, 237, 273, 303
Okhapkin, Sergei O., 211, 229
Omsk prison, 59-60, 61, 62
Operation Backfire, 72
OPO-4 (Fourth Experimental Section): Korolev at, 34-35
optics technology, 194-95. *See also* cameras
OR-1 engine, 36, 37
OR-2 engine, 37, 39-41
Orbiter program, 124, 251
orbiting mirror, 75
Ordzhonikidze, G. K., 22
Orieshkin, Yuri, 203, 212, 214, 233, 268-69
ORM-1 engine, 36
ORM-2 engine, 36
ORM-65 engine, 48
Osoaviakhim (aeronautical society), 32, 34, 36
Ostashev, Arkady, 111-12, 117-18, 144, 145
Otrieshka, Antonina, 1, 6, 199, 211, 214

Pallo, Arvid, 50, 141-42
Pashkov, Gyorgi, 210, 221
Patsaev, Viktor, 242, 285, 298, 302
Pchelka (dog), 168
Pedoseyev, P. N., 283
Peenemunde facility (Germany): construction of, 46; and ICBM development, 110; and satellite development, 122; and V-2 technology, 66, 68, 69, 72, 74, 75, 79, 83; von Braun's work at, 62
People's Commissariat of Aviation Industry, 62, 63
Perelman, Y. I., 8
Petlyakov 2 (Pe-2) dive bomber, 61, 62
Petrov, Boris, 52, 118, 283
Petrov, Gyorgi, 118
Petrovsky, Boris Vasilevich, 277-78, 279-80, 281
Phillips, Sam, 2
Pickering, William H., 2, 72, 133
Pilyugin, Nikolai: on Chelomei's influence, 258;

and ICBM development, 118; and Korolev as Communist Party member, 233; Korolev's relationship with, 282; and LK-1 project, 274; and manned space exploration, 167-68; and N-1 program, 266, 303; and N-1/L-3M program, 300; as NII-885 deputy chief designer, 78; and organizational structure, 203; and reasons for Soviet failures in manned space exploration, 303; reputation of, 2; and technology issues, 65, 69, 70, 71, 221

Pioneer program, 136, 137, 138, 147, 148, 154-55

Pivovarov, Sergei, 47

Platonov, Sasha, 221

Plesetsk launch center, 112, 152, 191, 217

Pobedonostsev, Yuri, 31, 37, 64-65, 72, 78

Podgorny, N. V., 284

Podlipki facility (Soviet Union). *See* NII-88

"pogo" effect, 107-8, 113, 291

Polaris program, 209

Polevoi, Gyorgi, 35

Poliatchenko, Vladimir, 262-63

Politburo, 216, 217, 232, 234, 235, 237

Polyansky, D. S., 284

Polyarny, A. I., 63

Polycarpov, Nikolai, 115

Polyot program, 262-63

Polytechnic Museum (Moscow), 241

Poniatowski, Stanislas, 29

Ponomarev, B. N., 284

Popovich, Pavel, 165, 170, 171, 177, 178, 188, 277, 283

Postyshev, Vladimir, 201, 203

Pravda: and *Apollo 11* mission, 295; and Communist Party paranoia, 242-43; and Gagarin flight, 160; Korolev's articles in, 3; Korolev's obituary in, 282; and *Luna 15* mission, 296; *Sputnik* program articles in, 121-22, 133; Van Allen belts pictured in, 136

Princeton, New Jersey: Soviet visitors in, 31-32

"Professor K. Sergeev" columns, 3, 241, 242

Profsoyuzny (Trade Union Doctorates), 98

Progress cargo vehicle, 320

propellants, 12. *See also* storable propellants; *type of propellant*

Prostreishiy Sputnik "PS," 126

Proton. *See* UR-500 (*Proton*) launch vehicle

Pustovoytenko, Alexandra, 236

quality control, 210-11, 224

R-1 missile: development of, 94-95, 97, 100, 101, 104, 109, 112, 214; and V-2 technology, 81, 83-85, 86

R-1A missile, 97, 99

R-2 missile, 84, 97-98, 100, 101, 104, 109, 112, 117, 215, 231

R-2E missile, 99

R-3 missile, 97, 99-101, 102, 104

R-5 (SS-3 Shyster) missile, 15, 84, 101-2, 104, 106, 109, 117, 215

R-7 launcher: *Atlas* compared with, 112-13; awards and honors for design of, 215; and communication satellites, 197; and Communist Party relations, 236, 238; and computer capabilities, 220; and cooperative U.S.-Soviet programs, 270; cost of site for, 113; deployment of, 112, 113; development of, 84, 91, 92-94, 103-9, 226, 251; engines for, 108, 113, 219, 249; and H-bomb, 113, 123, 127; as Korolev's responsibility, 251; launching of, 92, 109-10, 113; and manned space exploration, 161; and moon landings, 139; NATO designation for, 112; and "pogo" effect, 107-8, 113; problems with, 93-94, 291; and public relations, 236; and reconnaissance satellites, 192; reentry cone for, 106-7; and satellite development, 123-24; as Soviet threat to U.S., 92; and *Spiral* aerospace plane, 274-75; and *Sputnik* program, 126, 127, 220; stages in, 139; staging principle of, 92-93; technology for, 219, 220, 226; testing of, 109-10, 113; and unmanned space exploration, 149; and *Vostok* program, 249

R-7A missile, 109, 112

R-9 missile (SS-8 Sasin), 114, 115, 117-19, 120, 206

R-10 missile, 80, 84-85, 86, 87, 97

R-11 missile, 86

R-11FM, 208-9

R-12 missile (SS-4 Sandal), 86, 115, 116, 117

R-14 missile, 86-87, 117

R-15 missile, 86-87

R-16 missile, 117-20

R-56 booster, 265, 308

R-113 missile, 88

Rabfakists, 26, 30

radiation, 13, 248. *See also* radiation belts

radiation belts, 133, 135-36, 137, 138, 147, 148, 160

RAND Project, 122-23

Ranger program, 155, 156, 194, 251

Rashidov, Sh. R., 284

Rauschenbakh, Boris V.: and guidance and control systems, 47-48; imprisonment of, 56; and long-range missile development, 63; and manned space exploration, 160, 163, 170; and reconnaissance satellites, 194; and similarities between Korolev and von Braun, 1-2; and *Sputnik*, 121, 122; and unmanned space exploration, 143, 157-58

RCA, 190

RD-1 engine, 62

RD-100, RD-101, and RD-103 engines, 97, 100, 101

RD-107 and RD-108 engines, 108, 249

RD-110 engine, 101

RD-253 engine, 259, 303, 304

RD-270 engine, 265-66

Reaction Motors, Inc., 190

Reaction Propulsion Institute (RNII), 45-46, 47-48, 52, 53, 54

Reagan, Ronald, 259, 315

Rebrov, Mikhail, 129

reconnaissance satellites: and benefits of space exploration, 318; and *Discoverer* missions, 143-44, 190, 191, 192, 196; funding for, 318; and ICBM development, 108, 113; Korolev as developer of, 2, 212; and optics/camera technology, 192, 194-95, 196; recovery of, 192; and robotic spacecraft, 288; Soviet development of, 149, 170, 189, 190-96; U.S. development of, 124, 143-44, 190-92, 194, 196; and U.S. military, 318; and U.S.-Soviet competition, 124, 190; and *Vostok* program, 196

Red Star (glider), 34-35
Redstone Arsenal, 124
Redstone missiles, 2, 66, 159, 176, 177
reentry: and *Gemini 8* mission, 286-87; and manned space exploration, 162, 168, 169-71, 173-75, 177, 178, 181-82, 183, 184, 185, 186-87; and Moon flights, 247; and N-1 program, 289; and N-1/L-1 project, 294; and N-1/L-3 program, 309, 310; and N-1/L-3M program, 312; and *Saturn V/Apollo* mission, 290; and UR-500K/L-1 system, 290; and UR-700/LK-700 program, 312; and *Zond* program, 291, 292, 297
Rees, Eberhard, 2, 227
rendezvous in space: Houbolt's idea about, 9; and manned space exploration, 51, 96, 188, 189; and scenarios for manned Soviet Moon flights, 308-12; and SDI, 315; and Soviet space pioneers, 9; Soviet timetable for, 247-48; U.S., 128, 260. *See also specific program or mission*
rescue systems, 169, 185-86, 320. *See also specific program or mission*
Reshetnev, Mikhail, 2, 104, 198
Rhein-Metall Borsig, 78, 220
Richard, Paul, 34
RO-5 engine, 149
Rocketdyne Division. *See* North American Aviation
rockets: "closed loop," 149; development during World War II of, 44, 60; GDL development of, 37-38, 44; GIRD development of, 44; and Korolev's contributions, 15; multi-stage, 13-14; regenerative cooling of, 13; Stalin's purge as setback to development of, 54-55; Tsander's work on, 36, 37; Tsiolkovsky's work on, 12-14. *See also* V-2 technology; *specific rocket*
Rockwell International Corporation, 316, 321
Roosa, Stuart, 297
Roswell, New Mexico: Goddard's work at, 42, 49
rotor landings, 261-62
Rozhdestvensky, Mikhail, 103, 156, 157
RSC Energia, 64, 108, 165, 230, 237, 313, 316, 320, 321
RT-1 missile, 213
Rudakov, A. P., 284
Rudenko, Mikhail, 14
Rudnev, Konstantin, 115
Rudnitsky, Viktor, 71
Rudomino, Margarita Ivanova, 19, 23-24, 31
Rukavishnikov, Nikolai, 298
Rumer, Yuri, 60
Russian Revolution: anniversaries of, 132-33, 274
Ryabikov, Vassily Mikhailovich, 70-71, 139, 204
Ryazansky, Mikhail, 52, 69, 78, 79, 167-68, 203, 221
Rynin, Nikolai, 9, 10, 35

Sadovii, Viktor, 162
Sagdeev, Roald, 124, 135-36, 206, 306-7
Sakharov, Andrei, 4, 56, 58, 60, 91, 124, 205, 275
Salyut space stations, 205, 219, 242, 265, 298, 300, 320
Samoshina, Lydia, 276
Saturn IB, 293
Saturn program, 187, 226, 252, 264, 275, 298. *See also Saturn V* program; *Saturn V/Apollo* program
Saturn V program, 2, 12, 246, 257, 263-64, 265, 266-67
Saturn V/Apollo program: development of, 260, 264, 265; funding for, 267; in museums, 319; and N-1 program, 273; problems in, 291; purpose of, 302; unmanned, 290; and U.S.-Soviet competition, 263, 264, 265, 273
Schirra, Walter, 177, 188, 189, 293
Schmetterling (Butterfly) missile, 87-88
Schwinghamer, Robert, 226
Scientific Research Institute for Thermal Processe (NIITP), 107, 254, 305
Score Project, 138, 198
Scott, David, 286-87, 299
Seamans, Robert, 2, 176
Sedov, Leonid, 125, 147, 161, 283, 295
Seifert, Howard S., 72
Semenov, N. N., 283
Semenov, Yuri, 316
Semyonov, Gherman, 52
Semyorka project, 109, 113, 270
Serebrov, Alexander, 111
Serov, Ivan, 82
Sevastyanov, Vitaly, 197, 242, 296-97
Severin, Gai Ilyich, 162, 168-69, 181-82, 183
Sevruk, Dominic D., 62, 102
SFAF. *See* Society of Friends of the Air Force
Shakhurin, Alexei, 70
Shakmatov, Viktor, 206-8
sharaga: Korolev imprisoned in, 1, 57-62
Shatalov, Vladimir, 293-94, 296, 298
Shchetinkov, Yevgeni, 46, 47-48
Shelepin, A. N., 284
Shepard, Alan, 2, 159-60, 163, 177, 245, 250, 287-88, 297
Sheremetyevsky, Nikolai, 70, 237
Shevalyov, Vladimir, 160, 175, 199, 233, 234, 235, 243-45
Shevchenko, Andrei (grandson), 151, 278
Shevchenko, Gyorgi (son-in-law), 278
Shevchenko, Maria (granddaughter), 278
Shevchenko, Sergei Pavlovich Korolev (grandson), 278
Shevchenko, Taras: monument for, 24
Shlyapnikov, Alexander, 23
Shonin, Gyorgi, 296
Shustin, Konstantin, 212
Siemens Industries, 67
Sikorsky, Igor, 26
Sikorsky Aircraft Corporation, 115
Sisakian, N. M., 283
SK-4 light plane, 35
SK-9 glider, 48
Skylab program, 319-20
Slayton, Deke, 287-88
Smirnov Commission, 274
Smirnov, Leonid V., 151, 206, 283, 284, 299
Smirnov, Yuri, 240
Smith, Hedrick, 235
Smithsonian Institution, 42, 190
Society of Aviation and Aerial Navigation of Ukraine and the Crimea (OAVUK), 22, 25
Society of Friends of the Air Force (SFAF), 22-23
soft landings: lunar, 149, 156, 158, 310; and manned space exploration, 168, 169
Sokolov, A. I., 72

Sokolsky, Viktor, 9, 46
Solzhenitsyn, Aleksandr, 51, 57
Sommerda, Germany, 70, 71
Sonderhausen, Germany, 70
Sorokovka cruise missile, 103
Soyuz 1 mission, 242, 283, 289-90
Soyuz 2, 3, 4, and 5 missions, 289-90, 292, 293-94
Soyuz 6, 7, 8, and 9 missions, 296-97
Soyuz 10 mission, 298
Soyuz 11 mission, 242, 285, 298, 299, 302
Soyuz program, 265, 274, 300, 322; and *Apollo-Soyuz* program, 51, 96, 255, 256, 317; and Communist Party, 262; and computer capabilities, 219; development of, 178, 180, 254-56, 261-62, 288; and docking in space, 262, 286, 289-90, 292, 293-94, 320-21; and *Gemini* program, 255, 256; and Korolev's management style, 254-55; and Korolev's responsibilities, 255; and *Mercury* program, 255; Mishin as head of, 286; and N-1 program, 289; and N-1/L-3 program, 297; priority of, 292; and R-7 launcher, 219; and rotor landings, 261-62; spherical design of, 255-56; and T2K lunar lander, 297; and Ustinov-Korolev relationship, 262; and *Voskhod* program, 255-56; and *Vostok* program, 255-56, 262
Space Age: listing of attempts of first full year of, 137-38
space exploration: benefits of, 317; films about, 241-42; funding for Soviet, 149, 210, 253, 270, 298-99, 307, 322; Korolev's views about, 35-36, 139, 157, 247-48, 323; memorials to Soviet, 322; and organizational structure, 202, 206, 307; popularization of, 241-42; Soviet "firsts" in, 295; Soviet priority for, 260, 322; Soviet timetable for, 247-48; and technology issues, 229; U.S. funding for, 205-6, 248, 318; and U.S.-Soviet competition, 140, 147-48, 150, 152, 153-54, 190, 243, 248-50, 251, 252, 253, 313, 317. *See also* communications satellites; manned space exploration; Mars; Moon flights; reconnaissance satellites; Venus; *specific program or mission*
space industry, 78, 210, 248-49, 251
space medicine, 161, 162, 165, 187, 271-72, 281, 316, 318, 321
space organizations, 237, 321. *See also specific organization*
space rocket train. *See* rockets: multi-stage
space shuttle: Russian, 305, 313-15; U.S., 128, 313, 314, 315, 316, 319; and U.S.-Soviet competition, 315
space stations: Soviet, 13, 248, 293-94, 298, 315, 320; studies conducted from, 248; U.S., 319. *See also specific program*
space suits: on dogs, 161; and manned space exploration, 162-63, 168-69, 171, 180, 181, 183-84, 185; Soviet timetable for development of, 248; and space medicine, 271; and *Voskhod 2* mission, 269
Special Design Bureau Number 385, 125, 165
Spiral aerospace plane, 274-75
Spot Image satellite, 318
Sputnik 1 mission, 113, 121-22, 127, 128-29, 131-32, 133, 136, 143, 225-26, 227-28
Sputnik 2 mission, 129, 132, 133, 136, 161, 226
Sputnik 3 mission, 113, 126, 133, 134-36, 137, 205, 222-24
Sputnik 7 mission, 153
Sputnik 8 mission, 153
Sputnik program: and ballistic missiles, 131; and computer capabilities, 220; development of, 125, 233, 247, 249; dogs in, 125-26, 129, 132; effect on world of, 131-32; exhibit of, 322; and fortieth anniversary of Russian Revolution, 132-33; historical importance of, 128, 130; and ICBM development, 126, 127; Korolev as developer of, 2, 3, 120, 247; Korolev proposal for, 127; and manned space exploration, 166, 170; and Moon, 126, 131; and nuclear weapons, 132; *Pravda* articles about, 121-22, 133; and *Prostreishiy Sputnik* (PS), 126; and R-7 launcher, 220; and radiation belts, 133, 135-36, 137, 138; and reconnaissance satellites, 192; and shape of craft, 162; and Soviet military, 126; Soviets announce, 125; and space travel, 130; and unmanned space exploration, 139, 143; U.S. reaction to, 129-30, 252; and U.S.-Soviet competition, 126, 127, 131-33, 136-38, 252, 253; and Venus, 131; world reaction to, 122, 130-31
spy satellites. *See* reconnaissance satellites
SS-6 Sapwood. *See* R-7 missile
SS-8 Sasin. *See* R-9 missile
Stafford, Tom, 188, 189, 287
Stalin, Josef: and anti-Semitism, 115; death of, 105, 234; as hero of Bolshevism, 21; and Korolev's anonymity, 278; Korolev's relationship with, 97, 105-6, 202, 206, 233, 234; Korolev's views about, 234; and liquid rockets, 234; and missile development, 77, 114, 234; and NII-88 support, 96-97; purges of, 48, 49-63, 71, 93, 234, 237; and SFAF, 22; and U.S.-Soviet competition, 234; and V-2 technology, 64, 81; and von Braun's rocket experiments, 49
Stalin (Josef) directive (1946), 202-3
Stapp, John, 147
State Commission for Testing Boosters and Launches for the First Satellite, 204
State Commissions, 127, 184, 186, 204-5
State Gosplan, 139
State Union Design Bureau of Special Machine Building, 202
Stechkin, B. S., 62
Steinhoff, Ernst, 68
Stepanchonok, Vassily, 34-35
Stilianudi, A. N., 21
storable propellants, 113-15, 118-19, 208, 297
Strategic Defense Initiative (SDI), 257, 259, 314, 315
Strazheva, Irina (wife of Yangel), 116
Strekalov, Gennadi, 128
Strelka (dog), 167, 168
Stroyev, Nikolai Sergeevich, 167
Stroyprofshkola No. 1 (Professional Building Construction School), 21-22
Stuhlinger, Ernst, 74, 75, 133, 212
submarine delivery systems, 208-9, 246, 314
Summerfield, Martin, 130
Supreme Soviet, 217
surface, Moon, 157-58, 288
surface-to-air missiles, 80, 87. *See also specific missile*

Surveyor program, 158, 194, 251
Suslov, M. A., 284
Swigert, Jack, 296
Symington, Stuart, 191
Syncom communication satellite, 198
Syromiatnikov, Vladimir, 229, 241, 261, 320-21
systems analysis, 204-5, 301

T2K lunar lander, 297, 298, 299
T-1 study, 104
T-2 study, 104
Taran anti-ballistic missile system, 259
Tarasenko, Maxim, 78, 174-75, 191, 210, 241, 259, 266
TB-3 bomber, 37, 58, 59
technology: and basic research, 229; and computer capabilities, 219-21; and dedication and motivation of staff, 225; and electronic problems, 224-25; and experimentation, 226-28; and illustrations of complex designs, 228-29; and Korolev's management style, 221, 222-24, 225-29; and lint-free environments, 225; and practical manufacturing techniques, 225-29; and quality control, 224; and U.S.-Soviet competition, 219, 301. *See also* V-2 technology; *specific technology or program*
telephone tapping, 234
Telstar communication satellite, 198
Temtsunik, F. A., 21
Tereshkova, Valentina, 178, 179, 261, 283
test pilots, 163-64, 287. *See also specific person*
Thagard, Norman, 128
thermal regulation system, 145, 149, 289
thermonuclear warheads, 246
Thiel, Adolf, 74-75
Thiel, Walter, 65
Thiokol Chemical Corporation, 190
Thompson Ramo Wooldridge, 190
Thor-Able launcher, 147, 148
Thor-Agena booster, 196
Tikhomirov, Nikolai, 36, 37-38
Tikhonravov, Mikhail K., 2, 45, 253, 254; and GIRD-09 engine, 40, 41; as influence on Korolev, 139; Korolev's co-authorship with, 247-48; and Korolev's health, 277; and manned space exploration, 165, 170; and R-7 launcher, 92-93; and satellite development, 123; and *Sputnik*, 125; and V-2 technology, 64-65
Timchenko, Vladimir, 49, 255
Titov, Gherman Stepanovich, 165, 170, 171, 176, 177, 178, 283
Tokaty-Tokaev, Gregory A., 9, 14, 55
Tolstoy, Lev, 46
TOM-1 torpedo bomber, 34
Trunov, Konstantin Ivanovich, 98, 231
TsAGI. *See* Central Aerohydrodynamic Institute
Tsander, Friedrikh Arturovich, 2, 323; biographical information about, 36; death of, 40; at GIRD, 36, 37, 40, 41; honors for, 41; and jet propulsion theory, 36; Korolev's work with, 37; management style of, 40; and OR-2 engine, 39, 40; and rocket engines, 37; at TsAGI, 37
Tsblieyev, Vassily, 111
Tsiolkovsky, Konstantin Edvardovich: awards and honors for, 14; biographical information about, 10-11; contributions of, 44, 121-23, 306; early career of, 11-12; Goddard compared with, 43; health of, 11; influence on Glushko of, 38; influence on Korolev of, 8, 12-14, 35, 284, 323; Korolev's meeting with, 14; and satellite development, 123; as Soviet space pioneer, 8-9, 10-14, 35, 37, 43; and unmanned space exploration, 141
Tsiolkovsky Museum (Kaluga), 241
TsKBEM. *See* Central Design Bureau of Experimental Machine Building
TSNIIMash. *See* Central Scientific Research Institute for Machine Building
TsSBK. *See* Central Specialized Design Bureau (TsSKB)
TsUP. *See* Flight Control Center
Tsybin, Pavel, 186
Tsygan (dog), 161
Tsyganka (dog), 161
Tu-2 light bomber, 58, 59
Tukhachevsky, Mikhail, 44-45, 46, 48, 52, 54-55
Tumovsky, Yevgeni, 236
Tupolev, Andrei, 25, 30, 33-34, 57, 58-59, 112, 230. *See also* TB-3 bomber
TV-3 (*Test Vehicle 3*), 133
Tverdyy, V. P., 21
Twining, Nathan F., 192
Tyulin, Gyorgi, 140, 255; on Korolev, 71, 81; and Korolev-Glushko relationship, 114; and technology issues, 221; and V-2 technology, 72, 73, 81, 82

U-2 spy planes, 113, 124, 191, 196
UFOs (unidentified flying objects), 234
Ulam, Adam, 55, 56
Ulrich, Vassily, 50
United Aircraft, 190
United Nations, 123, 235, 270
United States: Korolev's views about, 234; ritual attitudes of Soviets about, 234. *See also specific person, topic, program, or mission*
unmanned space exploration. *See* Mars; Moon flights; Venus; *specific program*
UR-100 antiballistic missile, 259
UR-500 (*Proton*) launch vehicle: development of, 259, 260, 265, 267, 273; engines on, 259, 303; and ISS, 320; and *Kosmos* program, 198, 297; and LK-1 program, 273-74; as N-1 program rival, 259, 260, 266-67; and *Salyut* program, 298. *See also* UR-500K program
UR-500K program, 259, 274, 286, 289, 290, 294, 297, 309, 312
UR-700 program, 265, 266, 308, 312
USS George Washington, 209
Ustinov, Dimitri Fedorovich: character of, 206; and ICBM development, 96, 100, 114; and Korolev-Glushko relationship, 114; and Korolev's funeral, 284; Korolev's relationship with, 78, 115, 116, 202, 206, 209, 230, 258, 262, 276; management style of, 202; and manned space exploration, 167-68; and N-1 program, 272; and N-1/L-3 program, 288, 299; and organizational structure, 202, 205, 206, 209, 216; responsibilities of, 202; and satellite development, 123–24; and space shuttle, 314; and technology issues, 221; and U.S.-Soviet competition, 301; and V-2 technology, 70–71, 78, 82–83, 84, 86, 87; Yangel's relationship with, 115, 116

Utkin, Valdimir, 100
Utochkin, Sergi, 17–18, 19

V-1 buzz bomb, 64, 75
V-2 technology: and cruise missles, 86; Cuxhaven demonstrations of, 72; development in Soviet Union of, 79–80 Germans' accomplishments using, 85–86, 88–89; and Germans return to Germany, 89–90; and Germans in Soviet Union, 75–90; and ICBM development, 97, 108; and Korolev in post-war Germany, 60, 61, 63, 70-73; Korolev's views about, 71, 97; NII-88 development of, 78-90; obsolesence of, 81; and sabotage by German scientists, 83; Soviet reconstruction of, 67-75, 79-80; and Soviet scientists in Germany, 67-70, 73-74; Soviet test launches of, 73-74, 81-83; and U.S., 66, 67-68, 162; von Braun's development of, 44, 46, 63; World War II use of, 65-66
Vachnadze, Vahtang, 276
vacuum spray technique, 210
Vakhtangov, Vakhtang, 228
Valier, Max, 35
Van Allen, James, 133, 135, 136, 137, 138. *See also* radiation belts
Vandenberg Air Force Base, 314
Vanguard program, 124-25, 129, 130, 132-33, 137, 249
Venera program, 131, 152, 153, 154
Venus: Korolev's dreams about manned flights to, 247; and robotic program, 247, 248, 251; Soviet attempts to reach, 131, 148-49, 150, 152, 153-54, 166, 211-12, 248, 261, 291, 295; technology for flights to, 219, 221, 228; U.S. attempts to reach, 150, 152-54; and U.S.-Soviet competition, 138, 148, 150
Verne, Jules, 8
Vernov, Sergei, 136
Vetchinkin, V. P., 30, 35
Vetoshkin, Sergei, 84
Vetrov, Gyorgi Stepanovich, 16, 38, 112; awards and honors for, 214-15; and Communist Party decision-making, 237; and Communist Party membership, 230-32; and Germans' return to Germany, 89—90; and ICBM development, 93, 99, 104, 108; and Korolev-Glushko relationship, 114-15; and Korolev's arrest/imprisonment, 57, 61-62; and Korolev's desire to go to Moon, 139, 247; and Korolev's lifestyle, 200; and Korolev's management style, 98, 211, 213, 214-15; and manned space exploration, 180-81, 182-83, 252; and military needs, 253; and N-1 program, 257; and N-1/L-3 system, 265; and space shuttle, 313-14; and Tsiolkovsky-Korolev meeting, 14; and unmanned space exploration, 157; and V-2 technology, 64-65
Viking missions, 66, 194
Vilnitskii, Lev B., 186, 262
Vinogradov, Vladimir Nikitovich, 276-77
Vishnevsky, A. A., 279, 280, 281
Voinov, Lev, 274-75
Volchkov, Yevgeni, 281
Volkogonov, Dmitri, 51, 56, 57, 58
Volkov, Vladislav, 242, 285, 296, 298, 302
Volynov, Boris, 293-94
von Braun, Wernher, 62, 64, 67, 71, 258, 267; contributions to U.S. space program of, 2; death of, 1; imprisonment of, 1; Korolev compared with, 1-2, 212-13; liquid-fuel rockets work of, 43-44; and manned space exploration, 159; and Moon flights, 250, 251, 252; rocket development by, 49; and satellite development, 124, 129, 133; and space medicine, 161; and U.S.-Soviet competition, 251; and V-2 technology, 63, 66, 68, 71, 74
Vorobiev, Pavel, 7, 41, 224-25
Voronov, G. I., 284
Voroshilov, Kliment, 22, 52
Voskhod 1 mission, 128, 182, 189
Voskhod 2 ("walk in space") mission, 182-83, 184-86, 189, 212, 269, 277
Voskhod 3 concept, 186
Voskhod program, 165, 180-89, 205, 242, 243, 254, 255-56. *See also specific mission*
Voskresensky, Leonid A., 2, 73, 99, 104, 109, 118, 151
Vostok 1 mission, 177
Vostok 2 mission, 176, 177, 178
Vostok 3 mission, 177, 178, 188, 205, 277
Vostok 3A concept, 168
Vostok 4 mission, 177, 178, 188, 277
Vostok 5 mission, 178, 261
Vostok 6 mission, 178, 261
Vostok program: and Chelomei-Korolev relationship, 260; and cosmonaut deaths, 242; development of, 162-63, 165, 166-67, 168, 169, 178, 211-12; exhibit of, 322; and Korolev's management style, 166-67, 212; and Korolev's responsibilities, 255; and Moon flights, 260, 262; purpose of, 170; and R-7 launcher, 219, 249; and reconnaissance satellites, 170, 192, 193, 196; and *Soyuz* program, 254, 255-56, 262; and U.S.-Soviet competition, 249; and *Voskhod* missions, 180; *Zenit* program compared with, 193. *See also* Gagarin (Yuri) flight; *specific mission*
Voyager missions, 194
Vulkan program, 305

"walk in space": Soviet, 2, 182-83, 184-86, 189, 212, 269, 277, 295; U.S., 287, 295
Wallace, Donald Mackenzie, 9
Wasserfall (Waterfall) missile, 87, 89
weather satellites, 149, 269-70, 317
Webb, James, 2, 156, 176, 249, 251, 270
weightlessness: and manned space exploration, 161, 162, 172, 173, 176, 187, 188; space station development for studying, 248; and Tsiolkovsky's work, 13; and unmanned space exploration, 149
Weitz, Paul, 319
Wheelon, Albert D., 191
White, Edward, 163, 186-88, 189, 269, 289
Whitney, Thomas P., 57
Wiesner, Jerome, 250
wind tunnels, 12, 30, 48, 88
winged rockets, 46, 48, 50, 74-75, 230
Winkler, Johannes, 43
Witkin, Richard, 130
women: Korolev's relationships with, 199-200, 234; and space exploration, 178, 179; as translators, 244
World War I, 19

Yagoda, G. G., 57
Yakovchuk, Konstantin, 28
Yakovlev, N. D., 82, 98-99, 112
Yakovlev design bureau, 261-62
Yangel, Mikhail, 2, 258; awards and honors for, 14, 15; and Communist Party, 233; and ICBM development, 117 119; Korolev's relationship with, 14, 115-16, 233, 265, 272; and *Kosmos 379* mission, 297; and Moon flights, 265; and N-1 program, 272; at NII-88, 115-16; professional background of, 115; and R-12 missile, 115; reputation of, 14-15; and scenarios for manned Soviet Moon flights, 308; Ustinov's relationship with, 115, 116. *See also* NPO Yuzhnoye
Yangel design bureau. *See* NPO Yuzhnoye
Yazdovsky, Vladimir, 161, 165
Yeger, Sergei Mikhailovich, 60
Yegorov, Boris, 182, 189, 242, 243, 283
Yegorychev, N. G., 284
Yeliseev, Alexei, 185, 186, 289-90, 293-94, 296, 298
Yeltsin, Boris, 316
Yevpatoriya, Crimea: radio communications center at, 149-50
Yezhov, Nikolai, 21, 55, 57
Young Communist League: *See also* Komsomol
Young, John, 182, 189, 287, 299
Young, Laurence, 306

Zeldovich, Yaakov, 141
Zenchenko-Rshevskaya, Yevgeniya (stepsister), 16-17
Zenit reconnaissance satellite: and Communist Party paranoia, 241; development of, 191-92, 193-96, 212, 261; as Korolev's responsibility, 190, 251; and organizational structure, 205; and U.S.-Soviet competition, 191-92; and *Voskhod* program, 183; and *Vostok* program, 170, 193
Zhukov, Nikolai Nikolaevich, 98
Zhukovsky, Nikolai Yegorovich, 30
Zhukovsky Academic Circle (AKNEZH), 32
Zhukovsky Academy, 13, 25, 26
Zlotnikova, Antonina, 93-94, 199, 233, 277
Zond 1, 2, and 4 missions, 264-65, 291
Zond 5 and 6 missions, 292, 293
Zond 7 and 8 missions, 296, 297
Zond program, 152, 154, 155, 156, 221, 264, 291, 292-93. *See also specific mission*
Zudanov, Vladimir, 228
Zvezdochka (dog), 168
Zvyezda design bureau, 183-84. *See also* Severin, Gai Ilyich